Unsolvable Classes of Quantificational Formulas

Unsolvable Classes of Quantificational Formulas

Harry R. Lewis

Aiken Computation Laboratory
Harvard University
Cambridge, Massachusetts

1979

Addison-Wesley Publishing Company
Advanced Book Program
Reading, Massachusetts

London · Amsterdam · Don Mills, Ontario · Sydney · Tokyo

Library of Congress Cataloging in Publication Data

Lewis, Harry R
Unsolvable classes of quantificational formulas.

"Advanced Book Program."
Bibliography: p.
Includes indexes.
1. Unsolvability (Mathematical logic) 2. Combinatorial analysis. 3. Computable functions. I. Title.
QA9.63.L48 511'.3 79-17573
ISBN 0-201-04069-7

American Mathematical Society (MOS) Subject Classification Scheme (1970): 02B10, 02G05, 02F15, 68A25

Reproduced by Addison-Wesley Publishing Company, Inc., Advanced Book Program, Reading, Massachusetts, from camera-ready copy prepared at the office of the author.

Manufactured in the United States of America

ABCDEFGHIJK-AL-79

CONTENTS

PART II: FIRST-ORDER LOGIC

PREFACE

The last published survey of the theory of unsolvable classes of quantificational formulas is the 1959 treatise by János Surányi, Reduktionstheorie des Entscheidungsproblems. That work collected together most of the results then known on the reduction of the decision problem for all of first-order logic to that for syntactically specified classes of formulas. Its proofs are mainly of the following type: given some syntactic restriction on the structure of a formula, to present an effective method by which any quantificational formula can be reduced to one obeying that restriction. Unsolvability of the decision problem for each restricted class then follows from the unsolvability of quantification theory as a whole.

Shortly after the completion of Surányi's book, much stronger logical unsolvability results were obtained by methods not previously available. Büchi and Wang showed how simple computational or combinatorial systems--Turing machines or dominoes--could be succinctly described by simple quantificational formulas. Unsolvability results in the logical realm were then obtained from the unsolvability of extralogical decision problems. The subsequent growth of interest in the theory of computation has made available yet stronger tools, and has therefore made possible yet finer and sharper negative results on logical decision problems. The main purpose of this book is to provide a uniform treatment of the strong results these new tools have yielded.

The publication of this work is also occasioned by that of The Decision Problem: Solvable Classes of Quantificational Formulas, by Burton Dreben and Warren D. Goldfarb. Many of the results presented here precisely delimit the solvability results to which that book is devoted. As the approach and terminology of the two works are largely uniform, readers of each will find the other illuminating. Either book can, however, be read by itself.

This book will be of interest both to logicians, for its analysis of the fine structure of quantification theory, and to theoretical computer scientists and combinatorial mathematicians, for its treatment of decision problems outside as well as within logic. A level of mathematical sophistication is expected, and as a text the book is therefore most likely to be useful on the graduate level. It is, however, mostly self-contained, and is therefore accessible to students who have had only limited exposure to logic and to computability theory.

HARRY R. LEWIS

ACKNOWLEDGMENTS

I am deeply indebted to all those from whom I have learned, and that means, above all others, my mentor, Burton Dreben. To my collaborators and coauthors Warren Goldfarb and Stål Aanderaa I owe thanks not only for what they have taught me in their own writings and in our joint research, but also for their unpublished ideas appearing here for the first time. To Yuri Gurevich, to whom the program, and many of the specifics, of Chapter IIC are due, I am especially grateful for his willingness to let me see his unpublished survey article (1971). Egon Börger, too, has kept me abreast of the progress of his research. To my students in Mathematics 286 in 1975 a word of thanks for their tolerance while these ideas were in their formative stages. And finally, to my superb typists Renate D'Arcangelo and Karin Young, I am sincerely grateful for their long and uncomplaining labor at the preparation of this manuscript.

This research was supported by the National Science Foundation under Grant MCS76-09375.

INTRODUCTION

The subject of this monograph is the recursive unsolvability of a number of decision problems for (I) combinatorial systems such as automata and Wang's dominoes and (II) first-order logic. The main objective is to present the strongest known results on unsolvable classes of first-order formulas, that is, classes of formulas for which satisfiability cannot be effectively decided. To achieve this goal a number of combinatorial problems must be presented; since these problems are of interest in their own right, however, we pursue their ramifications somewhat beyond the point at which we might stop if our only purpose were to lay the foundations on which the logical results could stand. Still, no attempt is made to survey the multitude of related problems. Indeed, even in the logical realm we have not attempted to be exhaustive; the literature is too extensive and contains too many constructions that are technical variations on each other under different syntactic disguises to make such an effort rewarding. Instead, we have included only the hardest problems, those that took the development of new insights to resolve, plus whatever related problems are needed to complete a classification of formulas along some natural lines. In our exposition we have tried to separate the specific combinatorial designs that make the proofs succeed, from characterizations of those properties of the designs that are actually used in the proofs. This attempt has, we hope, both expository and mathematical merit. It tends to modularize the proofs, and to focus attention on the details of a construction when the details have to be checked; but it also leads to the development of general concepts and methods that may illuminate the underlying structures of these problems and prove useful in the analysis of related ones.

Aside from intrinsic interest, the principle used for selecting problems for inclusion is to consider for first-order logic without function signs or identity the most obvious syntactic features--number and type of the predicate letters and atomic formulas, structure of the quantifier prefix, and truth-functional form of the matrix. The combinatorial problems are mostly what we call ''tiling problems,'' a broad class that includes (slightly reformulated) Wang's domino problems and the so-called ''linear sampling'' problems. Before we move to a more detailed synopsis, it may be helpful to discuss what is omitted. On the combinatorial side, there is no mention of word problems for grammatical or algebraic systems (Thue systems, tag systems, groups, etc.), or of diophantine equation problems, and only incidental discussion of automata. For first-order formulas classified syntactically the presentation is rather more complete, although we do not discuss the situation that arises when function signs or identity are part of the language. We have also avoided discussion of computational notions either more general or more specific than mere unsolvability. Thus there is no discussion of degrees of unsolvability, even of the representation of r.e. degrees by the particular kinds of systems we discuss (see, for example,

Hughes, Overbeek, and Singletary (1971) and Börger (1976)). Nor in our unsolvability proofs for classes of logical formulas do we also establish conservative unsolvability (i.e., the unsolvability of the question of whether a formula has a finite model). In fact the constructions can generally be adapted to prove the stronger result, but our feeling has been that the proofs are complex enough as they are, and that no fundamental insights are to be gained by adding more technicalities.

Chapter IA presents notational conventions and preliminaries from computability theory. Turing machines are defined, so that the halting problem can be used as the basis for the rest of our unsolvability proofs, but we do not give a diagonalization argument for the unsolvability of the halting problem itself.

Chapter IB treats the origin-constrained tiling problem. This is the problem of determining whether the first quadrant of the plane can be tiled with copies of square tiles drawn from a given finite set of prototypes, subject to restrictions about which tiles may abut each other on the four sides, and which may appear at the origin. It is not a difficult matter to reduce the Turing machine halting problem to this problem; in a word, the ith row of a tiling $(i = 0, 1, \ldots)$ represents the state and the appearance of Turing machine tape after i steps in a computation, and the origin constraint specifies the starting configuration of the machine.

Chapters IC and ID treat two versions of the much more difficult linear sampling problem. To give the flavor, we now describe another version, not actually used in the monograph but the simplest of all to state. Imagine a two-way infinite tape; the tape is divided into cells, which have numerical coordinates $\ldots, -2, -1, 0, 1, 2, \ldots$. Each cell can be inscribed with a single symbol. The problem is: given a finite set of symbols, a finite set A of ordered pairs of those symbols, and a finite set B of ordered triples of those symbols, is there a way to write symbols in the cells so that (a) every adjacent pair of cells contains a pair of symbols in A and (b) every triple of cells whose coordinates sum to 0 contains a triple of symbols in B. This problem cannot be solved effectively in A and B, but there seems to be no direct proof of the unsolvability by a simple reduction of any other previously studied unsolvable problem. The versions of the linear sampling problem actually treated here have various numbers of tapes and rules about which combinations of tape cells must contain sequences in the A and B sets. (The unsolvability of the version just defined follows trivially from that presented on p. 48.) The proofs are long, but depend ultimately on the origin-constrained tiling problem.

The unsolvability of a certain linear sampling problem provides a simple and elegant proof of the unsolvability of two unconstrained tiling problems in Chapter IE. One version is like the origin-constrained tiling problem of Chapter IB, but without the origin constraint. The other version uses hexagonal tiles instead of square ones.

Finally, Chapter IF treats the Post correspondence problem and the halting problem for two-counter machines, which are used in Part II.

Before we present the particular logical decision problems considered in Part II, it is well to discuss what might be viewed as two methods by which a class $\mathscr{S}$ of logical formulas may be shown to be unsolvable. The first is to begin with some class $\mathscr{S}'$ that is known to be unsolvable, for example the class of all formulas, and to give an effective method of translating each formula in $\mathscr{S}'$ into a formula in the class $\mathscr{S}$, in such a way that the original formula has a model if and only if its translation has a model; then $\mathscr{S}$ is unsolvable as well. Proofs by such methods are called reduction proofs, and were used, even before the notion of computability was formally explicated by Turing, to show classes to be reduction classes, i.e. classes to which the decision problem for all of first-order logic could be reduced. On the other hand, some unsolvability proofs may be called encoding proofs, because they rely on the encoding of some class of automata or other combinatorial systems into formulas in $\mathscr{S}$, in such a way that satisfiability of a formula corresponds to some unsolvable property of the encoded system.

In terms of results the distinction between reduction and encoding proofs is insubstantial, in that any proof using either class of methods can, in principle, be translated into a direct encoding of the Turing machine halting problem (or whatever we wish to take as bedrock), by telescoping into a single mapping the series of reductions and encodings that leads from Turing machines to $\mathscr{S}$. But the distinction is also artificial on other grounds, namely that the dichotomy between combinatorial systems and logical formulas is rendered unconvincing when formulas are viewed as generators of their Herbrand expansions. The Herbrand expansion of a formula is a recursive set of quantifier-free instances of the formula; it can be generated mechanically by a simple class of substitutions, and it has the property that a formula is satisfiable, in the sense of model theory, if and only if the expansion is truth-functionally consistent, i.e. if and only if there is a truth-assignment that verifies the entire expansion. Since truth-functional consistency is decidable for any finite portion of the Herbrand expansion, and since the whole expansion is consistent if each finite portion is, use of the Herbrand expansion reduces unsatisfiability to a condition similar to the halting of a Turing machine: the existence of a certain type of finite set of instances of a formula is parallelled by the existence of a certain type of finite sequence of instantaneous descriptions for a machine. Indeed, and this is the crux of the matter for this monograph, the particularly simple structure of certain Herbrand expansions makes.

it possible to associate combinatorial systems with first-order formulas in so direct a way, that computations of a machine (or whatever is the analogue for some other kind of system) can be extracted from, or encoded in, the Herbrand expansion of a formula, almost verbatim. This line of attack was opened up by Büchi and by Wang in the early 1960's; the invention of Wang's domino problems was motivated by formal similarities between the set of lattice points of the plane and the Herbrand expansions of formulas with prefixes of the form $\forall\exists\forall$. All the stronger results obtained since then have relied on methods of the same kind.

What is attempted here is, first, to exploit these methods as fully as possible, by applying them to new combinatorial problems, and in new ways to old combinatorial problems, and, second, to extend them to handle reductions between classes of formulas, as well as encodings of combinatorial systems in logical formulas. Thus our reduction proofs often show how the Herbrand expansion of one formula can be found, in an encoded form, embedded in the Herbrand expansion of another. Much of the machinery we use to establish such relationships, and indeed our whole approach to these problems, is based on the development of the theory of Herbrand expansions by Dreben and Goldfarb for the analysis of solvable classes of formulas.

Chapter IIA, then, contains the basic definitions of the theory of Herbrand expansions. Chapter IIB is an extended example: the unsolvability of the decision problem for formulas having only one predicate letter, which is dyadic. This is a classical result; much stronger versions are proved in Chapter IIC, but this proof is included to illustrate the use of Herbrand expansions in reduction proofs and to motivate the development of the bigraph machinery used throughout Part II. A bigraph is essentially a partial interpretation for a predicate letter, i.e., a set of n-tuples of elements of which the predicate letter is to be true, together with another set of n-tuples of which it is to be false. The terminology developed in Chapter IIB permits clear statements of how such partial interpretations combine and interlock to produce truth-assignments with specified properties.

Chapter IIC treats classes of prenex formulas determined by their quantifier structure and the number of monadic, dyadic, ... predicate letters a formula may have. A complete classification along these lines is presented. The method is to reduce a linear sampling problem to a class of formulas whose Herbrand expansions have a closely related structure, and then to reduce this class to others by purely logical constructions.

Chapter IID considers formulas restricted according to the possible sequences of arguments an atomic subformula may have. This classification has been a fruitful source of solvable classes, so it is natural to look for simple unsolvable classes as well. Two prefix forms, $\forall\exists\forall$ and $\forall\ldots\forall\exists\ldots\exists$, are considered. The $\forall\exists\forall$ classification uses the linear sampling problem; the $\forall\ldots\forall\exists\ldots\exists$ classification uses the origin-constrained tiling problem. (The main results of Chapter IIC, but not these, could have been obtained by using the unconstrained tiling problem instead of the linear sampling problem.)

Chapter IIE is about Krom and Horn formulas, which are special kinds of formulas in conjunctive normal form. A Krom formula has at most two disjuncts per conjunct, a Horn formula at most one disjunct that is an atomic formula (rather than the negation of an atomic formula) per conjunct. Such formulas are classified by their prefixes and the number of argument-places of their predicate letters. The unsolvability proofs depend on the Post Correspondence Problem and the halting problem for two-counter machines.

Chapter IIF proves that formulas with four atomic subformulas form an unsolvable class. The strengthening of this theorem to the three atomic formula class is an open problem. The use of two-counter machines makes it possible to prove this theorem in a sharp form, for the $\forall\exists\forall\ldots\forall$ prefix.

Familiarity with first-order logic and the basics of the theory of computation is assumed, but there are no specific prerequisites. In particular, the only theorems referred to that are not stated and explained in the text are the Chinese Remainder Theorem and König's Infinity Lemma.

Unsolvable Classes of Quantificational Formulas

PART I: COMBINATORIAL SYSTEMS

Chapter IA

PRELIMINARIES

IA.1 Set Theory

If $S_1, \ldots, S_k$ are sets $(k \geq 1)$ then $S_1 \times \cdots \times S_k$ is the k-fold Cartesian product of $S_1, \ldots, S_k$. In particular, for any set S, $S^k = S \times \cdots \times S$ is the k-fold Cartesian product of S with itself. The members of a k-fold Cartesian product $S_1 \times \cdots \times S_k$ are called k-tuples, or simply sequences; 2-tuples are also called pairs, 3-tuples triples, 4-tuples quadruples; and 1-tuples are identified with the elements themselves, so that $S^1 = S$. Sequences are usually written using $\langle\ \rangle$ brackets, so that

$$S_1 \times S_2 = \{\langle s_1, s_2\rangle \mid s_1 \in S_1,\ s_2 \in S_2\} \quad ,$$

but ( ) parentheses are occasionally used for variety and clarity. If the domain of a mapping f is a set S of k-tuples, we usually write $f(s_1, \ldots, s_k)$ instead of $f(\langle s_1, \ldots, s_k\rangle)$, but we sometimes write $f(s)$ if we have let $s = \langle s_1, \ldots, s_k\rangle$. The ith projection function π_i selects the ith component of a k-tuple, when $k \geq i$. That is, $\pi_i(\langle s_1, \ldots, s_k\rangle) = s_i$ for any $s_1, \ldots, s_k$, $k \geq i$.

A relation R is a subset of S^k for some set S and some fixed k. The assertion that $\langle s_1, \ldots, s_k\rangle \in R$ is sometimes written $R(s_1, \ldots, s_k)$, and if R is binary, i.e., if $k = 2$, the assertion may also be written $s_1 R s_2$. In the latter case we may also write $s_1 R s_2 R s_3 \cdots R s_k$ if $s_i R s_{i+1}$ for $i = 1, \ldots, k-1$. The reflexive, transitive closure of a binary relation $R \subset S^2$ is that binary relation R^* such that sR^*s' if and only if $s = s' \in S$ or there is a sequence of $k \geq 2$ elements $s_1, \ldots, s_k \in S$ such that $s_i R s_{i+1}$ for $i = 1, \ldots, k-1$, $s = s_1$, and $s' = s_k$.

For any set S and mapping f,

$$f[S] = \{f(s) \mid s \in S\} \qquad \text{and} \qquad f^{-1}[S] = \{s \mid f(s) \in S\} \quad .$$

Also $f^{-1}(s) = f^{-1}[\{s\}]$; but if $f^{-1}(s)$ has only one member, we sometimes write $f^{-1}(s)$ for that unique member. The cardinality of a set S is denoted by $|S|$.

$\mathbb{Z} = \{\ldots, -2, -1, 0, 1, 2, \ldots\}$ is the set of integers and $\mathbb{N} = \{0, 1, 2, \ldots\}$ is the set of natural numbers.

An alphabet is a nonempty finite set, whose elements are called symbols. If Σ is an alphabet then Σ^* is the free monoid with generators in Σ; the monoid operation is called concatenation and is denoted by juxtaposition. Thus, Σ^* is the set of all finite sequences $a_1 \ldots a_k$ $(k \geq 0)$ with $a_1, \ldots, a_k \in \Sigma$, and if $u = a_1 \ldots a_k$ and $v = b_1 \ldots b_\ell$ are members of Σ^* then so is $uv = a_1 \ldots a_k b_1 \ldots b_\ell$. The members of Σ^* are called words over Σ, or simply words. The length $|u|$ of a word $u = a_1 \ldots a_k$, with $a_1, \ldots, a_k \in \Sigma$, is k $(k \geq 0)$. The identity of the monoid Σ^* is the word of length 0, also called the empty word and denoted by ϵ. Σ^k is the set of all words over Σ of length exactly k, so that $\Sigma^0 = \{\epsilon\}$, $\Sigma^1 = \Sigma$, and $\Sigma^* = \bigcup_{k=0}^{\infty} \Sigma^k$ (no confusion with the Cartesian product will arise). Also, $\Sigma^+ = \Sigma^* - \{\epsilon\} = \{w \mid w \in \Sigma^* \text{ and } |w| > 0\}$. If $S_1, \ldots, S_k$ are sets of words, then $S_1 \cdots S_k = \{s_1 \cdots s_k \mid s_1 \in S_1, \ldots, s_k \in S_k\}$. For example, $\Sigma^+ = \Sigma\Sigma^*$.

IA.2 Theory of Computation

Formally, a problem may be taken to be a set of words over an alphabet. In practice we may regard as a problem any set of finite objects that may be encoded as words in a natural manner. Thus, for example, if $S_0, S_1, \ldots$ are finite sets such that for each $i \geq 0$, $S_i \subset \Sigma_i^* \times \Sigma_i^*$ for some alphabet Σ_i, then $P = \{S_0, S_1, \ldots\}$ might be considered a problem; some code could be established for representing the symbols in the (possibly infinite) set $\bigcup_{i=0}^{\infty} \Sigma_i$ by using only a finite set of symbols, and this finite set could then be augmented with various brackets, braces, and commas so that certain words over the augmented alphabet could be interpreted as representations of the sets S_i.

Turing machines are defined formally in Section IA. 3. (We consider only deterministic machines.) An effective procedure is a Turing machine that halts on every input word (over some fixed alphabet Σ). If $\mathcal{M}$ is an effective procedure, then $\mathcal{M}$ computes a function from words to words, namely, the mapping φ such that for each word $w \in \Sigma^*$, $\varphi(w)$ is the word left on the finite nonblank portion of the Turing machine tape after $\mathcal{M}$ has halted, given that $\mathcal{M}$ was started in its designated initial state on word w. Any such function φ is said to be computable. A problem $P \subset \Sigma^*$ is solvable if there is a computable function $\chi: \Sigma^* \to \{0, 1\}$ such that $\chi(w) = 1$ if and only if $w \in P$; that is, P is solvable if its characteristic function is computable. In this case, an effective procedure that computes the characteristic function is called a decision procedure. A problem is unsolvable if it is not solvable.

In practice we use a more general, and more informal, way of describing problems. We often refer to "the problem of determining, given an X, whether Y". Here we are asking for an effective procedure that correctly indicates (by computing a 0 or 1 answer, for example) whether an input x has property Y, provided that x is an X. (If x is not an X, it does not matter what happens.) The problem is solvable if such an effective procedure exists, otherwise, unsolvable. Note that whether the set of all Xs is solvable is not an issue; it might be unsolvable, and yet the problem of determining, given an X, whether Y, could be solvable (e.g., if X is Y). In most cases of interest the set of Xs is itself solvable; then the problem of whether Y holds for an X, and the problem of whether Y holds for any given input, are either both solvable or both unsolvable.

It is well known that unsolvable problems come in a rich variety of degrees of unsolvability. However, every problem in this monograph falls on the same spot in this degree structure: all are 1-complete, i.e., recursively isomorphic to the Turing machine halting problem. Hence these problems cannot be distinguished from each other by the techniques of recursive function theory.

Formally, problem $P \subset \Sigma^*$ is 1-reducible to problem $Q \subset \Sigma^*$ if and only if there is a one-one computable function $\varphi: \Sigma^* \to \Sigma^*$ such that for each $w \in \Sigma^*$, $\varphi(w) \in Q$ if and only if $w \in P$. This is the narrowest notion of reducibility significant in recursive function theory; since it is the only notion of reducibility we use, we say under these circumstances simply that P is reducible to Q, and call φ a reduction of P to Q. Clearly, reducibility is transitive, and a problem to which an unsolvable problem is reducible is also unsolvable. From a recursion-theoretic standpoint, the only technique used here for showing problems unsolvable is the explicit presentation (and verification) of reductions of the Turing machine halting problem, or of other problems previously shown unsolvable by the same technique.

A problem $P \subset \Sigma^*$ is recursively enumerable if and only if there is a Turing machine $\mathcal{M}$ that halts when given members of P as input, but does not halt when given members of $\Sigma^* - P$ as input. A problem that every recursively enumerable problem is reducible to is said to be 1-complete. Since every problem discussed in this monograph is recursively enumerable, "unsolvable" may be taken throughout as synonymous with "1-complete".

IA.3 Turing Machines

For our purposes a Turing machine is a deterministic automaton with a finite-state control and a single one-way infinite tape on which symbols can be read and written by a single read-write head. The tape is initially blank, except for a finite initial portion called the input. After a machine has been started in a designated start state with its read-write head on the leftmost square of the tape, it runs in the usual way as determined by its transition function. One of two outcomes is possible: the machine may ultimately halt, i.e., reach a state-symbol configuration from which no transition is possible, or it may fail to halt and thus run forever. The outcome will in general depend on both the machine and the input. Note that there is no designated "final" state. We assume that the finite control is arranged in such a way that a machine never attempts to move its head left off the end of the tape.

We shall have recourse to the Turing machine halting problem in two versions.

Turing's Theorem, Inputless Version. The following problem is unsolvable: to determine, given an arbitrary Turing machine $\mathcal{M}$, whether or not $\mathcal{M}$ halts when given the empty word as input.

Turing's Theorem, Universal Version. For a certain "universal" Turing machine $\mathcal{U}$, the following problem is unsolvable: to determine, given an arbitrary word w over the alphabet of $\mathcal{U}$, whether $\mathcal{U}$ halts when given w as input.

Proofs of these theorems may be found in standard works on the theory of computation. (Our conditions for "halting" are slightly unusual, in order to simplify the definitions below.)

Formally, a Turing machine is a quintuple $\mathcal{M} = (K, \Sigma, B, q_0, \delta)$ where

> K is a finite set (of states) with $q_0 \in K$ (the start state)
>
> Σ is a finite set (of symbols), with $B \in \Sigma$ (the blank symbol)
>
> δ, the transition function, is a mapping from a subset of $K \times \Sigma$ to $K \times (\Sigma - \{B\}) \times \{-1, 0, +1\}$.

This definition has an oddity that will be of service in Chapter IB: a machine cannot write (or leave unchanged) its blank symbol. In practice, $\mathcal{M}$ would have a "pseudoblank" symbol, which it would use instead of the true blank, without distinguishing between them in any way. (The term "nonblank" used in the definition of "computable" on p. 2 ought then to be interpreted as "containing neither the blank nor the pseudoblank symbol".)

An instantaneous description of $\mathcal{M} = (K, \Sigma, B, q_0, \delta)$ is a member of $\Sigma^*(K\times\Sigma)\Sigma^*$; it specifies a state, a word on the tape, and a head position. The relation $\vdash_{\mathcal{M}}$ ("yields in one step") between instantaneous descriptions is defined as follows: if $w_1\langle q_1, a_1\rangle v_1$ and $w_2\langle q_2, a_2\rangle v_2$ are instantaneous descriptions, with $q_1, q_2 \in K$, $w_1, v_1, w_2, v_2 \in \Sigma^*$ and $a_1, a_2 \in \Sigma$, then $w_1\langle q_1, a_1\rangle v_1 \vdash_{\mathcal{M}} w_2\langle q_2, a_2\rangle v_2$ if and only if $\delta(q_1, a_1)$ is defined, $\delta(q_1, a_1) = (q_2, a, m)$ for some $a \in \Sigma$, $m \in \{-1, 0, 1\}$, and one of the following holds:

(1) $m = -1$ ("move the head left") and $v_2 = av_1$, $w_1 = w_2a_2$;

(2) $m = 0$ ("leave the head in place") and $w_1 = w_2$, $v_1 = v_2$, and $a_2 = a$;

(3) $m = +1$ ("move the head right"), $w_2 = w_1a$, and either

> (a) $v_1 \neq \epsilon$, and $v_1 = av_2$, or
>
> (b) $v_1 = \epsilon$, and $a_2 = B$, $v_2 = \epsilon$.

(The last clause corresponds to moving right onto blank tape.) Since $\mathcal{M}$ is deterministic, i.e., since δ is a function, the relation $\vdash_{\mathcal{M}}$ is single-valued. A halting instantaneous description I is one such that $I \vdash_{\mathcal{M}} J$ for no instantaneous description J; note that whether or not I is halting depends only on whether the transition function of $\mathcal{M}$ is defined on the unique pair in $K \times \Sigma$ that is a symbol of I. Now let $\overset{*}{\vdash}_{\mathcal{M}}$ be the reflexive, transitive closure of $\vdash_{\mathcal{M}}$. For $w \in \Sigma^*$ let

$$I_w = \begin{cases} \langle q_0, B\rangle & , \text{ if } w = \epsilon \\ \langle q_0, a_1\rangle a_2 \ldots a_k , & \text{ if } w = a_1 \ldots a_k \text{ for some } a_1, \ldots, a_k \in \Sigma , \\ & \qquad k \geq 1 . \end{cases}$$

Intuitively, I_w represents that configuration of $\mathcal{M}$ in which the read-write head of $\mathcal{M}$ is placed on the leftmost square of a tape inscribed with the word w. Then $\mathcal{M}$ <u>halts</u> when started with input w if and only if there is a halting instantaneous description I such that $I_w \overset{*}{\vdash}_{\mathcal{M}} I$. Since $\mathcal{M}$ is deterministic, there can be at most one such instantaneous description I, and in this case the uniquely determined sequence $I_w \vdash_{\mathcal{M}} I_1 \vdash_{\mathcal{M}} I_2 \cdots$ is finite and ends with I; otherwise, the sequence is infinite.

This formal definition of halting gives precise content to the two versions of Turing's theorem stated above.

<u>Historical References</u>. Turing machines are due to Turing (1937). The version described here is essentially that of Hopcroft and Ullman (1969), but without final states. For material on recursive function theory, see Rogers (1967).

Chapter IB

TILING PROBLEMS

IB. 1 Tiling Problems in General

Many unsolvable combinatorial problems may be described as tiling problems. A tiling problem has the following general form: A set $\mathbb{X}$, called the space, and a relation $R \subset \mathbb{X}^k$ for some k, called the spatial relation, are combined to form a pair $\Theta = (\mathbb{X}, R)$, called the tiling type. A Θ-system, or a system of type Θ, is then a pair $\mathscr{S} = (T, Q)$, where T is a finite set (of tiles) and $Q \subset T^k$. A T-tiling of $\mathbb{X}$, or simply a tiling, is a mapping $\tau: \mathbb{X} \to T$, and such a tiling τ is accepted by $\mathscr{S}$ if and only if $Q(\tau(s_1), \ldots, \tau(s_k))$ whenever $R(s_1, \ldots, s_k)$. The tiling problem for Θ-systems, or for a restricted class of Θ-systems, is the problem of determining, given an arbitrary Θ-system (or one in the restricted class) whether or not there is a tiling accepted by that Θ-system.

The term "accepted" is borrowed from automata theory and fits best if one imagines the Θ-system as an active agent checking a proposed tiling for violations of the rules specified by R and Q. The Θ-system has a severely limited mechanism for inspecting the space $\mathbb{X}$: Only certain combinations of k points may be examined at one time, and no control can be exerted over the order in which such finite complexes of points may be visited. Thus, a tiling system is a sort of 1-state automaton with k read-only heads, which may not move independently, but are coordinated by a mechanical linkage. A tiling is accepted if every legal configuration of the heads results in the reading of a satisfactory k-tuple of symbols.

The tiling problems that are easiest to visualize are those of type $\Theta = (\mathbb{X}, R)$, where the space $\mathbb{X}$ is $\mathbb{Z} \times \mathbb{Z}$ or $\mathbb{N} \times \mathbb{N}$, i.e., the lattice points of the plane or the first quadrant, and the spatial relation R holds between certain adjacent or distinguished points. Two examples of problems of this kind will be given, the first in Section IB. 2, the second in Chapter IE. In these cases, if (T, Q) is a Θ-system, a T-tiling is a tessellation of the plane or first quadrant, if the members of T are thought of as prototypes for an infinite supply of square plates to be used for the tiling. A tiling is accepted by (T, Q) if points that are adjacent or distinguished in the way specified by R are tiled with combinations of tiles permitted by Q.

Planar tiling problems of the sort just described are intimately connected to the domino problems introduced by Wang in his investigations of the $\forall\exists\forall$ case of the decision problem (1961, 1962). In a domino problem, however, the acceptability of a tiling is determined not by an arbitrary set Q of combinations of tiles, but instead by the requirement that the colors of abutting edges match, where a tile is uniquely determined by the colors of its four edges. It is a fairly simple maneuver to translate Wang's domino problems into our tiling problems and vice versa, but since the two approaches do not coincide exactly, we prefer to avoid the term "domino". The last section of this chapter (p. 14) shows how the translations may be carried out.

To return to tiling problems in general, let $\Theta = (\mathbb{X}, R)$ be a tiling type and let $\mathscr{S} = (T, Q)$ be a Θ-system. Note that whether $\mathscr{S}$ accepts a T-tiling τ depends on R, which is not specified as part of $\mathscr{S}$. That is, $\mathscr{S}$ accepts (or does not accept) τ as a Θ-system; if the same pair $\mathscr{S}$ is regarded as a system of some other type, it might accept different tilings. Since confusion will rarely arise in practice, we generally omit the qualifying phrase.

Note that as long as the space $\mathbb{X}$ is recursively enumerable and the spatial relation R is recursive, the class of $(\mathbb{X}, R)$-systems not accepting any tiling is recursively enumerable. For one can always generate larger and larger finite portions of $\mathbb{X}$ and exhaustively check the finite sets of sequences satisfying R for the existence of an acceptable partial tiling; if there is no acceptable tiling of all of $\mathbb{X}$, this fact will be discovered at some finite stage. As these conditions on $\mathbb{X}$ and R are met in all the cases discussed below, the proofs that the Turing machine halting problem is reducible to these problems suffice to show that these problems are 1-complete.

Next we slightly generalize our definitions. When $\Theta = (\mathbb{X}, R)$ is a tiling type and $\mathscr{S} = (T, Q)$ is a Θ-system, it often happens in practice that R and Q are the Cartesian products of other relations. Thus suppose there are an $\ell \geq 1$, numbers $k_1, \ldots, k_\ell \geq 1$, and relations $R_i \subset \mathbb{X}^{k_i}$ and $Q_i \subset T^{k_i}$ $(1 \leq i \leq \ell)$ such that $R = R_1 \times \cdots \times R_\ell$ and $Q = Q_1 \times \cdots \times Q_\ell$. Then we also refer to the $(\ell+1)$-tuple $(\mathbb{X}, R_1, \ldots, R_\ell)$ as a tiling type and $(T, Q_1, \ldots, Q_\ell)$ as a system of this type; the definition of acceptance is the same as for Θ and $\mathscr{S}$. The next section treats a problem of this kind.

IB.2 Origin-Constrained Tiling Problem

In this problem the space to be tiled is the first quadrant of the plane, or to be precise, the set of lattice points of the first quadrant. The spatial relation is decomposed into three parts, which pick out (1) the origin; (2) points horizontally adjacent to each other; and (3) points vertically adjacent to each other. The rules of the game are thus to put copies of a finite number of given prototype tiles down on the lattice points of the first quadrant so as to observe certain restrictions about which tiles may appear at the origin and which may abut each other horizontally or vertically.

Formally, let Δ_0 be the tiling type $(\mathbb{N}^2, R_0, R_H, R_V)$, where

$$R_0(\langle x, y \rangle) \qquad \text{if and only if} \quad \langle x, y \rangle = \langle 0, 0 \rangle$$

$$R_H(\langle x, y\rangle, \langle x', y'\rangle) \quad \text{if and only if} \quad \langle x', y'\rangle = \langle x, y\rangle + \langle 1, 0\rangle$$

$$R_V(\langle x, y\rangle, \langle x', y'\rangle) \quad \text{if and only if} \quad \langle x', y'\rangle = \langle x, y\rangle + \langle 0, 1\rangle \quad .$$

Thus a Δ_0-system is a quadruple $\mathscr{D} = (D, D_0, H, V)$, where D is a finite set, $D_0 \subset D$, and $H, V \subset D^2$. A D-tiling $\tau\colon \mathbb{N}^2 \to D$ is accepted by $\mathscr{D}$ if and only if $\tau(0,0) \in D_0$ and $\langle \tau(i,j), \tau(i+1, j)\rangle \in H$, $\langle \tau(i,j), \tau(i, j+1)\rangle \in V$ for each $i, j \in \mathbb{N}$. The <u>origin-constrained tiling problem</u> is the tiling problem for Δ_0-systems.

<u>Origin-Constrained Tiling Theorem</u>. The origin-constrained tiling problem is unsolvable. Moreover, it is unsolvable even for origin-constrained tiling systems (D, D_0, H, V) such that $|D_0| = 1$, i.e., such that the tile that may appear at the origin is unique.

<u>Proof.</u> We reduce the first version of the Turing machine halting problem, i.e., the halting problem for inputless Turing machines. The basic idea is for a tiling to correspond to a two-dimensional presentation of an infinite computation. The horizontal dimension is the space dimension; each row contains one instantaneous description, extended infinitely to the right with blank symbols. The vertical dimension represents time: successive rows contain successive instantaneous descriptions. As in the construction of instantaneous descriptions, a tile represents either a symbol or a state-symbol pair. For example, Figure 1 shows a portion of the tiling that would correspond to a Turing machine that starts in state q_0 on the blank tape, on the first step writes symbol a_2, moves its head right, and enters state q_1, and on the second step writes symbol a_1, moves its head left, and enters state q_3.

Formally, the reduction of the halting problem is easiest achieved in two steps (plus a third step to establish the stronger form of the theorem). We first reduce the halting problem to a special tiling problem in which the positioning of diagonally adjacent tiles is restricted, as well as that of vertically adjacent tiles, horizontally adjacent tiles, and the tile at the origin. We then show how the diagonal restrictions can be encoded into the horizontal and vertical restrictions.

Thus we consider the tiling type $\Delta_X = (\mathbb{N}^2, R_0, R_H, R_V, R_{X_1}, R_{X_{-1}})$, where R_0, R_H, R_V are as above and for $m = \pm 1$,

$$R_{X_m}(\langle x, y\rangle, \langle x', y'\rangle) \quad \text{if and only if} \quad \langle x', y'\rangle = \langle x, y\rangle + \langle m, 1\rangle \quad .$$

Thus, a Δ_X-system is a 6-tuple $(D, D_0, H, V, X_1, X_{-1})$, with D a finite set, $D_0 \subset D$, and $H, V, X_1, X_{-1} \subset D^2$.

To show that the tiling problem for Δ_X-systems is unsolvable, let $\mathscr{M} = (K, \Sigma, B, q_0, \delta)$ be any Turing machine, and construct a Δ_X-system $\mathscr{D}_{\mathscr{M}} = (D, D_0, H, V, X_1, X_{-1})$ as follows. D, the set of tiles, is $\Sigma \cup (K \times \Sigma)$, the set of symbols from which instantaneous descriptions of $\mathscr{M}$ are constructed.

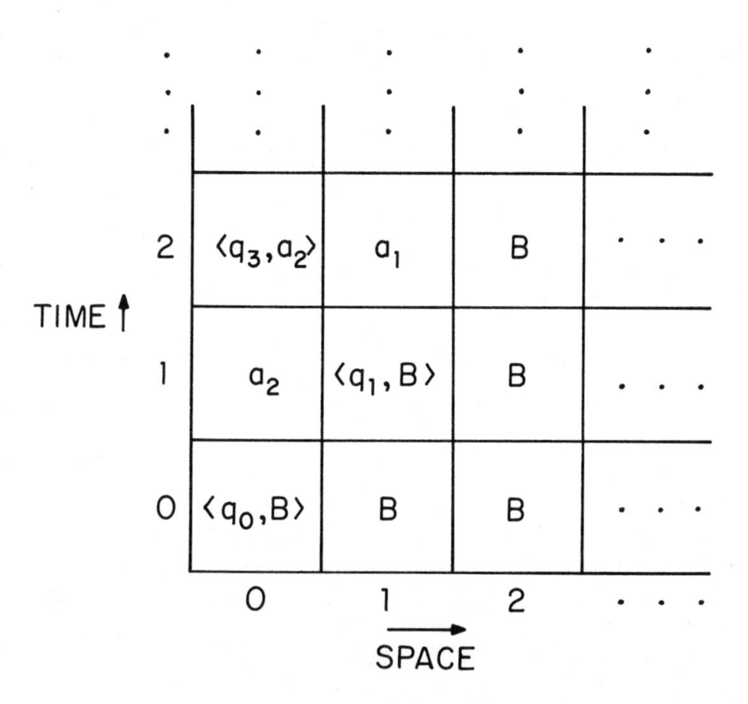

Figure 1

D_0, the set of tiles that may appear at the origin, is $\{\langle q_0, B\rangle\}$, indicating that $\mathcal{M}$ begins at the first instant of time in state q_0 scanning a blank on the leftmost tape square. H, V, X_1 and X_{-1} provide for the following:

(1) The tile to the right of one containing a blank also contains a blank. (This is where it is handy to assume that $\mathcal{M}$ cannot write B.) Hence the first row of an acceptable tiling consists entirely of blank squares.

(2) The tile above one not containing a state-symbol pair (indicating the head position) contains the same symbol as the tile. (The tile above may also contain a state, i.e., it may be a state-symbol pair.) Thus, unscanned squares do not change.

(3) The tiles directly above, and diagonally above to the left and right, a tile containing a state-symbol pair, require special attention to force the rewriting of the scanned symbol and the change of state to be represented correctly.

Formally,

H is the set $D \times D$, except that the tile to the right of one containing the blank must also contain the blank. That is, let $\alpha(\Sigma') = \Sigma' \cup (K \times \Sigma')$, for any $\Sigma' \subset \Sigma$; then

$$H = \alpha(\Sigma - \{B\}) \times \alpha(\Sigma) \cup \alpha(\{B\}) \times \alpha(\{B\})$$

$$V = \{\langle a, a\rangle \mid a \in \Sigma\} \cup \{\langle a, \langle q, a\rangle\rangle \mid a \in \Sigma,\ q \in K\}$$

(symbols that are not scanned do not change)

$$\cup\ \{\langle\langle q, a\rangle, \langle q', a'\rangle\rangle \mid \delta(q, a) = \langle q', a', 0\rangle\}$$

$$\cup\ \{\langle\langle q, a\rangle, a'\rangle \mid \delta(q, a) = \langle q', a', \pm 1\rangle\}$$

(symbols that are scanned are rewritten).

For $m = \pm 1$,

$$X_m = \Sigma \times \Sigma$$

$$\cup\ \{\langle\langle q, a\rangle, \langle q', a'\rangle\rangle \mid \delta(q, a) = \langle q', a'', m\rangle \text{ for some } a', a'' \in \Sigma\}$$

$$\cup\ \{\langle\langle q, a\rangle, a'\rangle \mid \delta(q, a) = \langle q', a'', m'\rangle \text{ for some } a', a'' \in \Sigma,\ m' \in \{0, -m\}\}.$$

If $\mathcal{M}$ does not halt, then let $I_0, I_1, \ldots$ be the sequence of instantaneous descriptions of $\mathcal{M}$ such that $I_i \vdash_{\mathcal{M}} I_{i+1}$ for each i, and let I'_i be for each i the one-way infinite sequence obtained by extending I_i infinitely to the right with blanks. Then the mapping

$$\tau(i, j) = i^{th} \text{ symbol of } I'_j$$

is a tiling of $\mathbb{N}^2$ accepted by $\mathcal{D}_{\mathcal{M}}$.

On the other hand, suppose there is an accepted tiling τ and let I'_j be the jth row of τ, i. e., the one-way infinite sequence whose i^{th} element is $\tau(i, j)$. Then the following may be shown by induction on j:

> For each j, there is an instantaneous description I_j of $\mathcal{M}$ such that $\langle q_0, B\rangle \vdash^*_{\mathcal{M}} I_j$, and I'_j is formed from I_j by extending infinitely to the right with blanks and then (possibly) replacing certain occurrences of symbols $a \in \Sigma$ by pairs $\langle q, a\rangle$ with $q \in K$.

But then there are instantaneous descriptions I_j such that $\langle q_0, B\rangle \vdash^j_{\mathcal{M}} I_j$ for arbitrarily large j, so $\mathcal{M}$ does not halt. (The possibility of more than one state-symbol pair occurring in the same row of τ can be excluded by refining the definition of $\mathcal{D}_{\mathcal{M}}$. We have chosen not to do so, since the outcome is the same with the simpler construction.)

Thus the tiling problem for systems of type Δ_X is unsolvable. To prove it unsolvable for Δ_0-systems as well, we must build into the tiles the capacity to communicate information between their diagonally adjacent neighbors on, say, the bottom and right, and the right and top. So let $\mathscr{D} = (D, D_0, H, V, X_1, X_{-1})$ be a Δ_X-system, and let d be a "dummy" tile not in D. Tile d plays the role of a tile that might have appeared in the row below the first, had this row been tiled. Let $X_1' = X_1 \cup \{d\} \times D$. We construct a Δ_0-system with tiles

$$D^* = \{\langle t, u, v, w\rangle \mid t \in D, \quad u \in D \cup \{d\}, \quad v, w \in D,$$

$$\text{and } \langle u, v\rangle \in X_1', \quad \langle v, w\rangle \in X_{-1}\} \quad .$$

Tile $\langle t, u, v, w\rangle$ represents tile t, together with the information that tile u is below it, v to its right, and w above it. Then let $\mathscr{D}^* = (D^*, D_0^*, H^*, V^*)$ where

$$D_0^* = D^* \cap (D_0 \times (D \cup \{d\}) \times D \times D)$$

$$\langle\langle t, u, v, w\rangle, \langle t', u', v', w'\rangle\rangle \in H^* \quad \text{if and only if} \quad \langle t, t'\rangle \in H \quad \text{and} \quad v = t'$$

$$\langle\langle t, u, v, w\rangle, \langle t', u', v', w'\rangle\rangle \in V^* \quad \text{if and only if} \quad \langle t, t'\rangle \in V$$

$$\text{and} \quad w = t', \quad t = u' .$$

Then if τ^* is any tiling of $\mathbb{N}^2$ accepted by $\mathscr{D}^*$, a tiling τ of $\mathbb{N}^2$ accepted by $\mathscr{D}$ may be obtained by taking the first component of each tile. Conversely, if τ is a tiling accepted by $\mathscr{D}$ then a tiling τ^* accepted by $\mathscr{D}^*$ may be obtained by adding as additional components the lower, righthand, and upper neighbors of each tile (taking d to be the lower neighbor of each tile in the bottom row). Figure 2 shows a portion of such a pair of tilings τ, τ^*.

This completes the proof of the unsolvability of the origin-constrained tiling problem; it remains to prove the stronger version in which D_0, the set of tiles that may appear at the origin, has only one member. Now clearly, there is a tiling accepted by (D, D_0, H, V) if and only if there is a tiling accepted by $(D, \{t\}, H, V)$ for some $t \in D_0$, so the stronger form of the problem is unsolvable. This proof, however, has the technical disadvantage of appealing to a weaker notion of reducibility (namely, truth-table reducibility) than is used in any of the proofs presented here (all of which involve only 1-reducibility). So we give an independent proof involving only 1-reducibility.

The idea is to replace the Δ_0-system $\mathscr{D} = (D, D_0, H, V)$ by a Δ_0-system $\mathscr{D}' = (D', \{t_0\}, H', V')$, such that a tiling is accepted by $\mathscr{D}$ if and only if the result of translating that tiling upwards by one row, and filling in the bottom row in a uniquely determined way, is accepted by $\mathscr{D}'$ (see Figure 3). Thus $D' = D \cup \{t_0, t_1\}$, where t_0

and t_1 are two new tiles; t_0 is to appear at position (0, 0) only, t_1 at all positions (n, 0), for $n > 0$. Next, $H' = H \cup \{\langle t_0, t_1\rangle, \langle t_1, t_1\rangle\}$; only t_1 can appear to the right of t_0, and only t_1 to the right of t_1, so that the first row of any tiling accepted by $\mathscr{D}'$ is $t_0, t_1, t_1, \ldots$ as shown. Finally $V' = (\{t_0\} \times D_0) \cup (\{t_1\} \times D) \cup V$; the tile above t_0 must be a member of D_0, and t_0 and t_1 may appear only in the first row. Then it is easy to check that $\mathscr{D}$ and $\mathscr{D}'$ are related as claimed, i. e. :

if τ is accepted by $\mathscr{D}$ then τ' is accepted by $\mathscr{D}'$, where

$$\tau'(0, 0) = t_0; \quad \tau'(n+1, 0) = t_1 \qquad \text{for} \quad n \geq 0;$$

and

$$\tau'(n, m+1) = \tau(n, m) \qquad \text{for} \quad n, m \geq 0;$$

and

if τ' is accepted by $\mathscr{D}'$ then τ is accepted by $\mathscr{D}$, where $\tau(n, m) = \tau'(n, m+1)$ for $n, m \geq 0$.

This completes the proof of the Origin-Constrained Tiling Theorem. ∎

t	v	u	
u	v	t	v
v	u	v	u
t	u	t	t

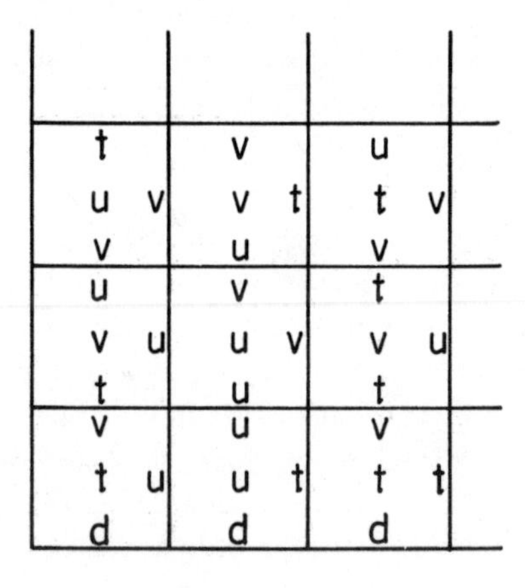

w t v u

$= \langle t, u, v, w\rangle$

Figure 2

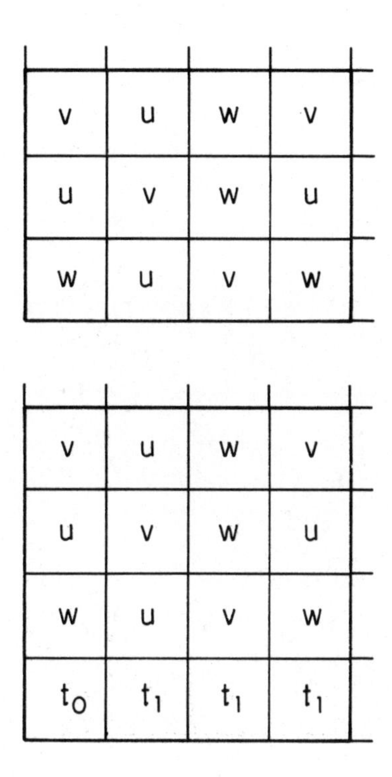

Figure 3

IB. 3 Planar Tiling Problems and Domino Problems

We have avoided the term "domino" coined by Wang to describe certain planar tiling problems because the correspondence between our formalism and his is not obvious. In this appendix we show that the two types of problems are indeed inter-translatable.

We consider only the origin-constrained domino problem, which corresponds to our origin-constrained tiling problem. This problem can be phrased as follows: Given a finite set C (of colors), a finite set $D \subset C^4$ (of dominoes) and a domino $t_0 \in D$ (the origin domino), determine whether there is a mapping $\tau: \mathbb{N} \times \mathbb{N} \to D$ (a tiling of the plane) such that the colors of abutting edges are the same, i. e., such that $\tau(0,0) = t_0$ and for all $m, n \in \mathbb{N}$, $\pi_2(\tau(m,n)) = \pi_4(\tau(m+1,n))$ and $\pi_1(\tau(m,n)) = \pi_3(\tau(m,n+1))$. (The π_i are the projection functions; the bottom, left, top, and right edges of domino $t \in D$ are its first, second, third, fourth components, respectively.)

Clearly the origin-constrained domino problem is reducible to the tiling problem for Δ_0-systems; one need merely consider $(D, \{t_0\}, H, V)$, where $H = \{\langle t_1, t_2\rangle \mid \pi_2(t_1) = \pi_4(t_2)\}$ and $V = \{\langle t_1, t_2\rangle \mid \pi_1(t_1) = \pi_3(t_2)\}$. It is the reverse reduction we wish to point out here. That is, the problem is to construct from a Δ_0-system (D, D_0, H, V) a corresponding set of dominoes with a designated origin domino. To carry through the construction it is convenient to make a few preliminary assumptions. First, assume that D_0 contains only one tile, say $D_0 = \{t_0\}$; if $|D_0| > 1$ then by introducing two new tiles as on p. 12 the number of origin tiles may be reduced to one. Also, assume of the origin tile t_0 that only one tile may appear to its right and only one above it; that is, there are unique tiles t_{0H}, t_{0V} such that $\langle t_0, t_{0H}\rangle \in H$ and $\langle t_0, t_{0V}\rangle \in V$. If this is not the case already, it can be effected by carrying out a second time the construction on p. 12 (note that t_0 as constructed there has a unique righthand neighbor t_1). Finally, assume that there is a "dummy" tile $d \in D$ such that $\langle d, t\rangle \in H$ and $\langle d, t\rangle \in V$ for each $t \in D$; we imagine d to occupy squares in the row below the first row and the column to the left of the first column of an accepted tiling of the first quadrant. If no such d is in D, one can be introduced without changing the set of tilings accepted by D by adding $\langle d, t\rangle$ (but not $\langle t, d\rangle$) to H and to V, for each $t \in D$.

The trick in the construction of the domino set D' is to color each edge of a domino so as to encode both the tile corresponding to that domino, and also the other tile adjacent to that edge. That is, let the set of colors be

$$C = D \times D \quad ;$$

let the set of dominoes be

$$D' = \{\langle\langle t_1, t\rangle, \langle t_2, t\rangle, \langle t, t_3\rangle, \langle t, t_4\rangle\rangle \mid t, t_1, t_2, t_3, t_4 \in D,$$

$$\langle t_1, t\rangle \text{ and } \langle t, t_3\rangle \in V, \quad \langle t_2, t\rangle \text{ and } \langle t, t_4\rangle \in H\} \ ;$$

and let the origin domino be

$$t_0' = \langle\langle d, t_0\rangle, \langle d, t_0\rangle, \langle t_0, t_{0V}\rangle, \langle t_0, t_{0H}\rangle\rangle \ ,$$

where t_0 is the origin tile, t_{0H} and t_{0V} are its unique horizontal and vertical neighbors, and d is the dummy tile.

Now if τ is a tiling of $\mathbb{N}^2$ accepted by $\mathscr{D}$, then extend τ to a mapping

$$\tau\colon (\mathbb{N} \cup \{-1\})^2 \to D$$

by letting $\tau(-1, n) = \tau(n, -1) = d$ for each $n \in \mathbb{N} \cup \{-1\}$. Then a tiling of the first quadrant consistent with the domino constraints is given by

$$\begin{aligned} \tau'(n,m) = \langle&\langle \tau(n,m-1), \tau(n,m)\rangle, \\ &\langle \tau(n-1,m), \tau(n,m)\rangle, \\ &\langle \tau(n,m), \tau(n,m+1)\rangle, \\ &\langle \tau(n,m), \tau(n+1,m)\rangle\rangle \end{aligned}$$

for each $n, m \in \mathbb{N}$. Conversely, if τ' is a tiling of the first quadrant of the plane with the dominoes, then the fact that the adjacency conditions are satisfied makes it possible to express τ' in terms of a mapping τ as shown, and then the construction of D' guarantees that τ is accepted by $\mathscr{D}$. Figure 4 illustrates a portion of a pair of correlated tilings.

Similar constructions work for the unconstrained problems treated in Chapter IE.

Historical References. Dominoes were introduced by Wang (1961, 1962). Our reduction of the Turing machine halting problem to the origin-constrained tiling problem is derived from Büchi's proof (1962) of the unsolvability of the $\exists \wedge \forall\exists\forall$ case of the decision problem for first-order logic. See also the references after Chapter IE.

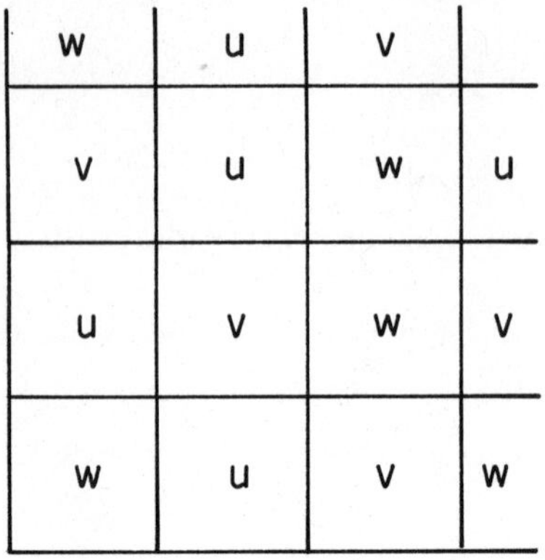

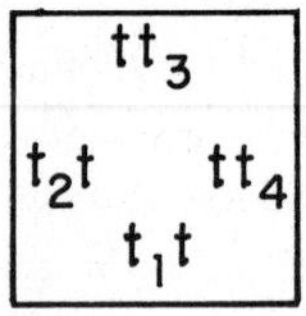

w	u	v	
v	u	w	u
u	v	w	v
w	u	v	w

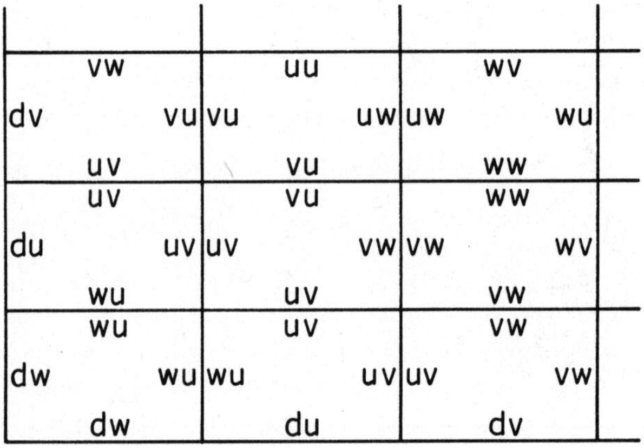

vw dv vu uv	uu vu uw vu	wv uw wu ww
uv du uv wu	vu uv vw uv	ww vw wv vw
wu dw wu dw	uv wu uv du	vw uv vw dv

tt_3
$t_2t \quad tt_4$
t_1t

$$= \langle \langle t_1, t\rangle, \langle t_2, t\rangle, \langle t, t_3\rangle, \langle t, t_4\rangle \rangle$$

Figure 4

Chapter IC

LINEAR SAMPLING PROBLEM, FIRST VERSION

This chapter and the next treat tiling problems in which the space to be tiled is $\mathbb{Z}$ (the set of integers) or several disjoint copies of $\mathbb{Z}$. Moreover the spatial relations that determine which points of the space can be simultaneously "inspected" or "sampled" are defined via linear equations. We therefore call these tiling problems <u>linear sampling problems</u>.

In the first version of the linear sampling problem the space to be tiled is

$$2\mathbb{Z} = \mathbb{Z} \times \{1,2\}$$

i.e., two disjoint copies of the integers. Thus a tiling may be viewed as a pair of two-way infinite <u>tapes</u>, each tape divided into <u>cells</u> in which are recorded symbols from a finite alphabet. Each tape cell has a numerical <u>coordinate</u>.

We are actually concerned with a class of tiling types $\Lambda_1^{(\theta)}$ for various natural numbers $\theta \geq 2$. Here $\Lambda_1^{(\theta)} = (2\mathbb{Z}, R_0^{(\theta)}, R_1)$, where $R_0^{(\theta)} \subset (2\mathbb{Z})^\theta$, $R_1 \subset (2\mathbb{Z})^4$, and

$$R_0^{(\theta)}(\langle n_1, e_1\rangle, \dots, \langle n_\theta, e_\theta\rangle) \quad \text{if and only if} \quad e_1 = e_2 = \dots = e_\theta$$
$$\text{and} \quad n_j = n_1 + j - 1 \quad (j = 2, \dots, \theta),$$

$$R_1(\langle n_1, e_1\rangle, \dots, \langle n_4, e_4\rangle) \quad \text{if and only if} \quad e_1 = e_3 = 1, \; e_2 = e_4 = 2,$$
$$\text{and} \quad n_2 - n_1 = n_4 - n_3 .$$

Thus $R_0^{(\theta)}$ relates a series of θ consecutive cells on the same tape, and R_1 a set of four cells, two from each tape, whose coordinates satisfy the given linear equation. Let us call the <u>distance</u> between two tape cells the difference of their coordinates (even if they are on different tapes). Then the restriction imposed by R_1 is that of the four cells simultaneously inspected, the distance between the cells in the first pair (one from each tape) is the same as the distance between the cells in the second pair (one from each tape).

A $\Lambda_1^{(\theta)}$-system is thus a triple $\mathscr{S} = (T, L, G)$, where T is a finite set, $L \subset T^\theta$, and $G \subset T^4$. The constraint on tilings imposed by $R_0^{(\theta)}$ and L is called the <u>local</u> condition, since $R_0^{(\theta)}$ permits direct comparison of tape cells of the same tape distant from each other by at most $\theta - 1$. The constraint imposed by R_1 and G is called the <u>global</u> condition, since it permits simultaneous inspection of cells at arbitrarily large distances from each other on different tapes.

IC. 1 Plan

The goal is to prove:

Linear Sampling Theorem, First Version. The tiling problem for $\Lambda_1^{(\theta)}$-systems is unsolvable for each $\theta \geq 2$.

We prove this theorem by proving the unsolvability of the tiling problem for $\Lambda_1^{(\theta)}$-systems for a certain θ (in fact, $\theta = 4$). Before proceeding to the main argument we note that this result will yield the unsolvability for each $\theta \geq 2$. Clearly it will suffice for the theorem to show the unsolvability of the tiling problem for $\Lambda_1^{(2)}$-systems. So suppose that the tiling problem for $\Lambda_1^{(\theta)}$-systems is unsolvable for some $\theta \geq 2$, and let $\mathscr{S} = (T, L, G)$ be any $\Lambda_1^{(\theta)}$-system. We construct a $\Lambda_1^{(2)}$-system $\mathscr{S}'$ corresponding to $\mathscr{S}$ by "blocking together" overlapping runs of θ consecutive tape squares. That is, let $\mathscr{S}' = (T', L', G')$, where $T' = T^{\theta}$, and

$\langle\langle s_1, \ldots, s_\theta\rangle, \langle s_{\theta+1}, \ldots, s_{2\theta}\rangle\rangle \in L'$ if and only if $s_{\theta+i} = s_{i+1}$ for $i = 1, \ldots, \theta-1$, and $\langle s_1, \ldots, s_\theta\rangle, \langle s_{\theta+1}, \ldots, s_{2\theta}\rangle \in L$;

$\langle\langle s_1, \ldots, s_\theta\rangle, \ldots, \langle s_{3\theta+1}, \ldots, s_{4\theta}\rangle\rangle \in G'$ if and only if $\langle s_1, s_{\theta+1}, s_{2\theta+1}, s_{3\theta+1}\rangle \in G$.

Then there is a tiling accepted by $\mathscr{S}'$ if and only if there is a tiling accepted by $\mathscr{S}$. Figure 5 illustrates a correlated pair of tilings of $2\mathbb{Z}$, for the case $\theta = 3$.

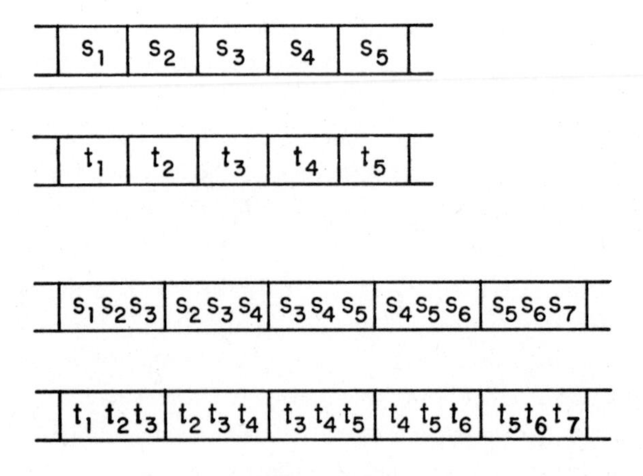

Figure 5

So we shall be content to prove the unsolvability of the tiling problem for $\Lambda_1^{(\theta)}$-systems for a certain $\theta \geq 2$. When the time comes to use the fact that $\theta = 4$ we shall mention it; but until then we simplify our notation by writing Λ_1 for $\Lambda_1^{(\theta)}$ and R_0 for $R_0^{(\theta)}$.

The basic plan is to reduce the tiling problem for Δ_0-systems, the origin-constrained tiling problem, to the tiling problem for Λ_1-systems. Thus we construct, for each Δ_0-system $\mathscr{D}$, a Λ_1-system $\mathscr{S}_{\mathscr{D}}$ such that $\mathscr{S}_{\mathscr{D}}$ accepts some tiling of $2\mathbb{Z}$ if and only if $\mathscr{D}$ accepts some tiling of $\mathbb{N}^2$. So far the reduction looks not unlike those of Chapter IB, but there are important differences, which we now discuss both to illuminate the general structure of the problem and to lay the groundwork for certain formal arguments that appear later.

The crucial difference between the kinds of combinatorial systems studied in Chapters IA and IB and the kinds studied here and in Chapters ID and IE is the <u>lack of a fixed point of reference</u> in the space to be tiled. In Chapter IB the origin is used to set an initial condition in tilings of $\mathbb{N}^2$, from which properties of parts of the tiling distant from the origin follow by induction. Here the space to be tiled is homogeneous; all points are treated alike, and so no point can play this special role. It will prove possible to use the local condition to encode the origin constraint, but only if the encoding of tilings of $\mathbb{N}^2$ into tilings of $2\mathbb{Z}$ is uniform, with no point of $2\mathbb{Z}$ intrinsically different from any other.

To be specific, one difference between the tilings accepted by a Λ_1-system and those accepted by a Δ_0-system is that a tiling of $2\mathbb{Z}$ accepted by a Λ_1-system can be shifted without spoiling its acceptability. That is, if α is a tiling of $2\mathbb{Z}$, call a <u>translation</u> of α any mapping α' such that for some $t_1, t_2 \in \mathbb{Z}$, $\alpha'(n, c) = \alpha(n+t_c, c)$ for $c = 1, 2$. Intuitively, a translation is obtained by <u>independently</u> shifting the two tapes.

<u>Translation Lemma.</u> If $\mathscr{S}$ is a Λ_1-system then any translation of a tiling accepted by $\mathscr{S}$ is also accepted by $\mathscr{S}$.

<u>Proof</u>. Immediate by inspection of the linear equations defining Λ_1. ■

Because of the Translation Lemma, in order to recover a tiling of $\mathbb{N}^2$ accepted by a Δ_0-system $\mathscr{D}$ from a tiling α of $2\mathbb{Z}$ accepted by the associated Λ_1-system $\mathscr{S}$, it will in general be necessary to discover a private, internal coordinate system for α, and to shift α so as to bring it into a "normal" alignment with the standard coordinate system, so that a direct correspondence between $\mathbb{N}^2$ and $2\mathbb{Z}$ can be established. Indeed, the exact relation between tilings of $\mathbb{N}^2$ and of $2\mathbb{Z}$ is more complicated than this: even in the normal alignment, we cannot hope for points of $2\mathbb{Z}$ to correspond to points of $\mathbb{N}^2$ by means of any simple 1-1 pairing function. For we now show that, because of the Infinity Lemma, the encoding of $\mathbb{N}^2$ in $2\mathbb{Z}$ must be such that each point of $\mathbb{N}^2$ is represented in $2\mathbb{Z}$

infinitely often and at bounded intervals along at least one of the tapes. Although we do not formalize this fact, we prove a lemma and then argue informally. The lemma will play an important role later on, in the proof of the Normalization Lemma.

First we need some definitions.

A <u>segment</u> of a tiling α of $2\mathbb{Z}$ is a pair of sequences $\underset{\sim}{\sigma} = \langle \sigma_1, \sigma_2 \rangle$, where for some k, t_1, and t_2, $\sigma_c = \langle \alpha(t_c, c), \alpha(t_c + 1, c), \ldots, \alpha(t_c + k - 1, c) \rangle$ for $c = 1, 2$. The <u>length</u> of the segment $\underset{\sim}{\sigma}$ is k, and it is a <u>central</u> segment of α if k is odd and $t_1 = t_2 = -(k+1)/2$ (so that the elements in the middle of σ_1 and σ_2 are $\alpha(0, 1)$ and $\alpha(0, 2)$). If $\langle \sigma_1, \sigma_2 \rangle$ is a pair of sequences, both of odd length k, then a <u>central subpair</u> is a pair of sequences formed by taking the middle ℓ elements of σ_1 and σ_2 for some odd $\ell < k$.

<u>Central Segment Lemma</u>. Let $\mathscr{S}$ be a Λ_1-system, let α be a tiling of $2\mathbb{Z}$ accepted by $\mathscr{S}$, and let Ξ be an infinite set of segments of α, each of odd length. Then there is a tiling accepted by $\mathscr{S}$, each of whose central segments is a central subpair of a segment in Ξ.

<u>Proof</u>. This lemma is a consequence of the Translation Lemma and the Infinity Lemma. Since Ξ is infinite there is an infinite sequence $\underset{\sim}{\sigma}_1, \underset{\sim}{\sigma}_2, \ldots$ of central subpairs of members of Ξ such that, for each i, $\underset{\sim}{\sigma}_i$ is a central subpair of $\underset{\sim}{\sigma}_{i+1}$. By the Infinity Lemma there is a tiling β such that each central segment of β is a central subpair of some $\underset{\sim}{\sigma}_i$. We need only show that $\mathscr{S}$ accepts β. But this follows immediately from the Translation Lemma and a kind of "compactness" for tiling problems: if arbitrarily large finite segments of β meet the acceptance conditions, then β is accepted.

This completes the proof of the Central Segment Lemma. ∎

To complete the argument about the embedding of $\mathbb{N}^2$ in $2\mathbb{Z}$: Suppose τ is a tiling of $\mathbb{N}^2$ accepted by $\mathscr{D}$, and α is a corresponding tiling of $2\mathbb{Z}$ accepted by $\mathscr{S}_{\mathscr{D}}$. Suppose, moreover, that for some $\langle i, j \rangle \in \mathbb{N}^2$, the points of $2\mathbb{Z}$ tiled with representations of the tile at position $\langle i, j \rangle$ of τ are separated by arbitrarily large intervals on both tapes. By collecting the intervening segments of α and applying the Central Segment Lemma we could construct a tiling accepted by $\mathscr{S}_{\mathscr{D}}$ with no representation of $\tau(i, j)$ at all.

With this much background we can describe the structure of $\mathscr{S}_{\mathscr{D}}$ and the tilings accepted by it in somewhat greater detail. Let $\mathscr{D} = (D, D_0, A_1, A_2)$ and $\mathscr{S}_{\mathscr{D}} = (T, L, G)$. (It will prove notationally convenient to use here subscripted variables A_1, A_2 rather than H and V, as were used in Chapter IB.) The set T is the threefold Cartesian product of two special sets Σ_{p_1} and Σ_{p_2} and the set D of tiles. Intuitively, the tapes of the Λ_1-system are ruled laterally into channels: two address channels, whose function we describe later, and a data channel, which contains tiles of the Δ_0-system being represented (Figure 6).

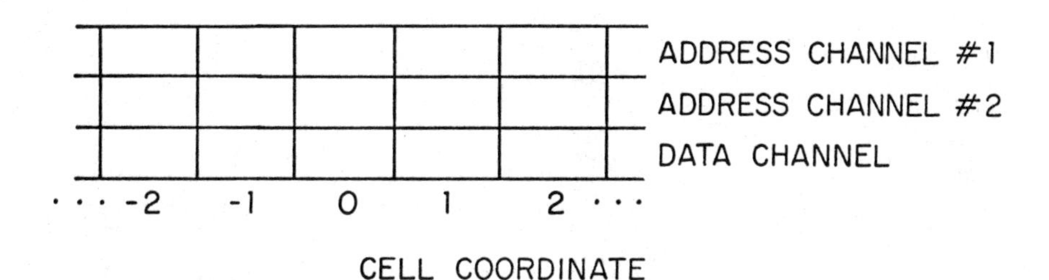

Figure 6

We are ready to consider how the tiles of $\mathscr{D}$ are distributed in the data channels, i.e., how the space $\mathbb{N}^2$ is correlated with $2\mathbb{Z}$. We need to introduce a special function:

> For any $p > 1$, if p^i divides n but p^{i+1} does not, then $\Omega_p(n) = i$. Also, let $\Omega_p(0) = \infty$, so that Ω_p maps $\mathbb{Z}$ onto $\mathbb{N} \cup \{\infty\}$. $\Omega_p(n)$ is the p-order of n.

Now if p_1 and p_2 are relatively prime natural numbers, then for each pair $\langle i,j\rangle \in \mathbb{N}^2$ there are infinitely many $n \in \mathbb{Z}$ with p_1-order i and p_2-order j. The essence of the encoding is to place the tile that would appear at position $\langle i,j\rangle \in \mathbb{N}^2$ in the data channel of any tape cell whose coordinate n satisfies $\Omega_{p_1}(n) = i$, $\Omega_{p_2}(n) = j$. Note that this encoding has the kind of "denseness" needed to avoid the consequences of the Central Segment Lemma.

For convenience, we sometimes refer to the "p_1-order of a tape cell" (or its p_2-order) instead of the p_1- (or p_2-) order of its coordinate.

To carry through the plan of distributing tiles in tape cells in this way, three things must be ensured:

Origin Constraint: every tape cell that has p_1-order 0 and p_2-order 0 contains in the data channel a tile of $\mathscr{D}$ that may occur at the origin;

Consistency Constraint: if two tape cells have the same p_1- and p_2-orders, then they contain in the data channel the same tile of $\mathscr{D}$;

Adjacency Constraint: two tape cells whose p_1- and p_2-orders are i,j and $i+1,j$ (or i,j and $i,j+1$) contain in the data channel two tiles that may be adjacent horizontally (or vertically).

To ensure that these three constraints are satisfied we record in the tape cells not only the tiles of $\mathscr{D}$ but also some information about the relative positions of those tape cells. Thus the first two channels of the tapes are called address channels because the information recorded in them is to help in determining the coordinates or "addresses" of tape cells. Of course, all the coordinates cannot be written out in full since they are unbounded in magnitude. Instead, the address channels help to determine relations between the orders of cells inspected at the same time. Specifically, the first address channel is for p_1-orders and the second is for p_2-orders; and for $i = 1$ or 2, the i-th address channel can be used to determine

(Z) whether the p_i-order of a cell coordinate is 0;

(E) whether the p_i-orders of the coordinates of two tape cells are equal;

(I) whether the p_i-order of one cell is one greater than that of another cell.

These three capabilities can be combined to ensure that the consistency, origin, and adjacency constraints are satisfied. Thus (Z) is used for $i = 1$ and 2 to yield the origin constraint; (E) for $i = 1$ and 2 for the consistency constraint; and (E) (for i) and (I) (for 3-i) to yield the adjacency constraint (horizontal for $i = 2$, vertical for $i = 1$). (Here p_1 and p_2 are some "sufficiently large" relatively prime numbers, and $p_1 < p_2$. Suitable values for p_1 and p_2 are 10 and 11.)

Thus if a pair of tapes correctly encodes some tiling of $\mathbb{N}^2$, then the Λ_1-system $\mathscr{S}_{\mathscr{D}}$ uses the information in the address channels to identify tape cells corresponding to the origin or to adjacent points in the first quadrant, so that the information in the data channel can be checked to ensure that the encoded tiling is one accepted by the Δ_0-system $\mathscr{D}$. There are two problems with this general view. The first is that, as suggested earlier, Λ_1-systems are too weak to accept only those tilings of $2\mathbb{Z}$ which are such tidy encodings of tilings of $\mathbb{N}^2$; for example any translation of an accepted tiling is accepted. We shall define two kinds of tilings, the normally and the admissibly addressed; the former are a subclass of the latter. Every tiling of $2\mathbb{Z}$ accepted by $\mathscr{S}_{\mathscr{D}}$ is admissibly addressed, whereas the preceding discussion applies only to normally addressed tilings (these are the ones referred to earlier as having a "normal" alignment). The gap between the two is bridged by the Normalization Lemma (p. 30), which will enable us to construct, from any admissibly addressed tiling of $2\mathbb{Z}$ accepted by $\mathscr{S}_{\mathscr{D}}$, a normally addressed tiling of $2\mathbb{Z}$ accepted by $\mathscr{S}_{\mathscr{D}}$; the argument just sketched will then show that this normally addressed tiling contains an encoded description of a tiling of $\mathbb{N}^2$ accepted by $\mathscr{D}$.

The other point is more troublesome. We have explained, in general terms, how the contents of the data channels of a tiling of $2\mathbb{Z}$ accepted by $\mathscr{S}_{\mathscr{D}}$ can be shown to encode a solution of the Δ_0-system $\mathscr{D}$, given that the address channels contain the correct addressing information. But this too must be checked, by a second induction on the orders of coordinates, which is contained in the proof of the Admissibility Lemma (p. 37). By means of the local condition the address channels of a tiling accepted by $\mathscr{S}_{\mathscr{D}}$ are shown to be correct for coordinates of order 0; then the global condition insures that if the addressing is correct for coordinates of order i it is also correct for coordinates of order $i+1$. The heart of the proof--the part that is combinatorially most difficult--deals only with the address channels; the correctness of the data channels will be established by the fairly direct argument given in the next section.

IC. 2 Reduction of the Origin-Constrained Tiling Problem

We now take as given two relatively prime natural numbers p_1, p_2 (in fact, $p_1 = 10$ and $p_2 = 11$), and for $p = p_1, p_2$ a fixed finite alphabet Σ_p (defined on p. 27). A p-tiling is a Σ_p-tiling of $2\mathbb{Z}$. We also assume that two classes of p-tilings, normal and admissible, have been specified; they are defined on pp. 28, 29.

In this section we state three crucial lemmata about normal and admissible p-tilings. The first is called the "Normalization Lemma" and states that any Λ_1-system accepting a p-tiling with admissible address channels also accepts one with normal address channels. The second is called the "Admissibility Lemma" and states that a certain Λ_1-system $\mathscr{I}_p$ constructed so as to accept every p-tiling with normal address channels accepts only p-tilings with admissible address channels. The third is called the "Addressing Lemma" and states that under certain circumstances, the address channels of a normal p-tiling can be used by $\mathscr{I}_p$ to yield information about the absolute or relative orders of tape cells. In this section these three lemmata are used to establish the unsolvability of the tiling problem for Λ_1-systems. The next three sections complete the proof by defining the undefined concepts and proving the three lemmata.

We first need a definition. A normally (admissibly) addressed tiling of $2\mathbb{Z}$ is a tiling α that consists of a pair of three-channel tapes, such that the first channels form a normal (admissible) p_1-tiling and the second channels form a normal (admissible) p_2-tiling. That is, the tilings

$$\alpha_1(n, c) = \pi_1(\alpha(n, c)) \qquad (c = 1, 2, \ n \in \mathbb{Z})$$

and

$$\alpha_2(n, c) = \pi_2(\alpha(n, c)) \qquad (c = 1, 2, \ n \in \mathbb{Z})$$

must both be normal (admissible). (The mapping π_i is the projection function on the i-th coordinate.)

Normalization Lemma. Any Λ_1-system accepting an admissibly addressed tiling accepts a normally addressed tiling.

In order to state the Admissibility Lemma we must define the Λ_1-system $\mathscr{I}_p$. However, this definition rests on the definition of normal, which is not given until later (p. 28). Let $p = p_1$ or p_2, and let

$$L_p = \{\langle\alpha(\nu_1), \ldots, \alpha(\nu_\theta)\rangle \mid \alpha \text{ is a normal p-tiling and } R_0(\nu_1, \ldots, \nu_\theta) \quad (\nu_1, \ldots, \nu_\theta \in 2\mathbb{Z})\}$$

$$G_p = \{\langle \alpha(\nu_1), \ldots, \alpha(\nu_4)\rangle \mid \alpha \text{ is a normal p-tiling and } R_1(\nu_1, \ldots, \nu_4) \quad (\nu_1, \ldots, \nu_4 \in 2\mathbb{Z})\}.$$

Thus L_p and G_p are simply the finite sets of all sequences of symbols observed by a Λ_1-system inspecting a normal p-tiling. (We use $\nu_1, \nu_2, \ldots$ as variables ranging over $2\mathbb{Z}$, n, m, n_2, etc. as variables ranging over $\mathbb{Z}$.)

Let $\mathscr{J}_p$ be the Λ_1-system (Σ_p, L_p, G_p).

Admissibility Lemma. Every p-tiling accepted by $\mathscr{J}_p$ is admissible.

Before stating the Addressing Lemma we extend to $2\mathbb{Z}$ the definition of Ω_p by

$$\Omega_p(n, c) = \Omega_p(n) \qquad (n \in \mathbb{Z},\ c = 1, 2) .$$

Addressing Lemma. For $p = p_1$ and p_2 there are relations $Z_p \subset \Sigma_p^\theta$, $E_p \subset \Sigma_p^4$, and $I_p \subset \Sigma_p^4$ such that

(Z) for $p = p_1$ or p_2, any normal p-tiling α, and any $\nu_1, \ldots, \nu_\theta \in 2\mathbb{Z}$, if $R_0(\nu_1, \ldots, \nu_\theta)$, then $Z_p(\alpha(\nu_1), \ldots, \alpha(\nu_\theta))$ if and only if $\Omega_p(\nu_1) = 0$;

(E-1) for $p = p_1$ or p_2, any normal p-tiling α, and any $\nu_1, \ldots, \nu_4 \in 2\mathbb{Z}$, if $R_1(\nu_1, \ldots, \nu_4)$ and $E_p(\alpha(\nu_1), \ldots, \alpha(\nu_4))$ then $\Omega_p(\nu_1) = \cdots = \Omega_p(\nu_4)$;

(E-2) for any normal p_1-tiling α_1 and p_2-tiling α_2, and for any $n_1, n_2 \in \mathbb{Z}$ such that $\Omega_{p_i}(n_1) = \Omega_{p_i}(n_2)$ for $i = 1, 2$, there are n, m, m_1, $m_2 \in \mathbb{Z}$ such that $m_j - m = n_j - n$ for $j = 1, 2$, and $E_{p_i}(\alpha_i(m, 1), \alpha_i(m_j, 2), \alpha_i(n, 1), \alpha_i(n_j, 2))$ and $E_{p_i}(\alpha_i(m_j, 1), \alpha_i(m, 2), \alpha_i(n_j, 1), \alpha_i(n, 2))$ for $i = 1, 2$, $j = 1, 2$;

(I-1) for $p = p_1$ or p_2, any normal p-tiling α, and any $\nu_1, \ldots, \nu_4 \in 2\mathbb{Z}$, if $R_1(\nu_1, \ldots, \nu_4)$ and $I_p(\alpha(\nu_1), \ldots, \alpha(\nu_4))$ then $\Omega_p(\nu_1) = \Omega_p(\nu_2) = \Omega_p(\nu_3) = \Omega_p(\nu_4) - 1$;

(I-2) For any normal p_1-tiling α_1 and p_2-tiling α_2, any $e_1, e_2 \in \mathbb{N}$ and for $i = 1, 2$, there are $\nu_1, \ldots, \nu_4 \in 2\mathbb{Z}$ such that $R_1(\nu_1, \ldots, \nu_4)$, $\Omega_{p_j}(\nu_1) = e_j$ for $j = 1, 2$, $I_{p_i}(\alpha_i(\nu_1), \ldots, \alpha_i(\nu_4))$, and $E_{p_{3-i}}(\alpha_{3-i}(\nu_1), \ldots, \alpha_{3-i}(\nu_4))$.

((Z), (E-1) and (E-2), and (I-1) and (I-2) are more precise versions of (Z), (E), (I) on p. 2?.) Note that, by definition of L_p and G_p, $Z_p \subset L_p$ and E_p, $I_p \subset G_p$.

Now let us assume the three crucial lemmata and prove from them the unsolvability of the tiling problem for Λ_1-systems. Let $\mathscr{D} = (D, D_0, A_1, A_2)$ be a Δ_0-system; we construct a Λ_1-system $\mathscr{S}_{\mathscr{D}} = (T, L, G)$, where $T = \Sigma_{p_1} \times \Sigma_{p_2} \times D$. As an intermediate stage we construct a Λ_1-system $\mathscr{K}$ that has the same set of tiles as $\mathscr{S}_{\mathscr{D}}$ but ignores the third or "data" channels. That is, the sole purpose of $\mathscr{K}$ is to combine $\mathscr{J}_{p_1}$ and $\mathscr{J}_{p_2}$; $\mathscr{S}_{\mathscr{D}}$ will then be derived from $\mathscr{K}$ by restricting the local and global conditions.

So we define $\mathscr{K} = (T, L', G')$, where as mentioned earlier $T = \Sigma_{p_1} \times \Sigma_{p_2} \times D$, and

$\langle s_1, \ldots, s_\theta \rangle \in L'$ if and only if $\langle \pi_i(s_1), \ldots, \pi_i(s_\theta) \rangle \in L_{p_i}$ for $i = 1, 2$;

$\langle s_1, \ldots, s_4 \rangle \in G'$ if and only if $\langle \pi_i(s_1), \ldots, \pi_i(s_4) \rangle \in G_{p_i}$ for $i = 1, 2$.

(Recall that $\mathscr{J}_p = (\Sigma_p, L_p, G_p)$, and $L_p \subset \Sigma_p^\theta$.)

It follows from the definition of $\mathscr{J}_p$ that $\mathscr{K}$ accepts every normally addressed T-tiling of $2\mathbb{Z}$, and it follows from the Admissibility Lemma that every T-tiling of $2\mathbb{Z}$ accepted by $\mathscr{K}$ is admissibly addressed. Next we construct $\mathscr{S}_{\mathscr{D}}$ from $\mathscr{K}$.

Let $\mathscr{S}_{\mathscr{D}} = (T, L, G)$, where

(1) $L \subset L'$ contains all those θ-tuples $\langle s_1, \ldots, s_\theta \rangle$ in L' such that, if $Z_{p_i}(\pi_i(s_1), \ldots, \pi_i(s_\theta))$ for $i=1$ and 2, then $\pi_3(s_1) \in D_0$.

(2) $G \subset G'$ contains all those quadruples $\langle s_1, \ldots, s_4 \rangle$ in G' such that

(a) if $E_{p_i}(\pi_i(s_1), \ldots, \pi_i(s_4))$ for $i=1$ and 2, then $\pi_3(s_3) = \pi_3(s_4)$;

(b) if, for $i=1$ or 2, $I_{p_i}(\pi_i(s_1), \ldots, \pi_i(s_4))$ and $E_{p_{3-i}}(\pi_{3-i}(s_1), \ldots, \pi_{3-i}(s_4))$, then $A_i(\pi_3(s_3), \pi_3(s_4))$.

Part (1) corresponds to the origin constraint of $\mathscr{D}$; (2a) to the requirement that tilings of $2\mathbb{Z}$ define tilings of $\mathbb{N}^2$, i.e., that the cells of an accepted tiling representing the same point on the plane contain the same member of D; and (2b) corresponds to the adjacency conditions of $\mathscr{D}$ (horizontal for $i = 1$, vertical for $i = 2$).

First suppose that there is a tiling $\tau : \mathbb{N} \times \mathbb{N} \to D$ accepted by $\mathscr{D}$. For $i = 1, 2$, let α_i be any normal p_i-tiling of $2\mathbb{Z}$. Define the D-tiling of $2\mathbb{Z}$ by $\alpha_3(n, c) = \tau(\Omega_{p_1}(n), \Omega_{p_2}(n))$ for $c = 1, 2$, if $n \neq 0$, and $\alpha(0,1) = \alpha(0,2) =$ any member of D. Finally, define $\alpha(n, c) = \langle \alpha_1(n,c), \alpha_2(n,c), \alpha_3(n,c) \rangle$; we claim that α is accepted by $\mathscr{S}_{\mathscr{D}}$.

By the definition of the $\mathscr{J}_p$, α_i is accepted by $\mathscr{J}_{p_i}$ for $i = 1, 2$; hence $\mathscr{K}$ accepts α. It remains to show that conditions (1) and (2) in the definition of $\mathscr{S}_{\mathscr{D}}$ do not result in α being unacceptable to $\mathscr{S}_{\mathscr{D}}$.

(1) Suppose $\nu_1, \ldots, \nu_\theta \in 2\mathbb{Z}$, $R_0(\nu_1, \ldots, \nu_\theta)$, and $Z_{p_i}(\alpha_i(\nu_1), \ldots, \alpha_i(\nu_\theta))$ for $i=1$ and 2. Then by (Z), $\Omega_{p_i}(\nu_1) = 0$ for $i=1$ and 2, and by the definition of α_3 and the fact that $\tau(0,0) \in D_0$, $\alpha_3(\nu_1) \in D_0$.

(2a) Suppose $\nu_1, \ldots, \nu_4 \in 2\mathbb{Z}$, $R_1(\nu_1, \ldots, \nu_4)$, and $E_{p_i}(\alpha_i(\nu_1), \ldots, \alpha_i(\nu_4))$ for $i=1$ and 2. Then by (E-1), $\Omega_{p_i}(\nu_3) = \Omega_{p_i}(\nu_4)$ for $i=1$ and 2, so $\alpha_3(\nu_3) = \alpha_3(\nu_4)$ by definition of α_3.

(2b) Suppose $\nu_1, \ldots, \nu_4 \in 2\mathbb{Z}$, $R_1(\nu_1, \ldots, \nu_4)$, and, for $i = 1$ or 2, $I_{p_i}(\alpha_i(\nu_1), \ldots, \alpha_i(\nu_4))$ and $E_{p_{3-i}}(\alpha_{3-i}(\nu_1), \ldots, \alpha_{3-i}(\nu_4))$. Then by (I-1), $\Omega_{p_i}(\nu_3) = \Omega_{p_i}(\nu_4)-1$ and by (E-1), $\Omega_{p_{3-i}}(\nu_3) = \Omega_{p_{3-i}}(\nu_4)$. Then by the definition of α_3 and the fact that τ satisfies the adjacency conditions of $\mathscr{D}$, $A_i(\alpha_3(\nu_3), \alpha_3(\nu_4))$.

Now suppose that $\mathscr{S}_{\mathscr{D}}$ accepts a tiling β. Since $\mathscr{K}$ also accepts β, β is admissibly addressed; then by the Normalization Lemma, $\mathscr{S}_{\mathscr{D}}$ accepts a normally addressed tiling α. For $i = 1, 2, 3$, let $\alpha_i = \pi_i \circ \alpha$.

The first step is to prove the consistency constraint, i.e., that cells whose coordinates have the same p_1- and p_2-orders contain the same member of D in the data channel. Thus we need to show that if $n_1, n_2 \in \mathbb{Z}$ and $\Omega_{p_i}(n_1) = \Omega_{p_i}(n_2)$ for $i = 1, 2$, then $\pi_3(\alpha(n_1, c_1)) = \pi_3(\alpha(n_2, c_2))$, where $c_1, c_2 \in \{1, 2\}$. For this we use (E-2). Thus there are n, m, m_1, m_2 such that $m_j - m = n_j - n$ for $j = 1, 2$, and $E_{p_i}(\alpha_i(m, 1), \alpha_i(m_j, 2), \alpha_i(n, 1), \alpha_i(n_j, 2))$ and $E_{p_i}(\alpha_i(m_j, 1), \alpha_i(m, 2), \alpha_i(n_j, 1), \alpha_i(n, 2))$ for $i = 1, 2$, $j = 1, 2$. Since $R_1(\langle m, 1\rangle, \langle m_j, 2\rangle, \langle n, 1\rangle, \langle n_j, 2\rangle)$ for $j = 1, 2$, by (2a) of the definition of $\mathscr{S}_{\mathscr{D}}$, $\alpha_3(n, 1) = \alpha_3(n_j, 2)$ for $j = 1, 2$, and similarly $\alpha_3(n_j, 1) = \alpha_3(n, 2)$ for $j = 1, 2$. Thus $\alpha_3(n_1, c) = \alpha_3(n_2, c)$ for $c = 1, 2$, and on either tape by itself the contents of the data channel depends only on the p_1- and p_2-orders. But since also $\Omega_{p_i}(n) = \Omega_{p_i}(n_j)$ for $j = 1, 2$ by (E-1), and $\alpha_3(n, 1) = \alpha_3(n_j, 2)$, the two tapes are also consistent with each other, i.e., $\alpha_3(n_1, 1) = \alpha_3(n_2, 2)$.

So we can define a mapping $\tau : \mathbb{N} \times \mathbb{N} \to D$ by $\tau(e_1, e_2) = \pi_3(\alpha(n, c))$, where $c = 1$ or 2, n is any integer such that $\Omega_{p_i}(n) = e_i$ for $i = 1, 2$. It remains to show that τ is accepted by $\mathscr{D}$. Clearly, by (Z), if $\Omega_{p_i}(\nu) = 0$ for $i = 1, 2$, then $\pi_3(\alpha(\nu)) \in D_0$, so $\tau(0, 0) \in D_0$, hence the origin constraint is satisfied. Now let $e_1, e_2 \in \mathbb{N}$, and let $i = 1$ or 2; by (I-2) there are $\nu_1, \ldots, \nu_4 \in 2\mathbb{Z}$ such that $R_1(\nu_1, \ldots, \nu_4)$, $\Omega_{p_j}(\nu_1) = e_j$ for $j = 1, 2$, $I_{p_i}(\alpha_i(\nu_1), \ldots, \alpha_i(\nu_4))$, and $E_{p_{3-i}}(\alpha_{3-i}(\nu_1), \ldots, \alpha_{3-i}(\nu_4))$. Then by (I-1) $\Omega_{p_i}(\nu_3) = \Omega_{p_i}(\nu_4) - 1 = e_i$ and by (E-1) $\Omega_{p_{3-i}}(\nu_3) = \Omega_{p_{3-i}}(\nu_4) = e_{3-i}$. Hence $\alpha_3(\nu_3) = \tau(e_1, e_2)$, and $\alpha_3(\nu_4) = \tau(e'_1, e'_2)$, where $e'_i = e_i + 1$ and $e'_{3-i} = e_{3-i}$. But then by clause (2b) of the definition of $\mathscr{S}_{\mathscr{D}}$, $A_i(\tau(e_1, e_2), \tau(e'_1, e'_2))$. Thus the adjacency constraint is satisfied.

Thus to prove the Linear Sampling Theorem (First Version) it suffices to define <u>normal</u> and <u>admissible</u> and to prove the Normalization, Admissibility, and Addressing Lemmata.

IC. 3 Normal and Admissible Tilings, and the Normalization Lemma

To complete the proof we must define normal and admissible tilings of $2\mathbb{Z}$ and prove the three crucial lemmata. Recall that the tapes used to encode tilings of $\mathbb{N}^2$ have three channels: a data channel and two address channels. The rest of the proof is concerned only with the address channels: definition of the special kinds of tilings that can occur in the address channels, and proof of the special properties asserted of such tilings by the crucial lemmata.

Let $\Sigma_p = \{q \mid 0 < q < p^2 \text{ and } q \not\equiv 0 \pmod p\}$. Σ_p is the finite alphabet for the p-tilings of $2\mathbb{Z}$ described in the three crucial lemmata. Let $\mathrm{rem}(n, k)$ (where $k > 0$) be the remainder of n when divided by k, i.e., the unique integer r such that $0 \le r < k$ and $n \equiv r \pmod k$. Finally, for $n \neq 0$ let

$$\lambda_p(n) = \mathrm{rem}(np^{-\Omega_p(n)}, p^2) .$$

Intuitively, $\lambda_p(n)$ is the result of writing out the value of n in p-ary notation, deleting any trailing zeroes, and then taking the last two digits of what is left. Any p-tiling α such that $\alpha(n, 1) = \alpha(n, 2) = \lambda_p(n)$ whenever $n \neq 0$, and $\alpha(0, 1) = \alpha(0, 2)$, will be called perfect (note that there are $p^2 - p$ such p-tilings, which differ from each other only at 0).

Before proceeding we make a notational convention. When a particular value of p is fixed we avoid mention of it wherever possible, writing, e.g., λ for λ_p, Ω for Ω_p, and tiling for p-tiling.

Note that for any $n_1, n_2 \in \mathbb{Z}$, at least two of n_1, n_2, and $n_2 - n_1$ must have the same order. The Perfect Tiling Lemma stated below implies that if n_1, n_2, and $n_2 - n_1$ all have the same order, then $\lambda(n_1)$ and $\lambda(n_2-n_1)$ determine $\lambda(n_2)$; if the order of n_2 is one greater than that of n_1 and $n_2 - n_1$, then $\lambda(n_1)$ and $\lambda(n_2-n_1)$ are complementary (mod p) and determine the second p-ary digit, but not the first p-ary digit, of $\lambda(n_2)$; and if the order of n_2 exceeds that of n_1 and $n_2 - n_1$ by 2 or more, then $\lambda(n_1)$ and $\lambda(n_2-n_1)$ are complementary (mod p^2) and give no information about $\lambda(n_2)$.

Perfect Tiling Lemma. Let $n_1, n_2 \in \mathbb{Z}$.

(a) If $\Omega(n_1) = \Omega(n_2-n_1) = \Omega(n_2)$ then $\lambda(n_1) + \lambda(n_2-n_1) \equiv \lambda(n_2) \pmod{p^2}$.

(b) If $\Omega(n_1) = \Omega(n_2-n_1) = \Omega(n_2) - 1$ then $\lambda(n_1) + \lambda(n_2-n_1) \equiv p\lambda(n_2) \not\equiv 0 \pmod{p^2}$.

(c) If $\Omega(n_1) = \Omega(n_2-n_1) \le \Omega(n_2) - 2$ then $\lambda(n_1) + \lambda(n_2-n_1) \equiv 0 \pmod{p^2}$.

Proof. We prove just part (c); proofs of the other parts are similar. The idea is that the p-ary notation for n_2 has at least two more trailing zeroes than that for n_1, so the last two nonzero digits of n_1 and $n_2 - n_1$ must be complementary. Formally, let $\Omega(n_1) = \Omega(n_2-n_1) = i < \infty$, so that $n_2 = n_2' p^{i+2}$ for some $n_2' \in \mathbb{Z}$ and $n_1 = n_1' p^i$ for some $n_1' \not\equiv 0 \pmod p$. Then

$$\lambda(n_1) + \lambda(n_2-n_1) = n_1 p^{-i} + (n_2-n_1)p^{-i}$$

$$= n_2 p^{-i} = n_2' p^2 \equiv 0 \pmod{p^2} .$$

This completes the proof of the Perfect Tiling Lemma. ∎

Note that no perfect tiling is periodic; that is, if α is perfect then there is no $d \in \mathbb{Z}$ such that $\alpha(n+d, c) = \alpha(n, c)$ for all $n \in \mathbb{Z}$, and $c = 1, 2$. For (if $p > 2$) either $d + p^{\Omega(d)}$ or $d - p^{\Omega(d)}$ has the same order as d; then letting n be whichever of $\pm p^{\Omega(d)}$ is such that $\Omega(n+d) = \Omega(d)$, it follows from (a) of the Perfect Tiling Lemma that $\lambda(n+d) \equiv \lambda(n) + \lambda(d) \equiv \lambda(d) \pm 1 \not\equiv \lambda(d) \pmod p$, defeating the possibility that $\alpha(n+d, c) = \alpha(n, c)$ for all n. Nevertheless, each perfect tiling is "almost-periodic" since it <u>is</u> true that $\lambda(n+d) = \lambda(n)$ whenever n has low order relative to d; for $\lambda(-n) \equiv -\lambda(n) \pmod{p^2}$ for all $n \neq 0$, so if $\Omega(d) \geq \Omega(n) + 2$ then by (c) of the Perfect Tiling Lemma

$$\lambda(n+d) - \lambda(n) \equiv \lambda(n+d) + \lambda(-n) \equiv 0 \pmod{p^2} .$$

And if d has very high order, then $\lambda(n+d)$ differs from $\lambda(n)$ for relatively few n, since numbers of high order are sparsely distributed.

Ideally we would design a Λ_1-system that would accept only perfect tilings, but we know by the Translation Lemma that this is impossible. In fact, the situation is worse than is indicated by the Translation Lemma alone. There are two degrees of similarity between the perfect tilings and the ones we can actually devise a Λ_1-system to accept: <u>normal</u> and <u>admissible.</u>

A tiling α is <u>normal</u> if and only if $\alpha(n, 1) = \alpha(n, 2)$ for all n, and there is a mapping $h : \mathbb{N} \to \{0, 1, \ldots, p-1\}$ such that for all $n \neq 0$, and for $c = 1, 2$,

$$\alpha(n, c) = \mathrm{rem}(\lambda(n) \cdot (1 + p \cdot h(\Omega(n))), p^2) .$$

Since a normal tiling α has the same values on both tapes, we often write simply $\alpha(n)$ for either $\alpha(n, 1)$ or $\alpha(n, 2)$.

Note that a normal tiling is identical to a perfect tiling, if first digits are ignored. The first digits differ from those of a perfect tiling in a systematic way, depending on the second digit and the order of the coordinate. To get a feeling for the structure of perfect and normal tilings, let $p = 4$ and consider the sequence of all cells of one of the tapes of a perfect or normal tiling whose coordinates have order i or greater, for some fixed i. This sequence is obtained by starting at 0 and writing down the contents of every p^i-th cell. Writing xx for cells whose coordinates have order exceeding i, the sequence for a perfect tiling would be (in base 4)

xx 01 02 03 xx 11 12 13 xx 21 22 23 xx 31 32 33 xx 01 02 ....

For a normal tiling there are four possible sequences. If $h(i) = 0$ the sequence would be just as above. If $h(i) = 1$ the sequence would be

xx 11 22 33 xx 21 32 03 xx 31 02 13 xx 01 12 23 xx 11 22 ... ;

if $h(i) = 2$ the sequence would be

xx 21 02 23 xx 31 12 33 xx 01 22 03 xx 11 32 13 xx 21 02 ... ;

and if $h(i) = 3$ the sequence would be

xx 31 22 13 xx 01 32 23 xx 11 02 33 xx 21 12 03 xx 31 22 ... .

We next define _admissible_ tilings, which form a yet broader class than the normal tilings; however they share with perfect and normal tilings the crucial property of being aperiodic, and can in a certain precise sense explained by the proof of the Normalization Lemma be "cut and pasted" to form normal tilings. They are of interest because they are all aperiodic, and because we can construct a Λ_1-system which accepts only these tilings; could we not find a class of tilings with these two properties the whole proof would founder.

A tiling β is i-_admissible_ $(i \geq 0)$ if and only if there are integers t_1, t_2 and a normal tiling α such that for $c = 1, 2$,

(a) if $\Omega(n) < i$ then $\beta(n+t_c, c) = \alpha(n)$; and

(b) if $\Omega(n) = i$ then $\beta(n+t_c, c) \equiv \alpha(n) \pmod{p}$.

The tiling β is _admissible_ just in case β is i-admissible for each $i \in \mathbb{N}$.

Admissible and i-admissible tilings differ from normal tilings in two ways. The first is due to the parameters t_1 and t_2, which simply translate a tiling. Consider the simple case in which $t_1 = t_2 = 0$. The second complication is that a normal and an i-admissible tiling may differ on cells whose coordinates are of order i or greater; they must agree on cells of order less than i. Moreover, on cells whose coordinates are of order exactly i the second digits must agree. The i-admissible tilings may be viewed as translations of approximations to normal tilings; the larger the value of i, the closer the approximation.

An admissible tiling might be viewed as the "limit" of a sequence of i-admissible tilings for increasing i. Indeed, this idea could be formalized, but one caution should be offered: an admissible tiling need not be the translation of a normal tiling, since the parameters t_1, t_2 for the defining sequence of i-admissible tilings may grow without bound.

We are nearly ready for the proof of the Normalization Lemma; we need one preliminary lemma. This lemma states, in essence, that if β is i-admissible then the result of shifting each tape of β by some (not necessarily the same) multiple of p^{i+1} is a tiling that agrees with β on p^i-1 of every p^i consecutive cells of either tape.

Agreement Lemma. Let β be an i-admissible tiling, and let α, t_1, t_2 be for β as in the definition of "i-admissible." Let $s_1, s_2 \in \mathbb{Z}$ be such that $\Omega(s_c) > i$ for $c = 1, 2$. Then for each n, if $\Omega(n) < i$ then $\beta(n+s_c+t_c, c) = \beta(n+t_c, c) = \alpha(n)$.

Proof. Since $\Omega(n) < i$ and $\Omega(s_c) > i$, $\Omega(s_c) - \Omega(n) \geq 2$ and $\Omega(n+s_c) = \Omega(n)$. Then by (c) of the Perfect Tiling Lemma $\lambda(n+s_c) = \lambda(n)$, and by the definition of "normal" $\alpha(n+s_c) = \alpha(n)$. Since $\beta(n+s_c+t_c, c) = \alpha(n+s_c)$ and $\beta(n+t_c, c) = \alpha(n)$, the result follows.

This completes the proof of the Agreement Lemma.

Normalization Lemma. Any Λ_1-system accepting an admissibly addressed tiling accepts a normally addressed tiling.

Proof. Say that a pair of finite sequences is normally addressed if it is a central segment (p. 20) of a normally addressed tiling (p. 23). We first show that if β is an admissibly addressed tiling accepted by a Λ_1-system $\mathscr{S}$, then β has arbitrarily long normally addressed segments. We shall then use the Central Segment Lemma to construct a normally addressed tape accepted by $\mathscr{S}$.

So let r be a positive integer; we show that β has a normally addressed segment of length $2r+1$. Let i be the least integer exceeding $\log_{p_1}(r)$; i has the property that if $0 < |n| \leq r$ then $\Omega_{p_k}(n) < i$ for $k = 1, 2$. For $k = 1, 2$, let $\beta_k = \pi_k \circ \beta$ be the admissible p_k-tape in the k-th address channel of β, and let α_k be a normal p_k-tiling and let t_{k1}, t_{k2} be integers such that $\beta_k(n+t_{kc}, c) = \alpha_k(n)$ for $c = 1, 2$ whenever $\Omega_{p_k}(n) < i$. (We are using here the i-admissibility of β_1 and β_2.) By the Chinese Remainder Theorem there are integers t_1, t_2 such that

$$t_c \equiv t_{kc} \pmod{p_k^{i+1}} \qquad \text{for} \qquad k = 1, 2, \quad c = 1, 2.$$

Then let $s_{kc} = t_c - t_{kc}$; then $\Omega_{p_k}(s_{kc}) > i$ and the Agreement Lemma applies, so that $\beta_k(n+t_c, c) = \alpha_k(n, c)$ for $k = 1, 2$, $c = 1, 2$, whenever $\Omega_{p_k}(n) < i$. We claim that the pair of sequences $\langle \sigma_1, \sigma_2 \rangle$ is normally addressed, where

$$\sigma_c = \langle \beta(t_c - r, c), \beta(t_c - r + 1, c), \ldots, \beta(t_c + r, c) \rangle .$$

For if $0 < |n| \leq r$ then $\Omega_{p_k}(n) < i$ by the choice of i; hence $\beta_k(n+t_c, c) = \alpha_k(n)$ for $c = 1, 2$.

Hence there is an infinite set of normally addressed segments of a tiling accepted by $\mathscr{S}$. By the Central Segment Lemma there is a tiling α accepted by $\mathscr{S}$ such that each central segment of α is normally addressed. But then α is normally addressed.

This completes the proof of the Normalization Lemma. ∎

IC.4 The Addressing Lemma

We begin the proof of the Addressing Lemma with part (Z).

Part (Z) of the Addressing Lemma. For $p = p_1$ or p_2 there is a relation $Z_p \subset \Sigma_p^\theta$ such that, for any normal p-tiling α and any $\nu_1, \ldots, \nu_\theta \in 2\mathbb{Z}$, if $R_0(\nu_1, \ldots, \nu_\theta)$, then $Z_p(\alpha(\nu_1), \ldots, \alpha(\nu_\theta))$ if and only if $\Omega_p(\nu_1) = 0$.

Proof. Here we fix the value of $\theta: \theta = 4$. (It would suffice to let $\theta = 3$ here; we shall need $\theta = 4$ in the proof of the Admissibility Lemma.) Define $Z_p(q_0, q_1, q_2, q_3)$ if and only if $\langle q_0, q_1, q_2, q_3 \rangle \in L_p$ and either $q_1 \equiv q_0 + 1 \pmod p$ or $q_2 \equiv q_0 + 2 \pmod p$.

Suppose α is a normal p-tiling. If $n \in \mathbb{Z}$ and $\Omega_p(n) = 0$, then $n \equiv \alpha(n) \not\equiv 0 \pmod p$. Let $q = \mathrm{rem}(n, p) = \mathrm{rem}(\alpha(n), p)$. There are three possibilities. If $1 \leq q \leq p-3$, then $\alpha(n+1) \equiv n+1 \equiv q+1 \pmod p$ and $\alpha(n+2) \equiv n+2 \equiv q+2 \pmod p$. If $q = p-2$, then $\alpha(n+1) \equiv n+1 \equiv q+1 = p-1 \pmod p$. And if $q = p-1$ then $\alpha(n+2) \equiv q+2 \equiv 1 \pmod p$. Thus in each case $Z_p(\alpha(n), \alpha(n+1), \alpha(n+2), \alpha(n+3))$.

Conversely, suppose $Z_p(\alpha(n), \alpha(n+1), \alpha(n+2), \alpha(n+3))$; we show that $n \not\equiv 0 \pmod p$. If $n \equiv 0 \pmod p$, then $\alpha(n+1) \equiv n+1 \equiv 1 \pmod p$ and $\alpha(n+2) \equiv n+2 \equiv 2 \pmod p$. Hence if $n \equiv 0 \pmod p$ and either $\alpha(n+1) \equiv \alpha(n) + 1 \pmod p$ or $\alpha(n+2) \equiv \alpha(n) + 2 \pmod p$, then (since $p > 0$) $\alpha(n) \equiv 0 \pmod p$, which is impossible. Therefore $\Omega_p(n) = 0$.

This completes the proof of Part (Z). ∎

Before completing the proof of the Addressing Lemma by proving parts (E) and (I) we must extend the Perfect Tiling Lemma.

<u>Normal Tiling Lemma</u>. Let α be a normal tiling and let $n_1, n_2 \in \mathbb{Z}$. Then

(a) If $\Omega(n_1) = \Omega(n_2-n_1) = \Omega(n_2)$ then $\alpha(n_1) + \alpha(n_2-n_1) \equiv \alpha(n_2) \pmod{p^2}$.

(b) If $\Omega(n_1) = \Omega(n_2-n_1) = \Omega(n_2)-1$ then $\alpha(n_1) + \alpha(n_2-n_1) \equiv p\alpha(n_2) \not\equiv 0 \pmod{p^2}$.

(c) If $\Omega(n_1) = \Omega(n_2-n_1) \leq \Omega(n_2)-2$ then $\alpha(n_1) + \alpha(n_2) \equiv 0 \pmod{p^2}$.

<u>Proof.</u> Let $\alpha(n) \equiv \lambda(n) \cdot (1+p \cdot h(\Omega(n))) \pmod{p^2}$ for each n, and use the Perfect Tiling Lemma. For example, to prove (a):

$$\alpha(n_1) + \alpha(n_2-n_1)$$

$$= \lambda(n_1) \cdot (1+p \cdot h(\Omega(n_1))) + \lambda(n_2-n_1) \cdot (1+p \cdot h(\Omega(n_2-n_1)))$$

by definition of α

$$= (\lambda(n_1) + \lambda(n_2-n_1)) \cdot (1+p \cdot h(\Omega(n_2)))$$

since $\Omega(n_1) = \Omega(n_2-n_1) = \Omega(n_2)$

$$\equiv \lambda(n_2) \cdot (1+p \cdot h(\Omega(n_2)))$$

by (a) of the Perfect Tiling Lemma

$$\equiv \alpha(n) \pmod{p^2}$$

by definition of α.

This completes the proof of the Normal Tiling Lemma. ∎

Now we are ready to define the finite relations E_p and I_p, which are subsets of G_p.

<u>Definition of I_p and E_p.</u> We let

$E_p(q_1, \ldots, q_4)$ iff $G_p(q_1, \ldots, q_4)$, $q_4 - q_3 \equiv q_2 - q_1 \pmod{p}$, and each of the following is incongruent to 0 (mod p): $q_2 - q_1$, $q_4 - q_3$, $q_3 - q_1$, $q_1 + q_4$, $q_2 + q_3$, $q_4 - q_2$;

$I_p(q_1, \ldots, q_4)$ iff $G_p(q_1, \ldots, q_4)$, $q_1 - q_2 - q_3 \equiv 0 \pmod{p}$, $q_1 - q_2 - q_3 \not\equiv 0 \pmod{p^2}$, and each of the following is incongruent to 0 (mod p): $q_4 \pm q_1$, $q_4 \pm q_2$, $q_4 \pm q_3$.

(These relations are not all independent; for example in the definition of E_p, from the facts that $q_4 - q_3 \equiv q_2 - q_1 \pmod{p}$ and that $q_2 - q_1 \not\equiv 0 \pmod{p}$ it follows that $q_4 - q_3 \not\equiv 0 \pmod{p}$.)

Now we resume our proof of the Addressing Lemma.

Part (E-1) of the Addressing Lemma. For $p = p_1$ or p_2, any normal p-tiling α, and $\nu_1, \ldots, \nu_4 \in 2\mathbb{Z}$, if $R_1(\nu_1, \ldots, \nu_4)$ and $E_p(\alpha(\nu_1), \ldots, \alpha(\nu_4))$, then $\Omega_p(\nu_1) = \ldots = \Omega_p(\nu_4)$.

Proof. Let $q_i = \alpha(\nu_i)$ for $i = 1, \ldots, 4$; let $n_i = \pi_1(\nu_i)$ for $i = 1, \ldots, 4$ (i.e., $\nu_1 = \langle n_1, 1 \rangle$, etc.). Assuming that $q_4 - q_3 \equiv q_2 - q_1 \pmod p$ and the six incongruences, we show that $\Omega_p(n_1) = \ldots = \Omega_p(n_4)$. Let ℓ, $1 \le \ell \le 4$, be such that n_ℓ has the lowest order among $n_1, \ldots, n_4$, and let $e_i = \Omega(n_i) - \Omega(n_\ell)$ for $i = 1, \ldots, 4$. Thus we wish to prove that $e_1 = \ldots = e_4 = 0$. Since by definition of λ,

$$\lambda(n_i) \equiv n_i p^{-\Omega(n_i)} = n_i p^{-(e_i + \Omega(n_\ell))} \pmod{p^2} ,$$

it follows that

$$n_i p^{-\Omega(n_\ell)} \equiv \lambda(n_i) p^{e_i} \pmod{p^2} .$$

Since $n_2 - n_1 = n_4 - n_3$, multiplying through by $p^{-\Omega(n_\ell)}$ we have

$$\lambda(n_2) p^{e_2} - \lambda(n_1) p^{e_1} \equiv \lambda(n_4) p^{e_4} - \lambda(n_3) p^{e_3} \pmod{p^2} .$$

Since $q_i = \alpha(n_i) \equiv \lambda(n_i) \pmod p$ for each i, it follows that $q_2 p^{e_2} - q_1 p^{e_1} \equiv q_4 p^{e_4} - q_3 p^{e_3}$ (mod p). It cannot be the case that $e_i > 0$ for three different i, for then $q_\ell \equiv 0$ (mod p). Nor can $e_i > 0$ for only one i, for then $q_2 - q_1 \not\equiv q_4 - q_3 \pmod p$. Finally, each of the six cases in which $e_i > 0$ for exactly two i contradicts one of the six incongruences.

This completes the proof of part (E-1). ■

Part (E-2) of the Addressing Lemma. For any normal p_1-tiling α_1 and p_2-tiling α_2, and for any $n_1, n_2 \in \mathbb{Z}$ such that $\Omega_{p_i}(n_1) = \Omega_{p_i}(n_2)$ for $i = 1, 2$, there are $n, m, m_1, m_2 \in \mathbb{Z}$ such that $m_j - m = n_j - n$ for $j = 1, 2$, and $E_{p_i}(\alpha_i(m, 1), \alpha_i(m_j, 2), \alpha_i(n, 1), \alpha_i(n_j, 2))$ and $E_{p_i}(\alpha_i(m_j, 2), \alpha_i(m, 1), \alpha_i(n_j, 2), \alpha_i(n, 1))$ for $i = 1, 2$, $j = 1, 2$.

Proof. Given n_1 and n_2, we can find, for $i = 1$ and 2, numbers N_i and M_i ($0 \le N_i$, $M_i < p_i$), such that

(i) $N_i \not\equiv 0 \pmod{p_i}$

(ii) $M_i \not\equiv 0 \pmod{p_i}$

(iii) $\lambda(n_1) - N_i + M_i \not\equiv 0 \pmod{p_i}$

(iv) $\lambda(n_2) - N_i + M_i \not\equiv 0 \pmod{p_i}$

(v) $\lambda(n_1) - N_i \not\equiv 0 \pmod{p_i}$

(vi) $\lambda(n_2) - N_i \not\equiv 0 \pmod{p_i}$

(vii) $N_i - M_i \not\equiv 0 \pmod{p_i}$

(viii) $\lambda(n_1) + M_i \not\equiv 0 \pmod{p_i}$

(ix) $\lambda(n_2) + M_i \not\equiv 0 \pmod{p_i}$.

Here we use the fact that $p_i \geq 10$; for there are p_i^2 possible choices of the pair $\langle N_i, M_i \rangle$, and each of the 9 incongruences eliminates at most p_i of them. By the Chinese Remainder Theorem there are N, M such that $N \equiv N_i \pmod{p_i}$ and $M \equiv M_i \pmod{p_i}$ for $i = 1, 2$. Now if $\Omega_{p_i}(n_1) = \Omega_{p_i}(n_2) = e_i$ for $i = 1, 2$, let

$$n = Np_1^{e_1} p_2^{e_2}, \qquad m = Mp_1^{e_1} p_2^{e_2}$$

$$m_j = m - n + n_j \qquad \text{for } j = 1, 2.$$

Obviously $m_j - m = n_j - n$ for $j = 1, 2$. To show that m, n, m_1, m_2 are as required, it suffices to show that

$$\alpha_i(n_j) - \alpha_i(n) \equiv \alpha_i(m_j) - \alpha_i(m) \pmod{p_i} \quad (i = 1, 2,\ j = 1, 2)$$

and that, for $i = 1, 2$, $j = 1, 2$, each of the following is incongruent to 0 (mod p_i):

$$\alpha_i(m_j) - \alpha_i(m),\ \alpha_i(n_j) - \alpha_i(n),\ \alpha_i(n) - \alpha_i(m),$$

$$\alpha_i(m) + \alpha_i(n_j),\ \alpha_i(n) + \alpha_i(m_j),\ \alpha_i(n_j) - \alpha_i(m_j)\ .$$

The congruence

$$\alpha_i(n_j) - \alpha_i(n) \equiv \alpha_i(m_j) - \alpha_i(m) \pmod{p_i}$$

follows from (a) of the Normal Tiling Lemma. For by (i), $\Omega_{p_i}(n) = e_i$ for $i = 1, 2$; by (ii), $\Omega_{p_i}(m) = e_i$ for $i = 1, 2$; by (iii) and (iv), $\Omega_{p_i}(m_j) = e_i$ for $i = 1, 2$, $j = 1, 2$; and by (v) and (vi), $\Omega_{p_i}(n_j - n) = e_i$ for $i = 1, 2$, $j = 1, 2$. Hence

$$\alpha_i(n_j) - \alpha_i(n) \equiv \alpha_i(n_j - n) = \alpha_i(m_j - m) \equiv \alpha_i(m_j) - \alpha_i(m) \pmod{p_i^2}\ .$$

It follows for the same reason that $\alpha_i(m_j) - \alpha_i(m)$ and $\alpha_i(n_j) - \alpha_i(n)$ are incongruent to 0 (mod p_i), since each is congruent to $\alpha_i(m_j - m) = \alpha_i(n_j - n)$. That $\alpha_i(n) - \alpha_i(m) \not\equiv 0 \pmod{p_i}$ follows from (vii); that $\alpha_i(m) + \alpha_i(n_j) \not\equiv 0 \pmod{p_i}$ follows from (viii) and (ix); and the incongruences $\alpha_i(n) + \alpha_i(m_j) \not\equiv 0 \pmod{p_i}$ and $\alpha_i(n_j) - \alpha_i(m_j) \not\equiv 0 \pmod{p_i}$ follow from the facts that $\alpha_i(n) - \alpha_i(m) \not\equiv 0 \pmod{p_i}$, $\alpha_i(m) + \alpha_i(n_j) \not\equiv 0 \pmod{p_i}$, and $\alpha_i(n_j) - \alpha_i(n) - \alpha_i(m_j) + \alpha_i(m) \equiv 0 \pmod{p}$.

This completes the proof of part (E-2). ∎

Part (I-1) of the Addressing Lemma. For $p=p_1$ or p_2, any normal p-tiling α, and any $\nu_1,\ldots,\nu_4 \in 2\mathbb{Z}$, if $R_1(\nu_1,\ldots,\nu_4)$ and $I_p(\alpha(\nu_1),\ldots,\alpha(\nu_4))$ then $\Omega_p(\nu_1) = \Omega_p(\nu_2) = \Omega_p(\nu_3) = \Omega_p(\nu_4)-1$.

Proof. As before, let $q_i = \alpha(\nu_i)$ and $n_i = \pi_1(\nu_i)$ for $i=1,\ldots,4$. Since $q_1 - q_2 - q_3 \equiv 0 \pmod p$, it follows from the fact that $q_i \not\equiv 0 \pmod p$ for each i that $q_3 - q_1$, $q_2 + q_3$, and $q_2 - q_1$ are all incongruent to 0 (mod p). Thus we may assume that all the values $q_3 - q_1$, $q_2 + q_3$, $q_2 - q_1$, $q_4 \pm q_1$, $q_4 \pm q_2$, $q_4 \pm q_3$ are incongruent to 0 (mod p). We prove that $\Omega(n_1) = \Omega(n_2) = \Omega(n_3) = \Omega(n_4) - 1$.

Let n_ℓ have the lowest order among $n_1,\ldots,n_4$, and for $i=1,\ldots,4$ let $e_i = \Omega(n_i) - \Omega(n_\ell)$. We shall show that $e_1 = e_2 = e_3 = 0$ and $e_4 > 0$; it will then follow by (a) and (c) of the Normal Tiling Lemma that $e_4 = 1$. For if $e_1 = e_2 = e_3 = 0$ and $e_4 \geq 2$, i.e., $\Omega(n_1) = \Omega(n_2) = \Omega(n_3) \leq \Omega(n_4) - 2$, then by (c) of the Normal Tiling Lemma $\alpha(n_3) + \alpha(n_4 - n_3) \equiv 0 \pmod{p^2}$. But $n_4 - n_3 = n_2 - n_1$ and $\Omega(n_4 - n_3) = \Omega(n_3) < \Omega(n_4)$ so

$$\alpha(n_3) + \alpha(n_4 - n_3) = \alpha(n_3) + \alpha(n_2 - n_1) \equiv \alpha(n_3) + \alpha(n_2) - \alpha(n_1)$$

$$= q_3 + q_2 - q_1 \not\equiv 0 \pmod{p^2}$$

by (a) of the Normal Tiling Lemma and the definition of I_p.

We have, once again, that

$$q_2 p^{e_2} - q_1 p^{e_1} \equiv q_4 p^{e_4} - q_3 p^{e_3} \pmod p .$$

As in the proof of (E-1), it is impossible that $e_i > 0$ for exactly three or exactly two i. Nor can all four e_i be zero, for then $q_1 - q_2 - q_3 \not\equiv 0 \pmod p$. If $e_1 > 0$ and $e_2 = e_3 = e_4 = 0$, then $q_2 - q_4 + q_3 \equiv 0 \pmod p$; since $q_1 - q_2 - q_3 \equiv 0 \pmod p$, we have $q_1 - q_4 \equiv 0 \pmod p$, a contradiction. If $e_2 > 0$ and $e_1 = e_3 = e_4 = 0$ then $-q_1 - q_4 + q_3 \equiv 0 \pmod p$, whence $q_2 + q_4 \equiv 0 \pmod p$, also a contradiction. Finally, if $e_3 > 0$ and $e_1 = e_2 = e_4 = 0$, then $q_2 - q_1 - q_4 \equiv 0 \pmod p$, whence $q_3 + q_4 \equiv 0 \pmod p$. The only remaining possibility is that $e_1 = e_2 = e_3 = 0$ and $e_4 > 0$, as was to be shown.

This completes the proof of part (I-1). ∎

Part (I-2) of the Addressing Lemma. For any normal p_1-tiling α_1 and p_2-tiling α_2, any $e_1, e_2 \in \mathbb{N}$, and for $i=1,2$, there are $\nu_1,\ldots,\nu_4 \in 2\mathbb{Z}$ such that $R_1(\nu_1,\ldots,\nu_4)$, $\Omega_{p_j}(\nu_1) = e_j$ for $j=1,2$, $I_{p_i}(\alpha_i(\nu_1),\ldots,\alpha_i(\nu_4))$, and $E_{p_{3-i}}(\alpha_{3-i}(\nu_1),\ldots,\alpha_{3-i}(\nu_4))$.

Proof. Given e_1, e_2, and i, let

$$n_1 = 2p_1^{e_1}p_2^{e_2}$$

$$n_2 = 4p_1^{e_1}p_2^{e_2}$$

$$n_3 = (p_i-2)p_1^{e_1}p_2^{e_2}$$

$$n_4 = p_i^{e_i+1}p_{3-i}^{e_{3-i}} ,$$

and let $\nu_1 = \langle n_1, 1\rangle$, $\nu_2 = \langle n_2, 2\rangle$, $\nu_3 = \langle n_3, 1\rangle$, $\nu_4 = \langle n_4, 2\rangle$. Clearly $n_2 - n_1 = n_4 - n_3$ so $R_1(\nu_1, \ldots, \nu_4)$; and clearly $\Omega_{p_j}(\nu_1) = e_j$ for $j = 1, 2$.

Next we show that $I_{p_i}(\alpha_i(\nu_1), \ldots, \alpha_i(\nu_4))$. That $\alpha_i(n_1) - \alpha_i(n_2) - \alpha_i(n_3) \equiv 0 \pmod{p_i}$ follows from (a) and (b) of the Normal Tiling Lemma, since

$$\alpha_i(n_3) + \alpha_i(n_4-n_3) \equiv 0 \pmod{p_i}$$

by (b) of that lemma, but $n_4 - n_3 = n_2 - n_1$ and $\Omega_{p_i}(n_2-n_1) = \Omega_{p_i}(n_1) = \Omega_{p_i}(n_3)$, so by (a) of the Normal Tiling Lemma

$$\alpha_i(n_3) + \alpha_i(n_4-n_3) = \alpha_i(n_3) + \alpha_i(n_2-n_1)$$

$$\equiv \alpha_i(n_3) + \alpha_i(n_2) - \alpha_i(n_1) \pmod{p_i} .$$

Similarly, it follows from (a) and (c) of the Normal Tiling Lemma that $\alpha_i(n_1) - \alpha_i(n_2) - \alpha_i(n_3) \not\equiv 0 \pmod{p_i^2}$. The proof that $\alpha_i(n_4) + \alpha_i(n_1)$, $\alpha_i(n_4) - \alpha_i(n_1)$, $\alpha_i(n_4) + \alpha_i(n_2)$, $\alpha_i(n_4) - \alpha_i(n_2)$, $\alpha_i(n_4) + \alpha_i(n_3)$, $\alpha_i(n_4) - \alpha_i(n_3)$ are all incongruent to 0 (mod p_i) comes down to the fact that

$$(1+2)p_{3-i}^{e_{3-i}},\quad (1-2)p_{3-i}^{e_{3-i}},\quad (1+4)p_{3-i}^{e_{3-i}},\quad (1-4)p_{3-i}^{e_{3-i}},\quad (1+p_i-2)p_{3-i}^{e_{3-i}},\ \text{and}\ (1-p_i+2)p_{3-i}^{e_{3-i}}$$

are all incongruent to 0 (mod p_i), which is easily seen to be true for $p_i = 10$, $p_{3-i} = 11$ and for $p_i = 11$, $p_{3-i} = 10$.

Finally we need to show that $E_{p_{3-i}}(\alpha_{3-i}(n_1), \ldots, \alpha_{3-i}(n_4))$. Clearly

$$\alpha_{3-i}(n_4) - \alpha_{3-i}(n_3) \equiv \alpha_{3-i}(n_2) - \alpha_{3-i}(n_1)$$

$$\equiv 2p_i^{e_i} \not\equiv 0 \pmod{p_{3-i}} .$$

Also

$$\alpha_{3-i}(n_3) - \alpha_{3-i}(n_1) \equiv (p_i-4)p_i^{e_i} \not\equiv 0 \pmod{p_{3-i}}$$

$$\alpha_{3-i}(n_1) + \alpha_{3-i}(n_4) \equiv (p_i+2)p_i^{e_i} \not\equiv 0 \pmod{p_{3-i}}$$

and the incongruence to 0 (mod p_{3-i}) of $\alpha_{3-i}(n_2) + \alpha_{3-i}(n_3)$ and $\alpha_{3-i}(n_4) - \alpha_{3-i}(n_2)$ follows from the facts already established.

This completes the proof of the Addressing Lemma. ∎

IC. 5 The Admissibility Lemma

The only step remaining in the proof of the theorem is the proof of the Admissibility Lemma. Because $\mathscr{J}_p$ is constructed from normal tilings, and because, by the proof of the Normalization Lemma, any admissible tiling can be turned into a normal one, this lemma may be taken as asserting that the only tilings accepted by $\mathscr{J}_p$ are, to a limited extent, similar to the ones from which $\mathscr{J}_p$ was constructed.

Admissibility Lemma. Every tiling accepted by $\mathscr{J}_p$ is admissible.

Proof. We prove that every tiling accepted by $\mathscr{J}_p$ is i-admissible for each i, by induction on i.

Basis: Every accepted tiling is 0-admissible.

Proof. Suppose β is accepted by $\mathscr{J}_p$. We need to show that there are t_1, t_2 such that $\beta(n+t_c, c) \equiv n \pmod p$ whenever $n \not\equiv 0 \pmod p$, for $c = 1, 2$. We proceed thus: For any $n \not\equiv 0 \pmod p$ let us write n' for the next larger number than n incongruent to $0 \pmod p$; that is, $n' = n+1$ if $1 \le \text{rem}(n, p) < p-1$ and $n' = n+2$ if $n \equiv p-1 \pmod p$. We also write n'' for $(n')'$, etc. We then show

Basis: For $c = 1, 2$, there is a $t_c \in \mathbb{Z}$ and an $n \in \mathbb{Z}$, $n \not\equiv 0 \pmod p$, such that $\beta(n+t_c, c) \equiv n \pmod p$ and $\beta(n'+t_c) \equiv n' \pmod p$;

Induction: For any n, t_c such that $n \not\equiv 0 \pmod p$, if $\beta(n'+t_c, c) \equiv n' \pmod p$ and $\beta(n''+t_c, c) \equiv n'' \pmod p$, then $\beta(n+t_c, c) \equiv n \pmod p$ and $\beta(n'''+t_c, c) \equiv n''' \pmod p$.

Note first that if $\langle q_0, \ldots, q_3 \rangle \in L_p$ and only the second p-ary digits of the q_i (i.e., the $\text{rem}(q_i, p)$) are considered, there are five possibilities:

	$\text{rem}(q_0, p)$	$\text{rem}(q_1, p)$	$\text{rem}(q_2, p)$	$\text{rem}(q_3, p)$
(a)	i	i+1	i+2	i+3
			(where $1 \le i \le p-4$)	
(b)	x	1	2	3
(c)	p-1	x	1	2
(d)	p-2	p-1	x	1
(e)	p-3	p-2	p-1	x

In each case x denotes an unspecified value.

From this the basis step follows almost trivially. The quadruple $\langle \beta(0, c), \beta(1, c), \beta(2, c), \beta(3, c) \rangle$ is of one of the five types (any other sequence of four tape cells would do just as well). If it is of type (a), let $n = i$, $t_c = -i$; if of type (b), let $n = 1$, $t_c = 0$; if of type (c), let $n = 1$, $t_c = 1$; if of type (d), let $n = p-2$, $t_c = -p+2$; and if of type (e), let $n = p-3$, $t_c = -p+3$.

Now assume that n and t_c satisfy the hypotheses of the induction step. We prove only that $\beta(n+t_c, c) \equiv n \pmod p$; the proof that $\beta(n'''+t_c, c) \equiv n''' \pmod p$ is similar. If $2 \le rem(n', p) \le p-3$ then $n' = n+1$, $n'' = n+2$ and the quadruple $Q = \langle \beta(n+t_c, c), \beta(n+1+t_c, c), \beta(n+2+t_c, c), \beta(n+3+t_c, c) \rangle$ must be of type (a) for $i = rem(n, p)$, so $\beta(n+t_c, c) \equiv i \equiv n \pmod p$. If $rem(n', p) = p-2$ then again $n' = n+1$, $n'' = n+2$, and the quadruple Q is of type (e), so $\beta(n+t_c, c) \equiv p-3 \equiv n \pmod p$. If $rem(n', p) = p-1$ then $n' = n+1$, $n'' = n+3 \equiv 1 \pmod p$ and Q is of type (d), so $\beta(n+t_c, c) \equiv p-2 \equiv n \pmod p$. And if $rem(n', p) = 1$ then $n' = n+2$, $n'' = n+3$ and Q is of type (c), so $\beta(n+t_c, c) \equiv p-1 \equiv n \pmod p$.

This completes the Basis Step of the proof of the Admissibility Lemma. ∎

Before beginning the quite technical proof of the induction step, we discuss in a general way how the definition of E_p and I_p (p. 32), parts (E-1) and (I-1) of the Addressing Lemma (p. 24), and the Normal Tiling Lemma fit together to complete the proof.

If a tiling is i-admissible, we know that, after a suitable translation, cells whose coordinates are of order i will be correct in their second digits. To prove (i+1)-admissibility we must show that the first digits are correct on cells whose coordinates are of order i, and we must also extend to order i+1 the correctness of second digits. The crux is that the quadruples in E_p are defined, among those in G_p, in terms of the second digits of the components; but by (E-1) of the Addressing Lemma the orders of the tape cells from which such quadruples arise are related in a fixed way; and then by (a) of the Normal Tiling Lemma the first digits are also related since the components are related by congruences $(\text{mod } p^2)$. Thus it is possible to draw conclusions about the first digits by means of observations about the second digits, provided that the second digits are related in certain ways and the tiling is accepted (so that all the appropriate quadruples are in G_p). Part (I) plays a similar role in establishing the correctness of second digits for order i+1.

The following lemma summarizes the information about G_p needed to complete the induction. It is called the "Extension Lemma", since it extends certain congruences from (mod p) to $(\text{mod } p^2)$.

<u>Extension Lemma</u>. Let $\langle q_1, \ldots, q_4 \rangle \in G_p$.

(E) Suppose $q_4 - q_3 \equiv q_2 - q_1 \pmod p$ and each of the following is incongruent to 0 (mod p): $q_2 - q_1$, $q_4 - q_3$, $q_3 - q_1$, $q_1 + q_4$, $q_2 + q_3$, $q_4 - q_2$. Then $q_4 - q_3 \equiv q_2 - q_1 \pmod{p^2}$.

(I) Suppose $q_3 + q_2 - q_1 \equiv 0 \pmod p$ but $q_3 + q_2 - q_1 \not\equiv 0 \pmod{p^2}$, and each of the following is incongruent to 0 (mod p): $q_4 \pm q_1$, $q_4 \pm q_2$, $q_4 \pm q_3$. Then $q_3 + q_2 - q_1 \equiv pq_4 \pmod{p^2}$.

<u>Proof.</u> Let α be a normal p-tiling and let $n_1, \ldots, n_4$ be such that $n_2 - n_1 = n_4 - n_3$ and $\langle q_1, \ldots, q_4 \rangle = \langle \alpha(n_1), \ldots, \alpha(n_4) \rangle$.

(E) By part (E-1) of the Addressing Lemma, $\Omega(n_1) = \ldots = \Omega(n_4) = e$, say; moreover $e = \Omega(n_2-n_1) = \Omega(n_4-n_3)$, since if $\Omega(n_2-n_1) > \Omega(n_1)$ then $q_1 \equiv q_3 \pmod{p}$. Hence by (a) of the Normal Tiling Lemma, $q_2 - q_1 \equiv q_4 - q_3 \equiv \alpha(n_2-n_1) = \alpha(n_4-n_3) \pmod{p^2}$.

(I) By (I-1) of the Addressing Lemma $\Omega(n_1) = \Omega(n_2) = \Omega(n_3) = \Omega(n_4)-1 = e$, say; and moreover $e = \Omega(n_2-n_1)$. Then by (b) of the Normal Tiling Lemma, $pq_4 \equiv q_3 + \alpha(n_4-n_3) \pmod{p^2}$, but by (a) of the Normal Tiling Lemma $\alpha(n_4-n_3) = \alpha(n_2-n_1) \equiv \alpha(n_2) - \alpha(n_1) = q_2 - q_1 \pmod{p^2}$, whence $pq_4 \equiv q_3 + q_2 - q_1 \pmod{p^2}$. This completes the proof of the Extension Lemma. ■

We can now complete the proof of the Admissibility Lemma.

<u>Induction:</u> <u>An accepted i-admissible tiling is (i+1)-admissible.</u>

<u>Proof.</u> Let $i \geq 0$, let β be an i-admissible tiling accepted by $\mathcal{J}_p$, and let α be a normal tiling and t_1 and t_2 be integers such that for $c = 1, 2$,

(a) if $\Omega(n) < i$ then $\beta(n+t_c, c) = \alpha(n)$;

(b) if $\Omega(n) = i$ then $\beta(n+t_c, c) \equiv \alpha(n) \pmod{p}$.

To show that β is (i+1)-admissible we need to show the existence of a normal tiling α' and integers t'_1, t'_2 such that for $c = 1, 2$,

(a') if $\Omega(n) \leq i$ then $\beta(n+t'_c, c) = \alpha'(n)$;

(b') if $\Omega(n) = i+1$ then $\beta(n+t'_c, c) \equiv \alpha'(n) \pmod{p}$.

We first show that

(A) for every $d \in \mathbb{Z}$ such that $\Omega(d) = i$, there is an $r(d)$ such that $0 < r(d) < p^2$, $r(d) \equiv \lambda(d) \pmod{p}$, and $\beta(n+d+t_2, 2) - \beta(n+t_1, 1) \equiv r(d) \pmod{p^2}$ whenever $\Omega(n) = \Omega(n+d) = i$.

It suffices to show that for any m_1, m_2 such that $\Omega(m_j) = \Omega(m_j+d) = i$ for $j = 1, 2$,

$$\beta(m_1+d+t_2, 2) - \beta(m_1+t_1, 1) \equiv \beta(m_2+d+t_2, 2) - \beta(m_2+t_1, 1) \pmod{p^2}.$$

Let n be any number such that $\Omega(n) = i$ and each of the following is incongruent to 0 (mod p): $\lambda(n) + \lambda(d)$, $\lambda(n) - \lambda(m_j)$ for $j = 1, 2$, $\lambda(n) + \lambda(m_j) + \lambda(d)$ for $j = 1, 2$. (We use the fact that $p > 6$.) Then for $j = 1$ and 2 the quadruple

$$\langle \beta(m_j + t_1, 1),\ \beta(m_j + d + t_2, 2),\ \beta(n + t_1, 1),\ \beta(n + d + t_2, 2) \rangle$$

is in G_p, since β is accepted; and it fulfills the hypotheses of (E) of the Extension Lemma, since $\beta(m+t_c, c) \equiv \lambda(m) \pmod{p}$ for all m of order i, by the i-admissibility of β. Hence by the Extension Lemma, $\beta(m_j+d+t_2, 2) - \beta(m_j+t_1, 1) \equiv \beta(n+d+t_2, 2) - \beta(n+t_1, 1) \pmod{p^2}$ for $j = 1, 2$, which yields the result.

We next show that

(B) for every $d \in \mathbb{Z}$ such that $\Omega(d) = i$, there is an $s(d)$ such that $0 < s(d) < p^2$, $s(d) \equiv \lambda(d) \pmod{p}$, and $\beta(n+d+t_c, c) - \beta(n+t_c, c) \equiv s(d) \pmod{p^2}$ for $c = 1, 2$, whenever $\Omega(n) = \Omega(n+d) = i$.

It suffices to prove (B) for $c = 2$; for then given n and d with $\Omega(n) = \Omega(n+d) = i$, we can find d' with $\Omega(d') = \Omega(n'+d) = \Omega(n'+d+d') = i$; then by (A), and (B) for $c = 2$,

$$\begin{aligned}\beta(n+d+t_1, 1) - \beta(n+t_1, 1) &\equiv (\beta(n+d+d'+t_2, 2) - r(d')) - (\beta(n+d'+t_2, 2) - r(d')) \\ &= \beta(n+d+d'+t_2, 2) - \beta(n+d'+t_2, 2) \equiv s(d) \pmod{p^2} .\end{aligned}$$

We claim that in fact $s(d)$ is the common residue $\pmod{p^2}$ of all the differences $r(d_2) - r(d_1)$, where $\Omega(d_1) = \Omega(d_2) = i$ and $d_2 - d_1 = d$. If all such $r(d_2) - r(d_1)$ have the same residue $\pmod{p^2}$, then their common value $s(d)$ satisfies (B); for given n with $\Omega(n) = i$, we can find d_1, d_2 such that $d_2 - d_1 = d$ and $\Omega(d_1) = \Omega(d_2) = \Omega(n-d_1) = \Omega(n+d_2-d_1) = i$. Then by (A)

$$\beta(n+t_2, 2) - \beta(n-d_1+t_1, 1) \equiv r(d_1) \pmod{p^2}$$

$$\beta(n+d_2-d_1+t_2, 2) - \beta(n-d_1+t_1, 1) \equiv r(d_2) \pmod{p^2}$$

whence

$$\beta(n+d_2-d_1+t_2, 2) - \beta(n+t_2, 2) \equiv r(d_2) - r(d_1) \pmod{p^2} .$$

Moreover $r(d_2) - r(d_1) \equiv \lambda(d_2) - \lambda(d_1) \equiv \lambda(d) \pmod{p}$.

Thus to prove (B) it suffices to show that if $d_2 - d_1 = d_2' - d_1'$ and $\Omega(d_1) = \Omega(d_2) = \Omega(d_1') = \Omega(d_2') = \Omega(d_2-d_1) = i$, then $r(d_2) - r(d_1) = r(d_2') - r(d_1')$. This is easily done; we need merely find an n such that $\Omega(n) = \Omega(n-d_1) = \Omega(n-d_1') = \Omega(n+d_2-d_1) = \Omega(n+d_2'-d_1') = i$; then repeating the most recent calculation twice yields

$$\begin{aligned}r(d_2) - r(d_1) &\equiv \beta(n+d_2-d_1+t_2, 2) - \beta(n+t_2, 2) \\ &= \beta(n+d_2'-d_1'+t_2, 2) - \beta(n+t_2, 2) \\ &\equiv r(d_2') - r(d_1') \pmod{p^2} .\end{aligned}$$

This completes the proof of (B).

From (B) it follows that $s(kd) \equiv ks(d) \pmod{p^2}$ whenever $\Omega(d) = \Omega(kd) = i$, and hence that $\beta(kp^i+t_c, c) \equiv \beta(p^i+t_c, c) + (k-1)s(p^i) \pmod{p^2}$ for $c = 1$ and 2, provided that $k \not\equiv 0 \pmod p$. Since $s(d) \equiv \lambda(d) \pmod p$ for each d of p-order i, there is for each $c = 1, 2$, a k_c, $0 < k_c < p^2$ and $k \equiv 1 \pmod p$, such that $\beta(k_c p^i + t_c, c) = s(p^i)$. Let $\ell_c = (k_c - 1)p^i$; then $\Omega(\ell_c) \geq i+1$ for $c = 1, 2$, and $\beta(\ell_c + kd + t_c, c) \equiv ks(d) \pmod{p^2}$ whenever $\Omega(d) = i$ and $k \not\equiv 0 \pmod p$. Therefore $\beta(\ell_c + n + t_c, c) \equiv \lambda(n) \cdot s(p^i) \pmod{p^2}$ for each n of order i. Let $t'_c = \ell_c + t_c$ for $c = 1, 2$, and let h be such that $\alpha(n) \equiv \lambda(n) \cdot (1 + p \cdot h(\Omega(n))) \pmod{p^2}$ for each n; define

$$h'(j) = \begin{cases} h(j), & \text{if } j \neq i \\ (s(p^i) - 1)/p, & \text{if } j = i \end{cases}$$

(recall that $s(p^i) \equiv 1 \pmod p$); and let α' be the normal p-tiling determined by h', i.e., $\alpha'(n) \equiv \lambda(n) \cdot (1 + p \cdot h'(\Omega(n))) \pmod{p^2}$. We claim that α', t'_1, t'_2 satisfy (a') and (b').

Proof of (a'). If $\Omega(n) < i$ then

$$\begin{aligned} \beta(n + t'_c, c) &= \beta(\ell_c + n + t_c, c) \\ &= \beta(n + t_c, c) \quad \text{by the Agreement Lemma} \\ &= \alpha(n) \\ &\equiv \lambda(n) \cdot (1 + p \cdot h(\Omega(n))) \\ &= \lambda(n) \cdot (1 + p \cdot h'(\Omega(n))) \\ &\equiv \alpha'(n) \pmod{p^2}. \end{aligned}$$

Also, if $\Omega(n) = i$ then

$$\begin{aligned} \beta(n + t'_c, c) &= \beta(\ell_c + n + t_c, c) \\ &\equiv \lambda(n) \cdot s(p^i) \\ &\equiv \lambda(n) \cdot (1 + p \cdot h'(i)) \quad \text{by definition of } h'(i) \\ &= \lambda(n) \cdot (1 + p \cdot h'(\Omega(n))) \\ &\equiv \alpha'(n) \pmod{p^2}. \end{aligned}$$

Proof of (b'). Now let n be a number of order $i+1$. To show that α', t'_1, t'_2 satisfy (b'), we find an n' and d such that n', $n'+d$, and $n-d$ are of order i and the quadruple

$$\langle \beta(n'+t'_1, 1),\ \beta(n'+d+t'_2, 2),\ \beta(n-d+t'_1, 1),\ \beta(n+t'_2, 2) \rangle$$

satisfies (I) of the Extension Lemma. For if we can find such n' and d, then

$$
\begin{aligned}
p \cdot \beta(n+t_2', 2) &\equiv \beta(n-d+t_1', 1) + \beta(n'+d+t_2', 2) - \beta(n'+t_1', 1) \\
&\equiv \alpha'(n-d) + \alpha'(n'+d) - \alpha'(n') \\
&\equiv p \cdot \alpha'(n) - \alpha'(d) + \alpha'(n') + \alpha'(d) - \alpha'(n') \\
&\qquad \text{by (a) and (b) of the Normal Tiling Lemma} \\
&= p \cdot \alpha'(n) \pmod{p^2}
\end{aligned}
$$

so that $\beta(n+t_2', 2) \equiv \alpha'(n) \pmod p$. Moreover by symmetry, it follows from consideration of

$$\langle \beta(n'+d+t_1', 1),\quad \beta(n'+t_2', 2),\quad \beta(n+t_1', 1),\quad \beta(n-d+t_2', 2) \rangle$$

that $\beta(n+t_1', 1) \equiv \alpha'(n) \pmod p$.

For n' we shall take one of the $p-1$ numbers $p^i, 2p^i, \ldots, (p-1)p^i$; and for d we shall also take one of these $p-1$ numbers. We require that $\lambda(n') + \lambda(d) \not\equiv 0 \pmod p$, so for each choice of n' there are only $p-2$ choices of d--a total of $(p-1) \cdot (p-2)$ choices for the pair $\langle n', d \rangle$. Some of these choices must also be disqualified. For in order to guarantee that the conditions in (I) of the Extension Lemma are met we must provide that $\beta(n+t_2', 2)$ be incongruent $\pmod p$ to each of the following: $\pm \beta(n'+t_1', 1)$, $\pm \beta(n'+d+t_2', 2)$, $\pm \beta(n-d+t_1', 1)$. The number of choices for $\langle n', d \rangle$ that must be eliminated for these reasons is no greater than $3 \cdot 2 \cdot (p-2)$: two choices for n', two for $n'+d$, and two for $n-d$ (i.e., for $-d$) must be eliminated, and each of these disqualifies $p-2$ pairs. (We are estimating conservatively.) But among the $(p-1) \cdot (p-2)$ possible choices for $\langle n', d \rangle$, at least one has not been eliminated, since $(p-1) \cdot (p-2) > 6 \cdot (p-2)$ when $p \geq 8$. So n' and d can be found as desired.

This completes the proof of the Admissibility Lemma. ■

Historical References. Linear sampling problems are first described by Aanderaa and Lewis (1974), but are derived ultimately from the automata of Aanderaa (1966). The 1-*systems* of Aanderaa and Lewis (1974) are single-tape versions of the $\Lambda_1^{(\theta)}$-systems of this chapter; the extension to two-tape systems here prepares for the three-tape systems of the next chapter, which are needed for the unsolvability of the unconstrained tiling problems of Chapter IE.

Chapter ID

OTHER VERSIONS OF THE LINEAR SAMPLING PROBLEM

The second version of the linear sampling problem, like the first, deals with a class of tiling types $\Lambda_2^{(\theta)}$ for $\theta \geq 2$. The space to be tiled is

$$3\mathbb{Z} = \mathbb{Z} \times \{1, 2, 3\} \quad ,$$

i.e., three "tapes" or disjoint copies of the integers. Then $\Lambda_2^{(\theta)} = (3\mathbb{Z}, Q_0^{(\theta)}, Q_1)$ where $Q_0^{(\theta)} \subset (3\mathbb{Z})^\theta$ and $Q_1 \subset (3\mathbb{Z})^3$ are as follows:

$$Q_0^{(\theta)}(\langle n_1, e_1\rangle, \ldots, \langle n_\theta, e_\theta\rangle) \quad \text{if and only if} \quad 1 \leq e_1 = e_2 = \cdots = e_\theta \leq 3$$
$$\text{and} \quad n_i = n_1 + i - 1 \quad \text{for} \quad i = 2, \ldots, \theta$$

$$Q_1(\langle n_1, e_1\rangle, \langle n_2, e_2\rangle, \langle n_3, e_3\rangle) \quad \text{if and only if} \quad e_i = i \quad \text{for} \quad i = 1, 2, 3,$$
$$\text{and} \quad n_3 = n_2 - n_1 \quad .$$

Thus $Q_0^{(\theta)}$ relates a sequence of θ consecutive cells from one of the three tapes, while Q_1 relates three cells, one from each tape, whose coordinates satisfy the given linear equation. As before, we refer to the constraint imposed by $Q_0^{(\theta)}$ as the local condition, and to that imposed by Q_1 as the global condition.

Most of this chapter is devoted to the proof that the tiling problem is unsolvable for $\Lambda_2^{(\theta)}$-systems. Section ID.4 considers a third version of the problem, closely related to this one but symmetric in its view of the three tapes.

ID.1 Plan

In order to obtain certain very sharp results on the $\forall\exists\forall$ case of the decision problem for first-order logic in Chapter IID (p. 140), it will be necessary to prove the unsolvability of the tiling problem for a narrow class of $\Lambda_2^{(\theta)}$-systems.

Linear Sampling Theorem, Second Version. For each $\theta \geq 2$, the tiling problem for $\Lambda_2^{(\theta)}$-systems is unsolvable, even for $\Lambda_2^{(\theta)}$-systems (T, L, G) satisfying the

0-1 restriction: $\{0, 1\} \subset T$, $\{0, 1\}^\theta \subset L$, and $G \subset (T - \{0, 1\}) \times (T - \{0, 1\}) \times \{0, 1\}$.

The 0-1 restriction implies that only the symbols 0 and 1 may appear on the third tape, and that no local condition is imposed on the third tape.

We first note that by blocking together sequences of θ tape cells, any $\Lambda_2^{(\theta)}$-system $\mathscr{T}$ can be reduced to a $\Lambda_2^{(2)}$-system $\mathscr{T}'$, such that $\mathscr{T}$ accepts a tiling if and only if $\mathscr{T}'$ accepts a tiling; moreover, the construction may be done in such a way that $\mathscr{T}'$ satisfies the 0-1 restriction if $\mathscr{T}$ does. The construction is much like that at the beginning of Chapter IC. Let $\mathscr{T} = (T, L, G)$; then $\mathscr{T}' = (T', L', G')$ where $T' = T^{\theta}$, and

$$\langle\langle s_1, \ldots, s_\theta\rangle, \langle s_{\theta+1}, \ldots, s_{2\theta}\rangle\rangle \in L' \text{ if and only if } s_{i+1} = s_{\theta+i}$$

$$\text{for } i = 1, \ldots, \theta-1, \text{ and } \langle s_1, \ldots, s_\theta\rangle, \langle s_{\theta+1}, \ldots, s_{2\theta}\rangle \in L \;;$$

$$\text{and } \langle\langle s_1, \ldots, s_\theta\rangle, \langle s_{\theta+1}, \ldots, s_{2\theta}\rangle, \langle s_{2\theta+1}, \ldots, s_{3\theta}\rangle\rangle \in G'$$

$$\text{if and only if } \langle s_1, s_{\theta+1}, s_{2\theta+1}\rangle \in G \;.$$

If $\mathscr{T}$ satisfies the 0-1 restriction, then modify the $\mathscr{T}'$ constructed in this way by replacing every member of $\{0, 1\}^{\theta}$ by its first component. We leave it to the reader to check that the $\Lambda_2^{(2)}$-system derived has the appropriate properties.

In order to establish the Linear Sampling Theorem (second version), our plan is to construct, for each $\Lambda_1^{(2)}$-system $\mathscr{S}$, a $\Lambda_2^{(\theta)}$-system $\mathscr{T}_{\mathscr{S}}$ satisfying the 0-1 restriction such that $\mathscr{S}$ accepts some tiling of $2\mathbb{Z}$ if and only if $\mathscr{T}_{\mathscr{S}}$ accepts some tiling of $3\mathbb{Z}$. In our construction the value of θ will depend on $\mathscr{S}$, but as just shown θ can be reduced to 2 without violating the 0-1 restriction. We now sketch the construction.

Let $\mathscr{S} = (T, L, G)$ be a $\Lambda_1^{(2)}$-system, and let $|T| = q$. Then $\mathscr{T}_{\mathscr{S}} = (T', L', G')$, where $T' = (\{0, 1, \ldots, q^2-1\} \times T) \cup \{0, 1\}$. The first two tapes of $\mathscr{T}_{\mathscr{S}}$ have two channels. The first channels contain numbers in the range $0, \ldots, q^2-1$; each such number will be correlated with a pair of symbols $\langle s, s'\rangle \in T^2$. The local condition will be used to guarantee that in any tiling accepted by $\mathscr{T}_{\mathscr{S}}$, these first channels must contain the periodic sequence $\ldots, 0, 1, \ldots, q^2-1, 0, 1, \ldots, q^2-1, 0, 1, \ldots$. The second channels of the first two tapes of $\mathscr{T}_{\mathscr{S}}$ correspond to the two tapes of $\mathscr{S}$, but with every cell repeated q^2 times. Figure 7 shows a portion of the first tape of $\mathscr{T}_{\mathscr{S}}$ that would correspond to the sequence $\ldots s_1 s_2 s_3 \ldots$ on the first tape of $\mathscr{S}$.

0	1	2	…	q^2-1	0	1	2	…	q^2-1	0	1
s_1	s_1	s_1	…	s_1	s_2	s_2	s_2	…	s_2	s_3	s_3

Figure 7

The third tape of $\mathscr{T}_{\mathscr{S}}$ contains only the symbols 0 and 1. It acts as a memory to compare different pairs of cells from the first two tapes that are the same distance apart (i.e., have the same difference between their coordinates) and thus would be compared directly by $\mathscr{S}$. Let $\alpha: 2\mathbb{Z} \to T$ be a tiling of $2\mathbb{Z}$ accepted by $\mathscr{S}$. Then a particular block of q^2 cells on the third tape of $\mathscr{T}_{\mathscr{S}}$ -- say, cells $dq^2, dq^2+1, \ldots, dq^2+q^2-1$ --is to be an encoding of the set of all pairs $\langle \alpha(n, 1), \alpha(n+d, 2)\rangle$ of symbols encountered in a cell of the first tape of $\mathscr{S}$ and a cell of the second tape whose coordinate is d greater. Let us fix some one-one correspondence $\gamma: \{0, \ldots, q^2-1\} \to T^2$. Then the presence of a 1 in cell dq^2+k of the third tape of $\mathscr{T}_{\mathscr{S}}$, where $0 \le k \le q^2-1$, signifies that for some integer n, $\langle \alpha(n, 1), \alpha(n+d, 2)\rangle = \gamma(k)$; the presence of a 0 indicates that the pair of symbols $\gamma(k)$ appears in no pair of cells of α (one from each tape) that are distance d apart.

ID. 2 The Construction

As stated, let $\mathscr{S} = (T, L, G)$ be a $\Lambda_1^{(2)}$-system and let $|T| = q$; then $\mathscr{T}_{\mathscr{S}} = (T', L', G')$ is a $\Lambda_2^{(\theta)}$-system, where $\theta = 2q^2$. Let $T' = (\{0, \ldots, q^2-1\} \times T) \cup \{0, 1\}$. Then $L' \subset (T')^{\theta}$ contains exactly those θ-tuples obtainable from L as follows. Let $\langle s_1, s_2\rangle$ and $\langle s_2, s_3\rangle \in T$; then for each j, $0 \le j \le q^2-1$, L' contains

$$\langle\langle j, s_1\rangle, \langle j+1, s_1\rangle, \ldots, \langle q^2-1, s_1\rangle, \langle 0, s_2\rangle, \langle 1, s_2\rangle, \ldots, \langle q^2-1, s_2\rangle, \langle 0, s_3\rangle, \langle 1, s_3\rangle, \ldots, \langle j-1, s_3\rangle\rangle .$$

Also, L' contains every member of $\{0, 1\}^{\theta}$.

Now let α be a tiling of $3\mathbb{Z}$ accepted by $\mathscr{T}_{\mathscr{S}}$. By means of the global condition we shall guarantee that the first two tapes contain only members of $\{0, \ldots, q^2-1\} \times T$ and the third tape contains only 0's and 1's. Moreover, the local condition guarantees that the periodic sequence $\ldots, 0, 1, \ldots, q^2-1, 0, 1, \ldots$ appears in the first channel of the first two tapes. Thus, there is a t, $0 \le t \le q^2-1$, with the following property: for any n_1, n_2, n_3, if $n_3 = n_2 - n_1$, and $\langle \alpha(n_1, 1), \alpha(n_2, 2), \alpha(n_3, 3)\rangle = \langle\langle j_1, s_1\rangle, \langle j_2, s_2\rangle, e\rangle$ (where $0 \le j_1, j_2 \le q^2-1$, $s_1, s_2 \in T$, $e \in \{0, 1\}$), then $n_3 \equiv j_2 - j_1 + t \pmod{q^2}$. Let us take the simple case in which $t = 0$, $j_2 > j_1$, $n_1 \equiv j_1$, $n_2 \equiv j_2$, and $n_3 \equiv j_2 - j_1 \pmod{q^2}$; say $n_1 = q^2 n + j_1$, $n_2 = q^2 n + q^2 d + j_2$, $n_3 = q^2 d + j_2 - j_1$. Then $\alpha(n_3, 3) = 1$ is to signify that in the correlated tiling β of $\mathbb{Z}_2$ accepted by $\mathscr{S}$, there is a pair of integers m_1, m_2 such that $m_2 - m_1 = d$ and $\langle \beta(m_1, 1), \beta(m_2, 2)\rangle = \gamma(j_2 - j_1)$, i.e., there is a pair of cells of β whose coordinates differ by d containing the pair of symbols associated with the number $j_2 - j_1$.

The triples in G' are of three kinds. The first kind guarantees (in the $t = 0$ case) that if $n_2 - n_1 = q^2 d + j_2 - j_1$ and $\langle \alpha(n_1, 1), \alpha(n_2, 2)\rangle = \langle\langle j_1, s_1\rangle, \langle j_2, s_2\rangle\rangle$ is such that $j_1 \le j_2$ and $\gamma(j_2 - j_1) = \langle s_1, s_2\rangle$, then a 1 appears in the cell of the third tape inspected at the same time, i.e., $\alpha(n_2 - n_1, 3) = 1$. Triples of the second kind guarantee that, under the same circumstances, if $j_2 - j_1 \geq 0$ but $\gamma(j_2 - j_1) = \langle s_3, s_4 \rangle \neq \langle s_1, s_2 \rangle$, then $\alpha(n_2 - n_1, 3) = 1$ if and only if the quadruple $\langle s_1, s_2, s_3, s_4 \rangle$ is in G. These two kinds of triples do the real work of imitating $\mathscr{S}$. The other triples are those $\langle \langle j_1, s_1 \rangle, \langle j_2, s_2 \rangle, e \rangle$ such that $j_1 > j_2$; G' contains all such triples.

Formally, G' is defined by the following three conditions:

(a) For any $s_1, s_2 \in T$, and any j_1, j_2, $0 \leq j_1 \leq j_2 \leq q^2 - 1$, such that $\gamma(j_2 - j_1) = \langle s_1, s_2 \rangle$,

(i) $\langle \langle j_1, s_1 \rangle, \langle j_2, s_2 \rangle, 0 \rangle \notin G'$; and

(ii) $\langle \langle j_1, s_1 \rangle, \langle j_2, s_2 \rangle, 1 \rangle \in G'$ if and only if $\langle s_1, s_2, s_1, s_2 \rangle \in G$.

(b) For any $s_1, s_2 \in T$, and any j_1, j_2, $0 \leq j_1 \leq j_2 \leq q^2 - 1$, such that $\gamma(j_2 - j_1) = \langle s_3, s_4 \rangle \neq \langle s_1, s_2 \rangle$,

(i) $\langle \langle j_1, s_1 \rangle, \langle j_2, s_2 \rangle, 0 \rangle \in G'$; and

(ii) $\langle \langle j_1, s_1 \rangle, \langle j_2, s_2 \rangle, 1 \rangle \in G'$ if and only if $\langle s_1, s_2, s_3, s_4 \rangle \in G$.

(c) For any $s_1, s_2 \in T$, any j_1, j_2 such that $0 \leq j_2 < j_1 \leq q^2 - 1$, and for $e = 0, 1$, $\langle \langle j_1, s_1 \rangle, \langle j_2, s_2 \rangle, e \rangle \in G'$.

ID. 3 <u>The Proof</u>

To prove that $\mathscr{T}_{\mathscr{S}}$ is as required, we introduce mappings between tilings of $2\mathbb{Z}$ and of $3\mathbb{Z}$. If α is a tiling of $3\mathbb{Z}$ such that $\alpha(n, i) \in \{0, \ldots, q^2 - 1\} \times T$ for $n \in \mathbb{Z}$, $i = 1, 2$, then the <u>contraction</u> of α is the tiling α' of $2\mathbb{Z}$ such that $\alpha'(n, i) = \pi_2(\alpha(q^2 n, i))$ for $n \in \mathbb{Z}$, $i = 1, 2$. (Here π_2 is the projection function on the second coordinate.) Also, if α is a tiling of $2\mathbb{Z}$ with $\alpha(n, i) \in T$ for $n \in \mathbb{Z}$, $i = 1, 2$, then the <u>dilation</u> of α is the tiling α' of $3\mathbb{Z}$ such that $\alpha'(q^2 n + j, i) = \langle j, \alpha(n, i) \rangle$ for $n \in \mathbb{Z}$, $0 \leq j \leq q^2 - 1$, $i = 1, 2$, and such that

$$\alpha'(q^2 d + j, 3) = \begin{cases} 1 & \text{if there is an } n \text{ such that } \langle \alpha(n, 1), \alpha(n+d, 2) \rangle = \gamma(j) \\ 0 & \text{otherwise} \end{cases}$$

($d \in \mathbb{Z}$, $0 \leq j \leq q^2 - 1$).

To complete the proof, we need two lemmata.

<u>Dilation Lemma</u>. The dilation of a tiling accepted by $\mathscr{S}$ is accepted by $\mathscr{T}_{\mathscr{S}}$.

<u>Proof</u>. Let α be a tiling of $2\mathbb{Z}$ accepted by $\mathscr{S}$, and let α' be the dilation of α. Clearly, $\langle \alpha'(n, i), \alpha'(n+1, i), \ldots, \alpha'(n+\theta-1, i) \rangle \in L'$ for $n \in \mathbb{Z}$, $i = 1, 2, 3$. Now consider a triple

$$\tau = \langle \alpha'(q^2 n_1 + j_1, 1), \alpha'(q^2 n_2 + j_2, 2), \alpha'(q^2(n_2 - n_1) + j_2 - j_1, 3)\rangle$$

where $n_1, n_2 \in \mathbb{Z}$, $0 \le j_1, j_2 \le q^2 - 1$; we wish to show that $\tau \in G'$. By the definition of dilation, $\pi_1(\alpha'(q^2 n_i + j_i)) = j_i$ for $i = 1, 2$; so let $s_1, s_2 \in T$ and $e \in \{0, 1\}$ be such that $\tau = \langle\langle j_1, s_1\rangle, \langle j_2, s_2\rangle, e\rangle$. If $j_1 > j_2$ then $\tau \in G'$ by (c) of the definition of G'. Now assume $j_1 \le j_2$; there are two cases to consider.

(i) If $\gamma(j_2 - j_1) = \langle s_1, s_2\rangle$, then by (aii) $\tau \in G'$ if $e = 1$ and $\langle s_1, s_2, s_1, s_2\rangle \in G$. But consider $\langle \alpha(n_1, 1), \alpha(n_2, 2), \alpha(n_1, 1), \alpha(n_2, 2)\rangle$; this quadruple is in G since $\mathscr{S}$ accepts α, and is equal to $\langle s_1, s_2, s_1, s_2\rangle$ by the definition of dilation. Moreover, $e = 1$ by the definition of dilation.

(ii) Suppose $\gamma(j_2 - j_1) = \langle s_3, s_4\rangle \ne \langle s_1, s_2\rangle$. If $e = 0$ then $\tau \in G'$ by (bi). If $e = 1$ then by the definition of dilation there are $n_1', n_2' \in \mathbb{Z}$ such that $n_2' - n_1' = n_2 - n_1$ and $\langle \alpha(n_1', 1), \alpha(n_2', 2)\rangle = \gamma(j_2 - j_1)$. Then the quadruple $\langle \alpha(n_1, 1), \alpha(n_2, 2), \alpha(n_1', 1), \alpha(n_2', 2)\rangle = \langle s_1, s_2, s_3, s_4\rangle$ is in G since $n_2 - n_1 = n_2' - n_1'$ and $\mathscr{S}$ accepts α. Hence $\tau \in G'$ by (bii).

This completes the proof of the Dilation Lemma. ■

Contraction Lemma. The contraction of a tiling accepted by $\mathscr{T}_{\mathscr{S}}$ is accepted by $\mathscr{S}$.

Proof. Let α be a tiling of $3\mathbb{Z}$ accepted by $\mathscr{T}_{\mathscr{S}}$, and let α' be the contraction of α. It is readily checked that $\langle \alpha'(n, i), \alpha'(n+1, i)\rangle \in L$ for all $n \in \mathbb{Z}$, and $i = 1, 2$. Moreover, there are t_1, t_2, $0 \le t_1, t_2 \le q^2 - 1$, such that for $i = 1, 2$, and for $0 \le j \le q^2 - 1$,

$$\pi_1(\alpha(q^2 n + j - t_i, i)) = j$$

and

$$\pi_2(\alpha(q^2 n + j - t_i, i)) = \pi_2(\alpha(q^2 n - t_i, i)) \quad ;$$

that is, the periodic sequence $\ldots, 0, 1, \ldots, q^2 - 1, 0, \ldots$ is in the first channels, and the cells in a contiguous block with $0, 1, \ldots, q^2 - 1$ in the first channel all have the same member of T in the second channel. Thus, $\alpha(q^2 n + j - t_i, i) = \langle j, \alpha'(n, i)\rangle$ for each $n \in \mathbb{Z}$, $0 \le j \le q^2 - 1$, $i = 1, 2$. Now let $n_1, n_2, d \in \mathbb{Z}$; we wish to show that $\tau = \langle \alpha'(n_1, 1), \alpha'(n_1 + d, 2), \alpha'(n_2, 1), \alpha'(n_2 + d, 2)\rangle$ is in G. Let $j_i = \gamma^{-1}(\alpha'(n_i), \alpha'(n_i + d))$ for $i = 1, 2$. Then for $i = 1, 2$, the triple

$$\tau_i = \langle \alpha(q^2 n_i - t_1, 1), \alpha(q^2(n_i + d) + j_i - t_2, 2), \alpha(q^2 d + j_i + t_1 - t_2, 3)\rangle$$

is in G' since $\mathscr{T}_{\mathscr{S}}$ accepts α. Since $\alpha(q^2 n + j - t_i, i) = \langle j, \alpha'(n, i)\rangle$ for any j, $0 \le j \le q^2 - 1$, and by the definition of the j_i and part (a) of the definition of G',

$$\tau_i = \langle\langle 0, \alpha'(n_i, 1)\rangle, \langle j_i, \alpha'(n_i + d, 2)\rangle, 1\rangle$$

for $i = 1, 2$. There are two cases to consider.

(i) If $\langle \alpha'(n_1, 1), \alpha'(n_1+d, 2)\rangle = \langle \alpha'(n_2, 1), \alpha'(n_2+d, 2)\rangle$, then $\tau \in G$ since by part (a) of the definition of G', $\langle \alpha'(n_i, 1), \alpha'(n_i+d, 2), \alpha'(n_i, 1), \alpha'(n_i+d, 2)\rangle \in G$ for $i = 1, 2$.

(ii) If $\langle \alpha'(n_1, 1), \alpha'(n_1+d, 2)\rangle \neq \langle \alpha'(n_2, 1), \alpha'(n_2+d, 2)\rangle$, then consider the triple

$$\langle \alpha(q^2 n_1 - t_1, 1), \alpha(q^2(n_1+d) + j_2 - t_2, 2), \alpha(q^2 d + j_2 + t_1 - t_2, 3)\rangle$$

which is in G' since $\mathscr{T}_{\mathscr{S}}$ accepts α, and is equal to

$$\langle\langle 0, \alpha'(n_1, 1)\rangle, \langle j_2, \alpha'(n_1+d, 2)\rangle, 1\rangle \ .$$

Since $\gamma^{-1}(\alpha'(n_1, 1), \alpha'(n_1+d, 2)) \neq j_2$, part (b) of the definition of G' applies. Hence

$$\langle \alpha'(n_1, 1), \alpha'(n_1+d, 2), \pi_1(\gamma(j_2)), \pi_2(\gamma(j_2))\rangle$$

is in G; but this is τ.

This completes the proof of the Dilation Lemma and the Linear Sampling Theorem, Second Version. ■

ID.4 A Refinement

The second version of the linear sampling problem can, by slight reformulation, be put in a simple and symmetric form $\Lambda_3 = (\mathbb{Z}_3, P_0, P_1)$, where $3\mathbb{Z} = \mathbb{Z} \times \{1, 2, 3\}$ as before,

$$P_0 = Q_0^{(2)}, \quad \text{i.e.,} \quad P_0(\langle n_1, e_1\rangle, \langle n_2, e_2\rangle) \text{ if and only if } 1 \leq e_1 = e_2 \leq 3 \text{ and } n_2 = n_1 + 1 \ ;$$

$$P(\langle n_1, e_1\rangle, \langle n_2, e_2\rangle, \langle n_3, e_3\rangle) \text{ if and only if } e_i = i \text{ for } i = 1, 2, 3 \text{ and } n_1 + n_2 + n_3 = 0 \ .$$

Thus the only difference between the tiling types $\Lambda_2^{(2)}$ and Λ_3 is that the linear relation $n_3 = n_2 - n_1$ is replaced by $n_1 + n_2 + n_3 = 0$; the global condition applies to three tape cells, one from each tape, whose coordinates sum to 0. (We do not use the 0-1 restriction here.)

The tiling problem for Λ_3-systems is shown unsolvable, in essence, by reversing the direction of the second tape of $\Lambda_2^{(2)}$-systems. Formally there is a slight technical problem because the local condition of a $\Lambda_2^{(2)}$-system is applied indifferently to all three tapes; but this is easily repaired by making the alphabets for the three tapes disjoint by means of the global condition. Thus, let

$\mathscr{T} = (T, L, G)$ be any $\Lambda_2^{(2)}$-system; form a Λ_3-system $\mathscr{T}' = (T', L', G')$ from $\mathscr{T}$ as follows:

$$T' = T \times \{1, 2, 3\} \quad ;$$

$$L' = \{\langle\langle s, e\rangle, \langle s', e\rangle\rangle \mid \langle s, s'\rangle \in L, \quad e = 1 \text{ or } 3\}$$

$$\cup \langle\langle s', 2\rangle, \langle s, 2\rangle\rangle \mid \langle s, s'\rangle \in L\}$$

$$G' = \{\langle\langle s, 1\rangle, \langle s', 2\rangle, \langle s'', 3\rangle\rangle \mid \langle s, s', s''\rangle \in G \quad .$$

Then α is a tiling of $3\mathbb{Z}$ accepted by $\mathscr{T}$ if and only if α' is a tiling of $3\mathbb{Z}$ accepted by $\mathscr{T}'$, where

$$\alpha'(n, e) = \langle \alpha(n, e), e\rangle \qquad (e = 1 \text{ or } 3)$$

$$= \langle \alpha(-n, e), e\rangle \qquad (e = 2) \quad .$$

This construction suffices to prove:

Linear Sampling Theorem, Third Version. The tiling problem for Λ_3-systems is unsolvable.

This version of the linear sampling problem is the most convenient for application to unconstrained tiling problems in the next chapter. To deduce from it the unsolvability of the one-tape version of the problem described in the introduction, simply regard the three tapes of a Λ_3-system as three independent channels of a single tape.

Historical Reference. The $\Lambda_2^{(\theta)}$-systems presented here are three-tape versions of the two-tape 2-*systems* of Aanderaa and Lewis (1974).

Chapter IE

UNCONSTRAINED TILING PROBLEMS

The tiling problems considered in this chapter, like the origin-constrained tiling problem, call for covering a planar surface with abutting plates in such a way that certain adjacency conditions are satisfied. These problems are "unconstrained" in that the entire plane is to be covered, and the plane is treated as a homogeneous space with no special status accorded the origin or any other point or set of points. Two versions of the problem are treated: that in which the plates to be used for tiling are hexagonal, and that in which the plates are square. The hexagonal version is easily shown unsolvable by reducing to it the tiling problem for Λ_3-systems; we then show how the hexagons can be encoded as complexes of squares so as to prove the square version unsolvable as well.

IE. 1 Hexagonal Version

Intuitively, the hexagonal tiling problem is to tell whether a honeycomb tiling of the plane can be made, the cells of the honeycomb to be filled with tiles from a finite set of prototypes and with restrictions to be observed on which tiles can neighbor each other on the six sides. Before formalizing the problem, we indicate why the tiling problem for Λ_3-systems is well-suited to reduction to the hexagonal tiling problem. Recall that the global condition for a Λ_3-system permits simultaneous inspection of any three tape cells, one from each tape, whose coordinates sum to 0. Our formulation of the hexagonal tiling problem is based on a coordinate system for the planar honeycomb under which each cell is assigned a unique triple of integers $\langle n_1, n_2, n_3 \rangle$ such that $n_1+n_2+n_3 = 0$ (Figure 8). Moreover, each neighbor of a cell has coordinates that differ from its own coordinates by at most one in each component. Thus there are three principal axes at angles of 120° to each other; all cells encountered by traveling in a direction perpendicular to any one of the three axes have the same coordinate in one of the three components.

Formally, then, the hexagonal tiling problem is the tiling problem for systems of type

$$\Upsilon_6 = (\mathbb{H}, P_1, P_2, P_3)$$

where $\mathbb{H} = \{\langle n_1, n_2, n_3 \rangle \mid n_1+n_2+n_3 = 0\}$ and for $i = 1, 2, 3$, $P_i(\langle n_1, n_2, n_3 \rangle, \langle m_1, m_2, m_3 \rangle)$ if and only if, for $j = 1, 2, 3$,

$$m_j = \begin{cases} n_j & (j = i) \\ n_j - 1 & (j \equiv i+1 \pmod 3) \\ n_j + 1 & (j \equiv i-1 \pmod 3) \end{cases}$$

The P_i represent the relations of adjacency in the directions perpendicular to the three axes. (The upsilon is for "unconstrained", the 6 for "hexagonal".)

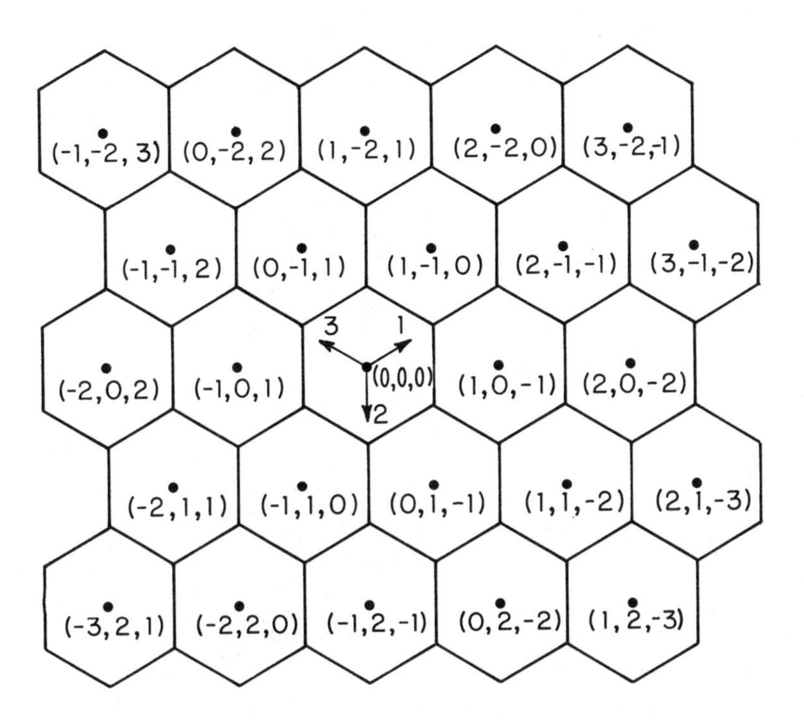

Figure 8

Now let $\mathscr{T} = (T, L, G)$ be any Λ_3-system; we show how to construct a Υ_6-system $\mathscr{U} = (U, A_1, A_2, A_3)$ such that there is a tiling of $3\mathbb{Z}$ accepted by $\mathscr{T}$ if and only if there is a tiling of $\mathbb{H}$ accepted by $\mathscr{U}$. We imagine the three tapes of $\mathscr{T}$ laid out on the plane perpendicular to the three axes of $\mathbb{H}$ and thus at angles of 120° to each other. Moreover, each tape is replicated infinitely, the copies laid out parallel to each other. Thus for each $i = 1, 2, 3,$ and for each $n \in \mathbb{Z}$, the set of honeycomb cells whose i^{th} coordinates are equal to n comprises a copy of the ith tape. Each honeycomb cell represents the intersection of copies of each of the three tapes; if the honeycomb cell has coordinate (n_1, n_2, n_3), it represents that intersection of the three tapes in which cell n_1 of the first tape, n_2 of the second, and n_3 of the third are brought into coincidence. Figure 9 illustrates the representation of three tapes $\ldots s_{-1}s_0s_1\ldots$, $\ldots t_{-1}t_0t_1\ldots$, and $\ldots u_{-1}u_0u_1\ldots$.

Thus the set of U of tiles of $\mathscr{U}$ will be a subset of T^3 --in fact, $U = G$, the set of triples of tiles of $\mathscr{T}$ that may lie on cells of the three tapes whose coordinates sum to 0. The adjacency conditions of $\mathscr{U}$ are used for two purposes:

first, to insure that in a tiling of $\mathbb{H}$ accepted by $\mathscr{U}$ all the copies of any one tape are indeed identical, and second, to encode the local condition of $\mathscr{T}$. Formally, then, $\mathscr{U} = (U, A_1, A_2, A_3)$ where

$$U = G$$

and for $i = 1, 2, 3$,

$$\langle\langle s_1, s_2, s_3\rangle, \langle s_1', s_2', s_3'\rangle\rangle \in A_i \quad \text{if and only if}$$

(1) $\langle s_j, s_j'\rangle \in L$, where $1 \leq j \leq 3$ and $j \equiv i-1 \pmod e$; and

(2) $s_i = s_i'$.

Then $\mathscr{T}$ and $\mathscr{U}$ are related in the following ways. First, if α is a tiling of $3\mathbb{Z}$ accepted by $\mathscr{T}$ then α' is a tiling of $\mathbb{H}$ accepted by $\mathscr{U}$, where if $n_1+n_2+n_3 = 0$ then

$$\alpha'(n_1, n_2, n_3) = \langle \alpha(n_1, 1), \alpha(n_2, 2), \alpha(n_3, 3)\rangle \quad .$$

And second, if α' is a tiling of $\mathbb{H}$ accepted by $\mathscr{U}$, then there is a tiling α of $3\mathbb{Z}$ related to α' as just stated, such that α is accepted by $\mathscr{T}$. We dispense with a formal proof.

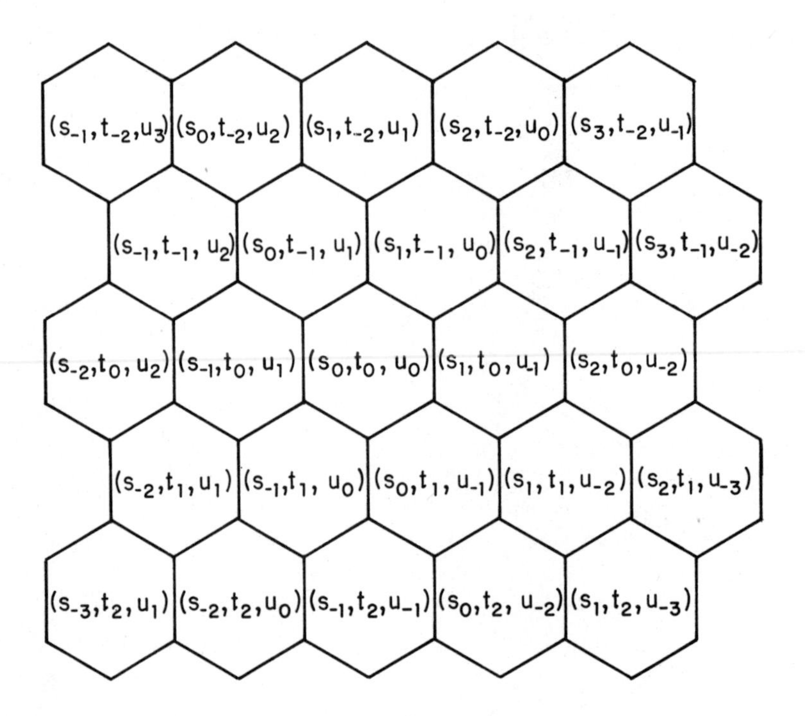

Figure 9

IE.2 Square Version

The square version of the unconstrained tiling problem is like the origin-constrained tiling problem Δ_0, but without the origin constraint. Formally, $\Upsilon_4 = (\mathbb{Z}^2, R_H, R_V)$, where $R_H(\langle n_1, n_2\rangle, \langle m_1, m_2\rangle)$ if and only if $m_1 = n_1 + 1$ and $m_2 = n_2$, and $R_V(\langle n_1, n_2\rangle, \langle m_1, m_2\rangle)$ if and only if $m_1 = n_1$ and $m_2 = n_2 + 1$. The crucial observation in the proof that Υ_6-systems can be reduced to Υ_4-systems is that a hexagonal tiling can be encoded as a square tiling by blocking together groups of eight squares as illustrated in Figure 10. That is, if $\mathcal{U} = (U, A_1, A_2, A_3)$ is an Υ_6-system, then we can construct an Υ_4-system $\mathcal{U}' = (U', H, V)$ such that there is a tiling of $\mathbb{H}$ accepted by $\mathcal{U}$ if and only if there is a tiling of $\mathbb{Z}^2$ accepted by $\mathcal{U}'$. Let $U' = U \times \{1, \ldots, 8\}$, so that eight square tiles associated with each hexagonal tile. Then arrange the horizontal and vertical adjacency conditions H and V so that

(1) the eight parts of a hexagonal tile must appear together as illustrated; and

(2) the three adjacency conditions for two hexagonal tiles t_1, t_2 of $\mathcal{U}$ are encoded into restrictions on the horizontal adjacency conditions between the corresponding square tiles $\langle t_1, 3\rangle$ and $\langle t_2, 6\rangle$; $\langle t_1, 8\rangle$ and $\langle t_2, 1\rangle$; and $\langle t_1, 2\rangle$ and $\langle t_2, 7\rangle$.

We omit the tedious technical details, and simply state the conclusion.

Unconstrained Tiling Theorem. The unconstrained hexagonal and square tiling problems Υ_6 and Υ_4 are unsolvable.

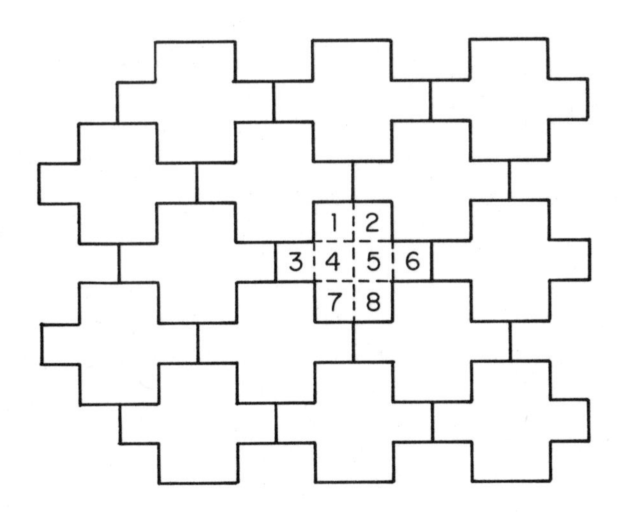

Figure 10

Historical References. The unconstrained domino problem was formulated by Wang (1961); it is interreducible with the tiling problem for Y_4-systems by the techniques explained in Section IB.3. Although the unconstrained domino problem was introduced in the context of Wang's investigations of the $\forall\exists\forall$ case, the $\forall\exists\forall$ case was settled (by Kahr, Moore, and Wang (1962)) with the aid of the diagonal-constrained domino problem, essentially the tiling problem for systems of type $(\mathbb{Z}^2, R_H, R_V, R_D)$, where R_H and R_V are as in this chapter and

$$R_D(\langle x, y\rangle) \qquad \text{if and only if} \quad x = y$$

i.e., D holds for points on the diagonal. The unconstrained problem was shown unsolvable by Berger (1966) by a direct encoding of the Turing Machine Halting Problem; the idea for the proof presented here is due to Aanderaa. Robinson (1971) greatly simplified Berger's proof. For an interesting variation on the unconstrained tiling problems of this chapter, see Martin Gardner's "Mathematical Games" column in the January, 1977 Scientific American.

Chapter IF

MISCELLANEOUS PROBLEMS

This chapter treats the Post Correspondence Problem and the halting problem for two-counter machines. The unsolvability of these problems is well known, and there is not much novelty in our presentation; we include these proofs for the sake of completeness, since they are used in Part II to establish the unsolvability of certain logical decision problems.

IF. 1 Post Correspondence Problem

A correspondence system is a finite subset $\mathscr{P}$ of $\Sigma^+ \times \Sigma^+$ for some finite alphabet Σ; i.e., a finite set of pairs of nonempty words. A presolution of $\mathscr{P}$ is a pair of words $\langle v_1 \cdots v_k, w_1 \cdots w_k \rangle$ such that $\langle v_i, w_i \rangle \in \mathscr{P}$ for $i = 1, \ldots, k$. This presolution is a solution of $\mathscr{P}$ provided that $k > 0$ and $v_1 \cdots v_k = w_1 \cdots w_k$. The Post correspondence problem is to determine, given a correspondence system $\mathscr{P}$, whether or not $\mathscr{P}$ has a solution.

Post Correspondence Theorem. The Post correspondence problem is unsolvable.

Proof. The first step is to define a modified Post correspondence problem in which one of the pairs in the correspondence system is distinguished. This pair must be the first used in the construction of a presolution. We show that the modified problem can be reduced to the unmodified problem. The second step is to show that the Turing machine halting problem (inputless version) can be reduced to the modified problem.

A modified correspondence system is a pair $\mathscr{P}' = (\mathscr{P}, \langle x, y \rangle)$, where $\mathscr{P}$ is a correspondence system and $\langle x, y \rangle \in \mathscr{P}$. A presolution of $\mathscr{P}'$ is a presolution of $\mathscr{P}$ of the form $\langle xv, yw \rangle$; that is, a pair $\langle v_1 \cdots v_k, w_1 \cdots w_k \rangle$, with $k \geq 1$, $\langle v_i, w_i \rangle \in \mathscr{P}$ for $i = 1, \ldots, k$, and $\langle v_1, w_1 \rangle = \langle x, y \rangle$. A solution of $\mathscr{P}'$ is a presolution of $\mathscr{P}'$ that is a solution of $\mathscr{P}$.

We now show how to construct, for each modified correspondence system $\mathscr{P}'_1$, a correspondence system $\mathscr{P}_2$ such that $\mathscr{P}'_1$ has a solution if and only if $\mathscr{P}_2$ has a solution. We omit a formal proof, providing instead a detailed commentary on the construction.

Let $\mathscr{P}'_1 = (\mathscr{P}_1, \langle x, y \rangle)$, where $\mathscr{P}_1 \subset \Sigma^+ \times \Sigma^+$, and let $*$ and \$ be new symbols not in Σ. If $\mathscr{P}_1$ contains k pairs, $\mathscr{P}_2$ will contain k+2: one for each pair in $\mathscr{P}_1$, and special pairs that can be used only at the beginning and end of a solution. The form of solutions of $\mathscr{P}_2$ is this: if $\langle a_1 \cdots a_n, a_1 \cdots a_n \rangle$ (where the a_i are symbols in Σ) is a solution of $\mathscr{P}'_1$, then

$\langle *a_1*a_2\cdots *a_n*\$, *a_1*a_2\cdots *a_n*\$\rangle$ will be a solution of $\mathscr{P}_2$; that is, the corresponding solution of $\mathscr{P}_2$ is obtained by placing a $*$ between each pair of symbols, and at the beginning and end, and then terminating with a \$. Moreover, every solution of $\mathscr{P}_2$ is of the form $\langle u_1\$u_2\$\ldots u_m\$, u_1\$u_2\$\ldots u_m\$\rangle$ for some $m \geq 1$, where each u_i is $*a_1*a_2*\ldots *a_n*$ for some string $a_1\ldots a_n$ such that $\langle a_1\ldots a_n, a_1\ldots a_n\rangle$ is a solution of $\mathscr{P}'_1$.

To force solutions of $\mathscr{P}_2$ to begin appropriately, if $\langle x, y\rangle$ (the starting pair for $\mathscr{P}'_1$) is $\langle a_1^0\cdots a_{m_0}^0, b_1^0\cdots b_{n_0}^0\rangle$, where $a_i^0, b_i^0 \in \Sigma$, then $\mathscr{P}_2$ contains the pair

$$p_0 = \langle *a_1^0*a_2^0\cdots *a_{m_0}^0*, *b_1^0*b_2^0\cdots *b_{n_0}^0\rangle \quad .$$

Moreover, every other pair $\langle v, w\rangle$ in $\mathscr{P}_2$ is such that w begins with $*$ but v does not. Thus p_0 must be the first pair used in the construction of a solution of $\mathscr{P}_2$.

For each pair $\langle a_1\cdots a_m, b_1\cdots b_n\rangle$ in $\mathscr{P}_1$, where $a_i, b_i \in \Sigma$, (including the pair $\langle x, y\rangle$), $\mathscr{P}_2$ contains the pair

$$\langle a_1*a_2*\cdots a_m*, *b_1*b_2\cdots *b_n\rangle \quad .$$

Thus, if $\langle c_1\cdots c_p, d_1\cdots d_q\rangle$ is any presolution of $\mathscr{P}'_1$, then

$$\langle *c_1*c_2\cdots *c_p*, *d_1*d_2\cdots *d_q\rangle$$

is a presolution of $\mathscr{P}_2$. Moreover, any presolution of $\mathscr{P}_2$ not containing \$ such that both strings begin with the same symbol corresponds to a presolution of $\mathscr{P}'_1$ in this way.

Finally, $\mathscr{P}_2$ contains the pair

$$\langle \$, *\$\rangle \ ,$$

which can be used to terminate solutions of $\mathscr{P}_2$. Moreover any solution of $\mathscr{P}_2$ is of the form $\langle u_1\$u_2\$\ldots u_m\$, u_1\$u_2\$\ldots u_m\$\rangle$ for some $m \geq 1$, where $\langle u_i\$, u_i\$\rangle$ is itself a solution for each i and no u_i contains \$. Thus the pair $\langle \$, *\$\rangle$ can be used to form a solution of $\mathscr{P}_2$ only if $\mathscr{P}'_1$ also has a solution.

This completes the construction of $\mathscr{P}_2$, and the sketch of the proof that $\mathscr{P}'_1$ has a solution if and only if $\mathscr{P}_2$ has a solution.

Thus to prove that the Post Correspondence Problem is unsolvable, it suffices to construct, for each Turing machine $\mathscr{M}$, a modified correspondence system $\mathscr{P}'$ such that $\mathscr{P}'$ has a solution if and only if $\mathscr{M}$ halts when given the empty word as input. As before, we provide a commentary on the construction rather than a formal proof.

The basic idea is that each of the (equal) strings comprising a solution of $\mathscr{P}' = (\mathscr{P}, p_0)$ will have the form of a sequence of instantaneous descriptions of $\mathscr{M}$, starting from the initial instantaneous description $\langle q_0, B\rangle$, separated by a special symbol # that cannot be part of any instantaneous description, and ending with a halting instantaneous description. Certain important pairs in $\mathscr{P}$ encode the transition function of $\mathscr{M}$. As a solution of $\mathscr{P}'$ is being constructed by concatenating the left and right members of a pair in $\mathscr{P}$ to the left and right words of a presolution, the left word must always be an initial subword of the right word; essentially, the left word always lags one instantaneous description behind the right, so that a step in the computation can be paralleled by concatenating the members of an appropriate pair.

Let $\mathscr{M} = (K, \Sigma, B, q_0, \delta)$. The alphabet of $\mathscr{P}'$ is $\Sigma \cup (K \times \Sigma) \cup \{\#\}$, i.e., the symbols appearing in instantaneous descriptions of $\mathscr{M}$, plus the special symbol #.

The first pair in $\mathscr{P}$ --in fact, the "starting" pair p_0 --is

$$\langle \#, \# \langle q_0, B \rangle \# \rangle \quad ,$$

which specifies the first instantaneous description.

The next pairs are those encoding the transition function of $\mathscr{M}$. As in the definition of the relation $\vdash_{\mathscr{M}}$, there are various cases depending on the direction of head movement and the position of the head. Suppose that $q_1, q_2 \in K$, $a_1, a_2 \in \Sigma$, and $m \in \{-1, 0, 1\}$, and $\delta(q_1, a_1)$ is defined and is equal to (q_2, a_2, m). Then $\mathscr{P}$ contains

$(\langle q_1, a_1 \rangle, \langle q_2, a_2 \rangle)$ if $m = 0$ (head stationary);

For each $a \in \Sigma$, $(a\langle q_1, a_1 \rangle, \langle q_2, a \rangle a_2)$ if $m = -1$ (head moves left);

For each $a \in \Sigma$, $(\langle q_1, a_1 \rangle a, a_2 \langle q_2, a \rangle)$ if $m = 1$ (head moves right);

$(\langle q_1, a_1 \rangle \#, a_2 \langle q_2, B \rangle \#)$ if $m = 1$ (head moves right onto blank tape).

Next we need pairs that permit symbols not near the read-write head to be propagated from one instantaneous description to the next. We also need a similar pair for the marker symbol #. Thus $(a, a) \in \mathscr{P}$, for each $a \in \Sigma - \{B\}$, and $(\#, \#) \in \mathscr{P}$.

Let us call a pair (v, w) a <u>partial solution</u> of $\mathscr{P}' = (\mathscr{P}, p_0)$ if (v, w) is a presolution that potentially could be completed to form a solution by the concatenation of more words from pairs from $\mathscr{P}$. That is, (v, w) is a partial solution if (v, w) is a presolution and for some word u, either $v = wu$ or $w = vu$, i.e., one of v, w is an initial subword of the other. The crucial properties of $\mathscr{P}'$ are:

If $I_0 \vdash_{\mathscr{M}} I_1 \vdash_{\mathscr{M}} \cdots \vdash_{\mathscr{M}} I_n$ $(n \geq 0)$, is a computation by $\mathscr{M}$, then

$$(\# I_0 \# I_1 \# \cdots \# I_{n-1} \#, \# I_0 \# I_1 \# \cdots \# I_n \#)$$

is a partial solution of $\mathscr{P}'$ constructible by using only the pairs presented thus far. Moreover, if (v, w) is a partial solution of $\mathscr{P}'$ constructed by using only the pairs presented thus far, then for some $u \in \Sigma^*$ and some instantaneous descriptions $I_0, \ldots, I_n$ of $\mathscr{M}$ $(n \geq 0)$, $v = \# I_0 \# I_1 \# \cdots \# I_{n-1} \# u$ and $w = \# I_0 \# I_1 \# \cdots \# I_n \# u$ and $I_0 \vdash_{\mathscr{M}} I_1 \vdash_{\mathscr{M}} \cdots \vdash_{\mathscr{M}} I_n$.

These properties are easily checked by an induction on n, starting from the basis case $n = 0$, $I_0 = \langle q_0, B \rangle$.

To complete the construction we need to add pairs with the aid of which such partial solutions can be turned into solutions, in case I_n is a halting instantaneous description. So suppose that I_n contains the state-symbol pair $\langle q, a \rangle$ where $\delta(q, a)$ is undefined, and suppose that the length of I_n is r. Then the corresponding solution of $\mathscr{P}'$ will be $\langle v, v \rangle$, where $v = \# I_0 \# I_1 \# \cdots \# I_{n-1} \# I_n^0 \# I_n^1 \# I_n^2 \# \cdots \# I_n^{r-1} \# \#$, and

$$I_n^0 = I_n \quad ,$$

$$I_n^{r-1} = \langle q, a \rangle \quad ,$$

and each I_n^j is derived from I_n^{j-1} by deleting one of the symbols in I_n^{j-1} adjacent to the occurrence of $\langle q, a \rangle$. These I_n^j are instantaneous descriptions of $\mathscr{M}$, but it is not true that $\langle q_0, B \rangle \overset{*}{\vdash}_{\mathscr{M}} I_n^j$ for $j > 0$. The only purpose of the I_n^j is to bring the two strings in the partial solution of $\mathscr{P}'$ close enough together in length so that they can be made to coincide exactly by using one additional pair from $\mathscr{P}$.

Thus $\mathscr{P}$ contains, for each $\langle q, a \rangle$ such that $\delta(q, a)$ is undefined, and each $a' \in \Sigma$, the pairs

$$(a' \langle q, a \rangle, \langle q, a \rangle) \quad \text{and} \quad (\langle q, a \rangle a', \langle q, a \rangle) \;;$$

these, together with the pairs (a, a) and $(\#, \#)$, make it possible to construct the I_n^j. Finally, $\mathscr{P}$ contains, for each q, a such that $\delta(q, a)$ is undefined, the pair

$$(\langle q, a \rangle \# \#, \#) \;;$$

one of these pairs would be the last used in the construction of a solution. We leave it to the reader to check that the only solutions constructible using these pairs are of the specified kind.

This completes the proof of the Post Correspondence Theorem. ∎

IF. 2 Two-Counter Machines

A counter machine (or register machine) is an automaton with a finite-state control and one or more counters, which are memory units capable of storing any natural numbers. The finite control may modify a counter as it changes state by adding or subtracting one, and it may test whether the number in a counter is zero. We here prove the unsolvability of the halting problem for machines with two counters. We then show that certain related machines, which we call special two-counter machines, also have an unsolvable halting problem; these special machines are needed for the proof of the unsolvability of certain classes of logical formulas (Chapters IIE and IIF). (We sometimes call the first kind standard two-counter machines for emphasis.) For the most part we merely sketch the construction; but we give the exact details for the reduction of standard two-counter machines to special ones.

Two-counter machines will be presented formally as sequences of instructions, like computer programs, rather than by use of the usual apparatus of automata theory. The state of a machine is thus the numerical position in the list of instructions of the instruction currently being executed. The instructions of the standard machines are one of five kinds:

H ("halt")

r' ("add one to the number in the first counter and proceed to the next instruction")

$r^-(j)$ ("if the number in the first counter is 0 then proceed to the next instruction, otherwise subtract one from it and go to instruction j")

s', $s^-(j)$ (same as r' and $r^-(j)$, but for the other counter) .

Formally, then, a two-counter machine is a sequence $\langle I_1, \ldots, I_n \rangle$ of such instructions. We assume that I_n, and only I_n, is H, and that if I_i is $r^-(j)$ or $s^-(j)$ then $1 \leq j \leq n$ (no jumping to instructions outside the program). We dispense with the formal semantics, though we shall be more precise in our discussion of special two-counter machines.

The halting problem for two-counter machines, like that for Turing machines, comes in two versions: inputless and universal.

Two-Counter Theorem, Inputless Version. The problem of determining, given a two-counter machine $\mathcal{M}$, whether or not $\mathcal{M}$ halts when started at the first instruction with both counters containing 0, is unsolvable.

Two-Counter Theorem, Universal Version. For a certain "universal" machine $\mathcal{U}$, the problem of determining, given a natural number n, whether

$\mathcal{U}$ halts when started at the first instruction with both counters containing n, is unsolvable.

These two theorems follow from the two versions of Turing's theorem (p. 4) by showing how a two-counter machine can simulate a Turing machine with the counters containing an encoded representation of the nonblank part of the Turing machine tape. We briefly describe the steps of this reduction: from Turing machine to two-pushdown machine to four-counter machine to two-counter machine.

First Step: Two-pushdown machines can simulate Turing machines.

Here a pushdown store is a memory device that, like a Turing machine tape, may contain any finite sequence of symbols from a finite alphabet, but that, unlike a Turing machine tape, may be examined or modified only at one end. (Think of the symbols as stacked up like plates, so that only the top is visible at any time.) A two-pushdown machine is then a finite control attached to two pushdown stores, which it may examine and change in the prescribed way. Although a single pushdown store is strictly weaker than a Turing machine tape, two pushdown stores can be used to simulate a single Turing machine tape: simply imagine the Turing machine tape to be cut or folded at the head position, so that the symbols just to the left and right of the head are on top of the pushdown stores and the symbols at the extreme ends of the finite nonblank portion of the Turing machine tape are at the bottoms of the pushdown stores. Motions of the Turing machine head back and forth within the nonblank part of the Turing machine tape are then simulated by "sloshing" symbols back and forth between the tops of the pushdown stores.

Second Step: Four-counter machines can simulate two-pushdown machines.

That is, two counters can be used instead of one pushdown. Suppose the finite alphabet from which the symbols in the pushdown store are drawn contains k symbols in all, $a_0, \ldots, a_{k-1}$. The trick is to regard the contents of a pushdown store as a number in base k. Thus if $a_{i_0} \cdots a_{i_p}$ is the word in the pushdown store, with the top (modifiable) end of the pushdown store shown on the left, then the corresponding number would be

$$\alpha = \sum_{j=0}^{p} i_j \cdot k^j \quad .$$

Determining the top symbol in the pushdown store then corresponds to finding the residue class of α (mod k). Removing the top symbol calls for subtracting this residue and dividing by k; placing a symbol on top of the store calls for multiplying by k and adding a number less than k. Since k is a fixed number, the operations of multiplying and dividing by k can be accomplished by repeated additions and subtractions by using the second counter in the pair to keep track of the result.

Third Step: Two-counter machines can simulate four-counter machines.

Again, the trick is to use an exponential encoding: four registers containing numbers p, q, r, s are encoded as the single number $2^p \cdot 3^q \cdot 5^r \cdot 7^s$. Adding or subtracting from one of the four counters corresponds to multiplying or dividing by 2, 3, 5, or 7, and checking whether a counter contains 0 is achieved by checking for congruence to 0 (mod 2, 3, 5, or 7). As before, the finite control and the second counter make these calculations possible.

This completes our sketch of the proof of the two versions of the Two-Counter Theorem; the details are a programming exercise. (For the universal version, note that an initial value for the counters may encode an input word for the Turing machine.)

Now our special two-counter machines are, like the standard ones, specified as sequences of instructions, but there are only three types of instruction:

H ("halt")

A(j) ("add one to the number in the second counter and go to instruction j")

S(j, k) ("if the number in the first counter is not zero, then subtract one from it and go to instruction j; otherwise, interchange the counters and go to instruction k") .

Formally, then, a special two-counter machine is a sequence $\langle I_1, \ldots, I_n \rangle$ of such instructions; as before we stipulate that I_n, and only I_n, is H, and that if I_i is A(j) or S(j, k), then $1 \le j, k \le n$.

The main purpose of introducing these special machines is to show that both the notion of computation implicit in the above paraphrases and also a peculiar variant notion of computation lead to unsolvable halting problems, in either the inputless or universal version. The variant notion of computation, in which the counters may contain negative as well as nonnegative numbers, is used in certain proofs involving formulas with prefixes of the form $\forall \exists \forall \ldots \forall$. As the distinction between the two notions of computation is somewhat delicate, we formalize both.

Let $\mathcal{M} = \langle I_1, \ldots, I_n \rangle$ be a special two-counter machine. An instantaneous description of $\mathcal{M}$ is a member of $\{1, \ldots, n\} \times \mathbb{N} \times \mathbb{N}$. The relation $\vdash_{\mathcal{M}}$ ("yields in one step") between instantaneous descriptions of $\mathcal{M}$ is the minimal relation satisfying, for each $i, j, k \in \{1, \ldots, n\}$ and $p, q \in \mathbb{N}$,

(1) $(i, p, q) \vdash_{\mathcal{M}} (j, p, q+1)$ if I_i is A(j) ;

(2) $(i, p+1, q) \vdash_{\mathcal{M}} (j, p, q)$ and
$(i, 0, q) \vdash_{\mathcal{M}} (k, q, 0)$ if I_i is S(j, k) .

The relation $\vdash^*_{\mathcal{M}}$ is the reflexive, transitive closure of $\vdash_{\mathcal{M}}$. $\mathcal{M}$ is said to <u>$\vdash$-halt</u> when started from instantaneous description (i, p, q) if and only if $(i, p, q) \vdash^*_{\mathcal{M}} (n, 0, 0)$.

Now let $\{1, \ldots, n\} \times \mathbb{Z} \times \mathbb{Z}$ be the set of <u>extended instantaneous descriptions</u> of $\mathcal{M}$. The relations $\Vdash_{\mathcal{M}}$ and $\Vdash^*_{\mathcal{M}}$ between extended instantaneous descriptions of $\mathcal{M}$ are defined exactly as the relations $\vdash_{\mathcal{M}}$ and $\vdash^*_{\mathcal{M}}$ are defined above, except that in (1) and (2) the variables p and q range over $\mathbb{Z}$ instead of $\mathbb{N}$. Unlike $\vdash_{\mathcal{M}}$, $\Vdash_{\mathcal{M}}$ is not single-valued. For example, if I_4 is $S(5, 6)$, then both $(4, 0, 10) \Vdash_{\mathcal{M}} (6, 10, 0)$ and also $(4, 0, 10) \Vdash_{\mathcal{M}} (5, -1, 10)$. The definition of $\Vdash$-halting is like that of $\vdash$-halting, but using $\Vdash_{\mathcal{M}}$ instead of $\vdash_{\mathcal{M}}$.

We can now state the results we need.

<u>Special Two-Counter Theorem, Inputless Version.</u> The problems of determining, given a special two-counter machine $\mathcal{M}$, whether $\mathcal{M}$

a) $\vdash$-halts

or

b) $\Vdash$-halts

when $\mathcal{M}$ is started from $(1, 0, 0)$, are both unsolvable.

<u>Special Two-Counter Theorem, Universal Version.</u> For a certain "universal" special two-counter machine $\mathcal{U}$, the problems of determining, given a number $p \in \mathbb{N}$, whether $\mathcal{M}$

a) $\vdash$-halts

or

b) $\Vdash$-halts

when $\mathcal{M}$ is started from $(1, p, p)$, are both unsolvable.

These theorems are proved from the two versions of the Two-Counter Theorem. There are two parts to the argument:

- Proof that special two-counter machines, under the $\vdash$ semantics, can simulate standard two-counter machines;
- Proof that $\vdash$-halting and $\Vdash$-halting from an instantaneous description are equivalent.

The second part may be argued very simply, because of the restricted types of instructions special machines may have. Let $\mathcal{M}$ be a special two-counter machine; let (i, p, q) be an instantaneous description of $\mathcal{M}$, and let I_n be the halt instruction of $\mathcal{M}$. Clearly, if $(i, p, q) \vdash^*_{\mathcal{M}} (n, 0, 0)$ then $(i, p, q) \Vdash^*_{\mathcal{M}} (n, 0, 0)$. For the converse, suppose

$$(i, p, q) \Vdash_{\mathcal{M}} (i', p', q') \Vdash_{\mathcal{M}} \cdots \Vdash_{\mathcal{M}} (n, 0, 0)$$

but not $(i, p, q) \vdash^*_{\mathcal{M}} (n, 0, 0)$. Let (i'', p'', q'') be the first extended instantaneous description in the sequence that is not an instantaneous description. Then $p'' < 0$. But for any extended instantaneous descriptions (i_1, p_1, q_1) and (i_2, p_2, q_2), if $p_1 < 0$ and $(i_1, p_1, q_1) \Vdash_{\mathcal{M}} (i_2, p_2, q_2)$, then $p_2 < 0$ (see the definition of $\Vdash_{\mathcal{M}}$ above). Hence not $(i'', p'', q'') \Vdash^{*}_{\mathcal{M}} (n, 0, 0)$, a contradiction.

Thus, it remains only to show that special two-counter machines can simulate standard ones under the $\vdash$ semantics. The idea is a familiar one: the contents p and q of the two counters of the standard machine are encoded as the single number $2^p \cdot 3^q$, and addition and subtraction in the standard machine are simulated by multiplication and division by 2 or 3 in the special machine.

Thus, let $\mathcal{M} = \langle I_1, \ldots, I_n \rangle$ be a standard two-counter machine. We construct a special two-counter machine $\mathcal{M}' = \langle J_1, \ldots, J_{t_{n+1}} \rangle$ by replacing each instruction I_i of $\mathcal{M}$ by a sequence of instructions $J_{t_i}, \ldots, J_{t_i + s_i}$ of $\mathcal{M}'$, called the ith <u>block</u>. Here $t_1 = 3$ and $t_{i+1} = t_i + s_i + 1$ for $i = 1, \ldots, n$, where

if I_i is H then $s_i = 0$ and the ith block is

$S(t_{n+1}, t_{n+1})$;

if I_i is r' then $s_i = 2$ and the ith block is

$S(t_i + 1, t_{i+1})$
$A(t_i + 2)$
$A(t_i)$;

if I_i is s' then $s_i = 3$ and the ith block is

$S(t_i + 1, t_{i+1})$
$A(t_i + 2)$
$A(t_i + 3)$
$A(t_i)$;

if t_i is $r^-(j)$ then $s_i = 5$ and the ith block is

$S(t_i + 1, t_j)$
$S(t_i + 2, t_i + 4)$
$A(t_i)$
$A(t_i + 4)$
$A(t_i + 5)$
$S(t_i + 3, t_{i+1})$;

and if I_i is $s^-(j)$ then $s_i = 7$ and the ith block is

$S(t_i+1, t_j)$

$S(t_i+2, t_i+6)$

$S(t_i+3, t_i+5)$

$A(t_i)$

$A(t_i+5)$

$A(t_i+6)$

$A(t_i+7)$

$S(t_i+4, t_{i+1})$.

Moreover, the first instructions J_1, J_2 are

$A(2)$

$S(2, 3)$

and the last instruction $J_{t_{n+1}}$ is

H .

The crucial property of $\mathcal{M}'$, which can be proved by a tedious induction, is that

$$\text{for any } p, q \in \mathbb{N}, \quad (1, p, q) \vdash^*_{\mathcal{M}} (n, 0, 0) \text{ if and only if}$$

$$(1, 2^p \cdot 3^q - 1, 2^p \cdot 3^q - 1) \vdash^*_{\mathcal{M}'} (t_{n+1}, 0, 0)$$

Thus the two versions of the Special Two-Counter Theorem follow from the two versions of the Two-Counter Theorem.

Historical References. The Post Correspondence Theorem is due to Post (1946). The proof we present here is adapted from Hopcroft and Ullman (1969, pp. 212-218).

Two-Counter machines are due to Minsky (1961, 1967). The unsolvability proof presented here is derived from Hopcroft and Ullman (1969, pp. 98-100), who took it from Fischer (1966). Special two-counter machines were used by Lewis and Goldfarb (1973).

PART II: FIRST-ORDER LOGIC

Chapter IIA
LOGICAL PRELIMINARIES

IIA.1 Solvability and Unsolvability

The *decision problem* for a class of formulas of first-order logic is the problem of determining, given a formula in the class, whether or not it is satisfiable. A class of formulas will be called *solvable* if its decision problem is solvable, otherwise *unsolvable*. (The *membership* problem for the classes we consider, i.e., the problem of determining whether a formula is in the class or not, is usually solvable, since the classes are defined by simple syntactic criteria.) In the narrow technical sense of Chapter IA, the *problem* that is solvable or unsolvable is the set of encodings, as words over an alphabet, of unsatisfiable formulas in the class. A class of formulas will be said to be *reducible* to another class of formulas if the decision problem for the one is reducible to that for the other.

All the general remarks made in Chapter IA about the terms "reducible" and "unsolvable" still apply here: the decision problem for each of the unsolvable classes presented is 1-complete. Thus each of the classes is what is called a *reduction class* in the literature: a class of formulas such that for each first-order formula F, we can effectively find a formula G in the class, such that G is satisfiable if and only if F is satisfiable. Moreover, the computable function that yields G from F may be seen in each case to be one-one.

IIA.2 Syntax

The formal language in which our logical formulas are written has the following primitive signs:

quantifiers $\forall$ and $\exists$;
truth-functional connectives $\vee$, $\neg$;
a countably infinite supply of variables;
for each $k > 0$ a countably infinite supply of k-place predicate letters;
for each $k \geq 0$ a countably infinite supply of k-place function signs.

(Note in particular that the identity sign is omitted.)

These symbols have an ordering, called the lexical order; thus we may refer to the ith k-place predicate letter in lexical order. This lexical ordering of the individual symbols induces a lexical ordering of all expressions built up from those symbols.

The degree of a predicate letter or function sign is its number of argument-places. A 1-place predicate letter or function sign is monadic, a 2-place one dyadic, a three-place one triadic; a 0-place function sign is a constant. Variables of the metalanguage that range over predicate letters (typically P, Q, R), function signs (typically f, g, h), or variables of the object language (typically u, v, x, y, z) are also sometimes used as particular object-language signs (specific predicate letters, function signs, or object-language variables); context makes clear what is intended.

Terms are constructed from variables and constants by iterated application of the function signs. Thus each variable and each constant is a term, and if $t_1, \ldots, t_k$ are terms and f is a k-place function sign, $k \geq 1$, then $f(t_1 \ldots t_k)$ is a term. If f is a monadic function sign and t is a term, we write $f^n(t)$ for the term $f(f(f(\ldots(f(t))\ldots)))$, with n occurrences of f $(n \geq 0)$. If $t_1, \ldots, t_k$ are terms and P is a k-place predicate letter then $Pt_1 \ldots t_k$ is an atomic formula. The terms $t_1, \ldots, t_k$ are the arguments of $f(t_1 \ldots t_k)$ or $Pt_1 \ldots t_k$. Formulas are built up from atomic formulas by means of the truth-functional connectives, quantifiers, and parentheses. A formula occurring in another formula is a subformula of it. While $\vee$ and $\neg$ are the only primitive truth-functional connectives, $\wedge$, $\rightarrow$, and $\leftrightarrow$ are defined in the usual way. Also, infix notation will be used, and referred to, freely. Thus if $F_1, \ldots, F_k$ are formulas then $(F_1 \vee \ldots \vee F_k)$ is a formula, called the disjunction of $F_1, \ldots, F_k$, and $(F_1 \wedge \ldots \wedge F_k)$ is a formula, called the conjunction of $F_1, \ldots, F_k$. The F_i are called the disjuncts in the first case and the conjuncts in the second. A signed atomic formula is either an atomic formula or an atomic formula preceded by the negation sign $\neg$. The matrix of a formula F is the formula F^M obtained from F by deleting each occurrence of a quantifier as well as the occurrence of the variable immediately to its right. The notions of scope of an occurrence of a quantifier or truth-functional connective, free and bound occurrences of a variable, quantifier prefix, and so on, are defined in the usual way.

Now let F be a quantifier-free formula, i.e., a formula with no occurrence of $\forall$ or $\exists$. For certain words w over the alphabet $\{0, 1, 2\}$, a subformula $\langle F \rangle_w$ is defined by the following rules.

$\langle F \rangle_\varepsilon = F$ (ε is the empty word);

if $\langle F \rangle_w$ is defined then $\langle \neg F \rangle_{w0} = \langle F \rangle_w$;

if i is 1 or 2, and F_1 and F_2 are quantifier-free formulas such that $\langle F_i \rangle_w$ is defined, then $\langle F_1 \vee F_2 \rangle_{wi} = \langle F_i \rangle_w$.

Thus each occurrence of a subformula of F has a unique "structural address"

w which encodes not only the position of that occurrence of the subformula within F but also the nature of the truth-functional connectives within whose scopes that occurrence lies. But $\langle F \rangle_w$ itself is a subformula, not an occurrence of one; thus it may be that $\langle F \rangle_w = \langle F \rangle_{w'}$, even though $w \neq w'$. We write $T(F)$ for the set of all w such that $\langle F \rangle_w$ is defined, and $\overline{T}(F)$ for the set of maximal members of $T(F)$, i.e., the set of all w such that $\langle F \rangle_w$ is defined but $\langle F \rangle_{wi}$ is not defined for any i. Thus $\langle F \rangle_w$ is atomic for each $w \in \overline{T}(F)$. Also, if F is a formula containing occurrences of quantifiers, let $T(F) = T(F^M)$, $\overline{T}(F) = \overline{T}(F^M)$, and $\langle F \rangle_w = \langle F^M \rangle_w$ for each $w \in T(F)$.

A formula is closed if it contains no free variables, and is rectified if no variable occurs both free and bound, and no two quantifier-occurrences bind the same variable. By renaming variables each formula can be transformed into an equivalent rectified formula. The rest of this monograph is concerned with decision problems for classes of what we call schemata, which are closed, rectified formulas not containing function signs. However, the theory of Herbrand expansions presented below applies to the broader class of closed, rectified formulas, including formulas with occurrences of function signs, and for technical reasons it is convenient to develop the theory in this broader setting.

Let F be a closed, rectified formula. Then the variables of F are in one-to-one correspondence with the occurrences of quantifiers in F. Thus we say that a variable v governs a variable w in F if and only if the scope of the (unique) quantifier-occurrence in F that binds v properly includes the scope of the (unique) quantifier-occurrence that binds w. A variable v is an x-variable of F if and only if the quantifier-occurrence binding v is either an occurrence of $\exists$ within the scopes of an even number (possibly 0) of occurrences of negation signs, or else an occurrence of $\forall$ within the scopes of an odd number. Otherwise, a variable v of F is a y-variable. Thus the x-variables of F are precisely those variables that become existential variables in any prenex equivalent of F obtained by the usual prenexing rules, and the y-variables of F are those that become universal variables.

Let F be a formula, let $v_1, \ldots, v_k$ be distinct variables, and let $t_1, \ldots, t_k$ be (not necessarily distinct) terms. Then $F[v_1/t_1, \ldots, v_k/t_k]$ is the result of replacing simultaneously for each i, $1 \leq i \leq k$, the variable v_i at all its free occurrences in F by t_i. The variable v_i need not actually occur in F for this notation to be defined. If, for example, v_1 does not occur free in F then $F[v_1/t_1, \ldots, v_k/t_k]$ is just $F[v_2/t_2, \ldots, v_k/t_k]$. We also have a similar notation for substitution of formulas for atomic formulas: if $A_1, \ldots, A_k$ are distinct atomic formulas and $G_1, \ldots, G_k$ are any formulas, then $F\{A_1/G_1, \ldots, A_k/G_k\}$ is the result of replacing simultaneously for each i, $1 \leq i \leq k$, the atomic formula A_i by the formula G_i. We also write $F\{A/G(A = \ldots)\}$ for the result of replacing each atomic subformula A by G, where the relation between A and G is described by the parenthesis; and use other self-explanatory abbreviations where convenient.

An *instance* of a formula F is any formula resulting from the matrix F^M by substituting, for each variable, some term. That is, an instance of F is any formula $F' = F^M[v_1/t_1, \ldots, v_k/t_k]$, where $v_1, \ldots, v_k$ are all the (distinct) variables of F^M and $t_1, \ldots, t_k$ are any (not necessarily distinct) terms. In most cases the terms will be variable-free, so that F' will be a variable-free formula. The instance F' is said to be an instance of F *over* D if and only if D is any set of terms including $t_1, \ldots, t_k$. The *substituent in* F' *for the variable* v_i is the term t_i; of course this phrase implicitly regards F' as an instance of the particular formula F.

IIA.3 Herbrand Theory

Let S be a finite set of function signs. Then the set of all variable-free terms constructible from the function signs in S alone is said to be a *domain*, specifically, the *domain generated by* the function signs in S. Thus the domain generated by the monadic function sign f and the constant c is $\{c, f(c), f(f(c)), \ldots\} = \{f^n(c) \mid n \geq 0\}$.

If D is a domain and F is a quantifier-free formula, then $E(F, D)$, the *expansion of* F *over* D, is the set of all instances of F over D. Thus $E(F, D)$ is a set of variable-free formulas; the terms appearing in these formulas need not be restricted to the terms in D. (For example, if F contains an occurrence of a function sign not occurring in any member of D, then $E(F, D)$ will contain occurrences of terms not in D.)

We next extend the notion of expansion to closed, rectified formulas. We do so by associating with each such formula F a unique quantifier-free formula F^* and domain $D(F)$, and then defining $E(F) = E(F^*, D(F))$. $E(F)$ will be the Herbrand expansion of F.

We first associate with each variable v and each number $k \geq 0$ a k-place function sign called the k-*place indicial correlate* of v; we also pick out a special constant, that is, a 0-place function sign, ł. No distinct variables have indicial correlates in common, and ł is not an indicial correlate of any variable. An indicial correlate of any variable is called an *indicial function sign.*

Let F be a closed, rectified formula. The *functional form* F^* of F is the result of replacing in the matrix F^M each x-variable x of F by the term $f(y_{i_1} \ldots y_{i_r})$, where $r \geq 0$ is the number of y-variables governing x in F, f is the r-place indicial correlate of x, and $\langle y_{i_1}, \ldots, y_{i_r} \rangle$ is the sequence of y-variables of F that govern x, in left-to-right order of their first occurrence in F. Thus F^* is a quantifier-free formula whose free variables are precisely the y-variables of F. The term $f(y_{i_1} \ldots y_{i_r})$ is called the *indicial term supplanting* x *in* F^*. If x is governed by no y-variable then the indicial term supplanting x in

F^* is just the 0-place indicial correlate of x.

The <u>Herbrand domain</u> D(F) of a closed, rectified formula F is the domain generated by the indicial function signs occurring in F^* and the special constant ɨ. (Alternatively, in case a constant occurs in F^*, D(F) may be defined as the domain generated by the indicial function signs of F^*, without the constant ɨ. The Expansion Theorem stated below holds under either definition.) For example, if F is a prenex schema and has prefix $\forall y_1 \exists x \forall y_2$, then D(F) = {ɨ, f(ɨ), f(f(ɨ)), . . . }, where f is the monadic indicial correlate of x, but if F is a prenex schema and has prefix $\exists z \forall y_1 \exists x \forall y_2$, then D(F) = {ɨ, c, f(ɨ), f(c), . . . }, where c is the 0-place indicial correlate of z ({c, f(c), f^2(c), . . . } under the alternative definition). On the other hand, if F has a quantifier prefix containing only universal quantifiers and a matrix containing the dyadic function sign f and no other function signs, then F^* contains no constants and D(F) = {ɨ, f(ɨ ɨ), f(ɨ f(ɨ ɨ)), . . . }.

If x is an x-variable of F then an <u>x-term</u> in D(F) is any term whose outermost function sign is the indicial correlate of x that occurs in F^*. Thus each term in D(F), except ɨ, is an x-term for a unique x-variable x of F.

An <u>Herbrand instance</u> of F is any instance of the functional form F^* over the Herbrand domain D(F). That is, an Herbrand instance is any formula $H = F^*[y_1/t_1, \ldots, y_k/t_k]$, where $y_1, \ldots, y_k$ are all the y-variables of F (and hence all the free variables of F^*), and $t_1, \ldots, t_k$ are any (not necessarily distinct) terms in D(F). Thus each Herbrand instance is a variable-free formula. Moreover, each Herbrand instance F' of F is an instance of F fulfilling the following constraints:

(1) The substituents in F' for the y-variable of F are terms in D(F);

(2) If the substituents in F' for the y-variables $y_{i_1}, \ldots, y_{i_r}$ of F are $t_1, \ldots, t_r$, respectively, and $f(y_{i_1} \ldots y_{i_r})$ is the indicial term supplanting the x-variable x of F in F^*, then the term $f(t_1 \ldots t_r)$ is the substituent in F' for x.

In particular, the substituent in any Herbrand instance F' of F for any x-variable x of F is an x-term.

We also speak of Herbrand instances of subformulas of the matrix of a formula F: if $G = \langle F \rangle_w$, for some $w \in T(F)$, then an Herbrand instance of G is $\langle F' \rangle_w$, for some Herbrand instance F' of F. Note that an Herbrand instance of G as a subformula of F need not be an Herbrand instance of G as a subformula of some other formula; but as confusion will not arise in practice, we omit the phase "as a subformula of F". (Since F is rectified, an Herbrand instance of a subformula G at one of several occurrences of G in F is also an Herbrand instance of G at any other occurrence of G in F, so there is no need to specify which occurrence of G is meant.)

The <u>Herbrand expansion</u> $E(F)$ is the set of all Herbrand instances of F, that is, the expansion $E(F^*, D(F))$ of the functional form over the Herbrand domain.

A model $\mathfrak{A}$ for a closed formula F consists of a nonempty universe U together with interpretations of the predicate letters and function signs of F--each k-place predicate letter P is assigned a k-ary relation $P^{\mathfrak{A}}$ on U, i.e., a subset of U^k, and each k-place function sign f is assigned a mapping $f^{\mathfrak{A}}$ from U^k to U--such that F is true in U under these interpretations. $\mathfrak{A}$ is a model <u>over</u> the universe U; we abbreviate "element of the universe of $\mathfrak{A}$" to "element of $\mathfrak{A}$." If F has a model then F is said to be <u>satisfiable</u>.

Let S be a set of atomic formulas, and let S' be the set of all quantifier-free formulas constructible from members of S by the truth-functional connectives. A <u>truth-assignment on</u> S' is a function $\mathscr{A}$ from S' into $\{-1, 1\}$ uniquely determined by its values on S by means of the following inductive rules:

$\mathscr{A}(\neg F) = 1$ if and only if $\mathscr{A}(F) = -1$

$\mathscr{A}(F \vee G) = 1$ if and only if $\mathscr{A}(F) = 1$ or $\mathscr{A}(G) = 1$

$\mathscr{A}(F \wedge G) = 1$ if and only if $\mathscr{A}(F) = 1$ and $\mathscr{A}(G) = 1$.

A truth-assignment $\mathscr{A}$ is said to <u>verify</u> a formula F if $\mathscr{A}(F) = 1$ and to <u>falsify</u> F if $\mathscr{A}(F) = -1$. (The use of -1, instead of 0 as is usual, for falsehood has a minor technical advantage for us, which will be exploited in Section IIB. 2.) We write $\mathscr{A} \models F$ in case $\mathscr{A}$ verifies F. If E is any set of quantifier-free formulas, then a truth-assignment on E is just a truth-assignment on the set E', where E' is the set of all quantifier-free formulas constructible from atomic subformulas of members of E; and E is <u>truth-functionally consistent</u> if and only if there is a truth-assignment on E that verifies each member of E. We say that $\mathscr{A}$ verifies E, and write $\mathscr{A} \models E$, if $\mathscr{A}$ is a truth-assignment on the set E of quantifier-free formulas such that $\mathscr{A}$ verifies each member of E.

<u>Expansion Theorem</u> (Skolem-Herbrand-Gödel). A closed, rectified formula F is satisfiable if and only if its expansion $E(F)$ is truth-functionally consistent.

The utility of the Expansion Theorem in unsolvability proofs is that it enables us to treat first-order formulas as combinatorial systems of the type studied in Part I. Thus the model-theoretic notion of satisfiability plays a very minor role here; we find it possible to obtain sharper results, and to prove them more explicitly, by referring nearly always to the expansion of a formula rather than to models of it. We do so both when reducing an unsolvable combinatorial problem to a logical decision problem, and when reducing an unsolvable logical decision problem to another logical decision problem. As the bulk of our proofs are of the latter kind, we close this chapter with a general account of how the Expansion Theorem is typically used to show that a certain effective procedure does in fact reduce the decision problem for a class of formulas $\mathscr{S}_1$ to that for a class $\mathscr{S}_2$.

Such a proof naturally takes the following form: Starting from a typical formula $F \in \mathscr{S}_1$, we show how to construct a formula $G \in \mathscr{S}_2$, and then prove that G is satisfiable if and only if F is satisfiable. (We use the rubric $(F \triangleright G)$ for the proof that, if F is satisfiable, then G is satisfiable, and $(G \triangleright F)$ for the proof of the converse.) In many cases, especially those of Chapter IIC, the construction of G from F has the property that for some subformula $\langle G \rangle_w$ of G^M, $\langle G \rangle_w$ is obtained from F^M by substituting quantifier-free formulas for atomic formulas. That is, for some $w \in T(G)$, and some quantifier-free formulas $B_1, \ldots, B_n$,

$$\langle G \rangle_w = F^M\{A_1/B_1, \ldots, A_n/B_n\},$$

where $A_1, \ldots, A_n$ are the distinct atomic subformulas of F. Thus $T(F) \subset T(\langle G \rangle_w)$. Then to show that G is satisfiable if and only if F is satisfiable, it suffices to show that

<u>$(F \triangleright G)$</u> if $\mathscr{A}$ is a truth-assignment on E(F) such that $\mathscr{A} \models E(F)$, then there is a truth-assignment $\mathscr{B}$ on E(G) such that

(A) for each Herbrand instance G' of G, there is an Herbrand instance F' of F such that $\mathscr{A} \models F'$ if and only if $\mathscr{B} \models G'$; and

<u>$(G \triangleright F)$</u> if $\mathscr{B}$ is a truth-assignment on E(G) such that $\mathscr{B} \models E(G)$, then there is a truth-assignment $\mathscr{A}$ on E(F) such that

(B) for each Herbrand instance F' of F, there is an Herbrand instance G' of G such that $\mathscr{B} \models G'$ if and only if $\mathscr{A} \models F'$.

Because of the structure of G, it is possible to show (A) by

(a) showing, for each $G' \in E(G)$, that $\mathscr{B} \models G'$ provided that $\mathscr{B} \models \langle G' \rangle_w$, and then

(b) finding an $F' \in E(F)$ such that $\mathscr{A}(\langle F' \rangle_{w'}) = \mathscr{B}(\langle \langle G' \rangle_w \rangle_{w'})$ for each $w' \in \overline{T}(F)$.

Conversely, to show (B) we can show that for any $F' \in E(F)$ there is a $G' \in E(G)$ such that

(a) $\mathscr{B} \models \langle G' \rangle_w$ and such that

(b) $\mathscr{A}(\langle F' \rangle_{w'}) = \mathscr{B}(\langle \langle G' \rangle_w \rangle_{w'})$ for each $w' \in \overline{T}(F)$.

The next section gives an extended example of a proof using these techniques.

Historical References. Versions of the expansion theorem were given by Skolem (1922, 1928, 1929), Herbrand (1930), and Gödel (1930). Herbrand's concern was, however, not the Expansion Theorem as we have stated it, but to establish by constructive means the equivalence of the truth-functional consistency of each finite portion of E(F) on the one hand, and the irrefutability of F in some standard system of quantification theory on the other. For a more complete discussion, see the introduction to Herbrand (1971) and the introductions to Skolem (1928) and Gödel (1930) in van Heijenoort (1971).

Chapter IIB

A MOTIVATING EXAMPLE

In this chapter we motivate, develop, and then illustrate the use of general machinery for reducing classes of schemata to other classes of schemata. This machinery is especially useful for the classification by prefix and similarity type (Chapter IIC) but certain elements are used in later proofs as well.

IIB.1 The Example: Schemata with One Dyadic Predicate Letter

To show what kind of structures we shall need to describe, we prove the unsolvability of the class of schemata having only a single, dyadic, predicate letter. To simplify the proofs we assume the unsolvability of the class of schemata having only dyadic predicate letters (but any number of them); a straightforward generalization of the construction would reduce arbitrary schemata to schemata with only one dyadic predicate letter.

Thus the problem is: Given a schema F with $q \geq 2$ dyadic predicate letters $P_1, \ldots, P_q$, to construct a schema G with one dyadic predicate letter P, such that G is satisfiable if and only if F is satisfiable. The basic idea is to construct G so that every model of F will be correlated with some model of G, and vice versa. (The formal proof will deal with expansions, rather than models.) However, if we take this approach, we must immediately recognize that the correlated models cannot have the same universes: for example, F may have as many as 2^{qn^2} different models over a fixed universe of cardinality n, whereas G can have only 2^{n^2}. And since we cannot assume a priori any understanding of the structure of models of F, there can be no effective method for correlating several models of F with one of G.

So if $\mathfrak{A}$ is a finite model for F, the correlated finite model $\mathfrak{A}'$ of G must have more elements than $\mathfrak{A}$ has. How will these elements be used? One way (by no means the only way, but typical of the methods used below) is to introduce q distinct new elements $e^1_{ab}, \ldots, e^q_{ab}$ for each pair $\langle a, b\rangle$ of (not necessarily distinct) elements of $\mathfrak{A}$ and to use the truth or falsehood of $P^{\mathfrak{A}'}(e^i_{ab}, e^i_{ab})$ to represent the truth or falsehood of $P^{\mathfrak{A}}_i(a, b)$ $(i = 1, \ldots, q)$. Thus q relations on a set of n elements would be replaced by one relation on a set of $qn^2 + n$ elements.

The idea of introducing these new elements presents two problems in the construction of G:

(1) introducing them, i.e., forcing any model for G to contain such new elements;

(2) using them, i.e., replacing the atomic formulas constructed from the q dyadic predicate letters and pairs of variables of F by atomic formulas constructed from one dyadic letter and variables ranging over the new elements.

The first problem is simple enough in principle; G will have a conjunct

$$G_0 = \forall u_1 \forall u_2 \exists x_1 \cdots \exists x_q G_0^M$$

where G_0^M is some matrix stating the relation between the new elements and the old. Of course, the relationship has to be expressible by means of the one dyadic predicate letter. Also, the new elements, when substituted for u_1 and u_2, will generate more new elements; but this problem we put aside for the moment.

As for (2), $x_1, \ldots, x_q$ cannot be used in the revised matrix, since these variables stand for the new elements associated with the particular pair $\langle u_1, u_2 \rangle$, and the matrix of F contains many variables. Assume that F is prenex, with prefix $Q_1 v_1 \cdots Q_m v_m$. Then we need q new variables for each pair of old variables: for example, y_{jk}^i might correspond to the ith new element associated with the pair of old elements v_j, v_k. Thus an atomic formula $P_i v_j v_k$ in F^M would be replaced by $Py_{jk}^i y_{jk}^i$, giving a new matrix G_1. So to a first approximation, G might look like

$$\forall u_1 \forall u_2 \exists x_1 \cdots \exists x_q G_0^M \wedge Qv_1 \cdots Qv_m \forall y_{11}^1 \cdots \forall y_{mm}^q G_1 \quad ,$$

the idea being that the u_i and v_i range over old elements and the x_i and y_{jk}^i over new elements--specifically, x_i is the ith new element for $\langle u_1, u_2 \rangle$, and y_{jk}^i the ith new element for $\langle v_i, v_j \rangle$.

Naturally, these restrictions cannot be accomplished by fiat; they have to be built into G. The way to arrange them is to provide an antecedent to G_1, stating that the relationship of y_{jk}^i to v_j and v_k is the same as that stated in G_0 of x_i to u_1 and u_2. Thus, the formula G will look like

$$\forall u_1 \forall u_2 \exists x_1 \cdots \exists x_q G_0^M \wedge Q_1 v_1 \cdots Q_m v_m \forall y_{11}^1 \cdots \forall y_{mm}^q$$

$$\left(\left(\bigwedge_{1 \le j,k \le m} G_0^M[u_1/v_j, u_2/v_k, x_1/y_{jk}^1, \ldots, x_q/y_{jk}^q]\right) \rightarrow G_1\right)$$

where

$$G_1 = F^M\{P_i v_j v_k / Py_{jk}^i y_{jk}^i \ (1 \le j,k \le m, \ 1 \le i \le q)\} \quad .$$

But this will accomplish the desired purpose only if the relationship of old elements to new is unambiguous in the context of the complete model $\mathfrak{A}'$: if $q+2$ elements of $\mathfrak{A}'$ stand in the stated relationship, then two must be old elements, and the other q must be the corresponding new elements. The construction of an appropriate G_0 and an antecedent to G_1 requires both exploitation of the limited expressiveness of the logical language and the design of a suitable combinatorial structure.

Since only one dyadic predicate letter is used in G, we can describe these combinatorial structures using directed graphs: draw an arrow from element a to element b if $P^{\mathfrak{A}'}(a,b)$, and not otherwise. Thus if G_0^M were

$$Px_1x_2 \wedge Px_2x_3 \wedge \cdots \wedge Px_{q-1}x_q \wedge Px_1u_1 \wedge Px_2u_2$$

then the corresponding graph would be as shown in Figure 11. Would this graph satisfy the unambiguity condition? By this we mean that if we used as the antecedent to G_1 the conjunction of formulas

$$Py_{jk}^1y_{jk}^2 \wedge Py_{jk}^2y_{jk}^3 \wedge \cdots \wedge Py_{jk}^{q-1}y_{jk}^q \wedge Py_{jk}^1v_j \wedge Py_{jk}^2v_k$$

for $1 \leq j,k \leq m$, could we assume in the conclusion that v_j and v_k were old elements, and that y_{jk}^i was the ith associated new element? The answer is no. Suppose, for example, that a, b is a pair of elements of $\mathfrak{A}$ of which $P_1^{\mathfrak{A}}(a,b)$ is to hold, so that $P^{\mathfrak{A}'}(e_{ab}^1, e_{ab}^1)$ is to hold. Then the antecedent to G_1 could be made true by taking e_{ab}^1 for the value of each of the variables y_{jk}^i $(i = 1,\ldots,q)$, v_j, and v_k! (If v_j or v_k is an x-variable this might not be possible; but in general F will contain some atomic formulas $P_iv_jv_k$ such that both v_j and v_k are y-variables.) If the identity sign were part of the language, we could add to the antecedent assertions $y_{jk}^i \neq y_{jk}^{i+1}$ for $i = 1,\ldots,q-1$; but there is no direct way to make these assertions without use of the identity sign.

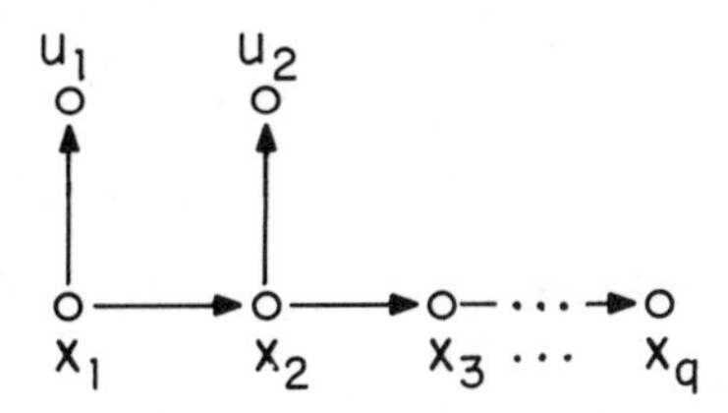

Figure 11

Instead we revise G_0^M to be

$$Px_1x_2 \wedge \neg Px_2x_1 \wedge Px_2x_3 \wedge \neg Px_3x_2 \wedge \cdots \wedge Px_{q-1}x_q \wedge \neg Px_qx_{q-1}$$

$$\wedge Px_1u_1 \wedge \neg Pu_1x_1 \wedge Px_2u_2 \wedge \neg Pu_2x_2 \quad .$$

By using a dashed arrow to represent a pair $\langle a, b\rangle$ of which $P^{\mathfrak{A}'}(a, b)$ is denied, this design could be illustrated as shown in Figure 12. The corresponding antecedent formula, the conjunction of

$$Py^1_{jk}y^2_{jk} \wedge \neg Py^2_{jk}y^1_{jk} \wedge Py^2_{jk}y^3_{jk} \wedge \neg Py^3_{jk}y^2_{jk} \wedge \cdots \wedge Py^{q-1}_{jk}y^q_{jk} \wedge \neg Py^q_{jk}y^{q-1}_{jk}$$

$$\wedge\, Py^1_{jk}v_j \wedge \neg Pv_jy^1_{jk} \wedge Py^2_{jk}v_k \wedge \neg Pv_ky^2_{jk} \quad ,$$

for $1 \le j, k \le m$, cannot "mistakenly" be made true, as before, by taking e^1_{ab} to be the value for each of the variables, for some a, b such that $P^{\mathfrak{A}'}(e^1_{ab}, e^1_{ab})$ holds; for example, $\neg Py^2_{jk}y^1_{jk}$ would fail. But the formula is still not "unambiguous" since, for example, it could be made true by taking as values for $y^1_{jk}, \ldots, y^q_{jk}$, v_j, and v_k, the elements $e^1_{ab}, \ldots, e^q_{ab}$, e^2_{ab}, and b, for some old elements a, b. That is, e^2_{ab} and a cannot be distinguished.

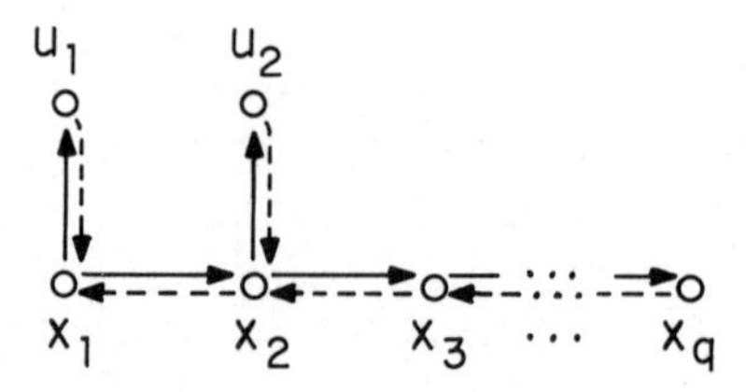

Figure 12

We might try to repair this flaw by reversing the direction of the arrows between u_1 and x_1, as shown in Figure 13. We might then try to argue that in any model $\mathfrak{A}'$ for G constructed in the way we have described, there can be a "chain" of $q+1$ elements $a_0, \ldots, a_q$ such that $P^{\mathfrak{A}'}(a_i, a_{i+1})$ but not $P^{\mathfrak{A}'}(a_{i+1}, a_i)$ for each i only if, for $i = 1, \ldots, q$, $a_i = e^i_{a_0a'_0}$ for some a'_0. Not so: if, for example, the values of u_1 and u_2 are the same element of $\mathfrak{A}'$, a "cycle" is created with no distinguishable beginning or end.

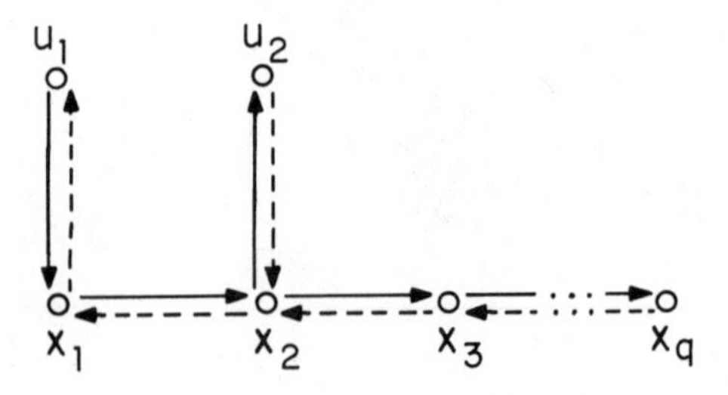

Figure 13

Many other subtle problems arise, for example, the fact that additional "new" elements are needed when u_1 or u_2 has a new element as its value. Setting aside these problems for the time being, we illustrate in Figure 14 a design which actually fills the bill--for $q \geq 4$. We give a formal proof of this in Section IIB. 3, after introducing certain definitions which will make it possible to discuss more easily exactly what properties of Figure 14 are being used in the proof.

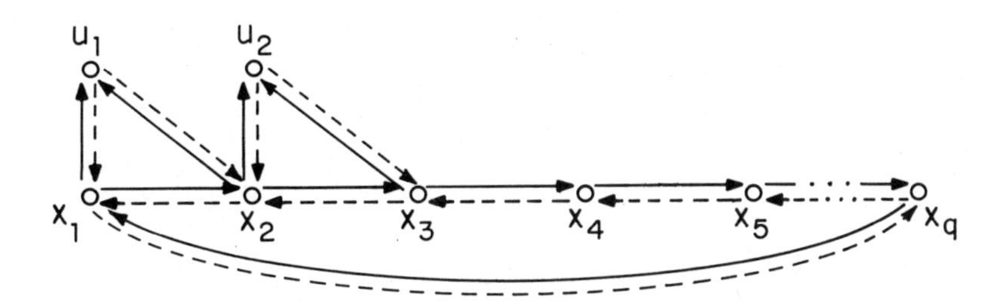

Figure 14

(Note that Figure 14 becomes Figure 15 in the $q = 3$ case, which is "ambiguous" because the beginning of the cycle cannot be distinguished. But this is not a serious problem, since we can apply the construction for the $q = 4$ case to schemata that actually have fewer than four predicate letters.)

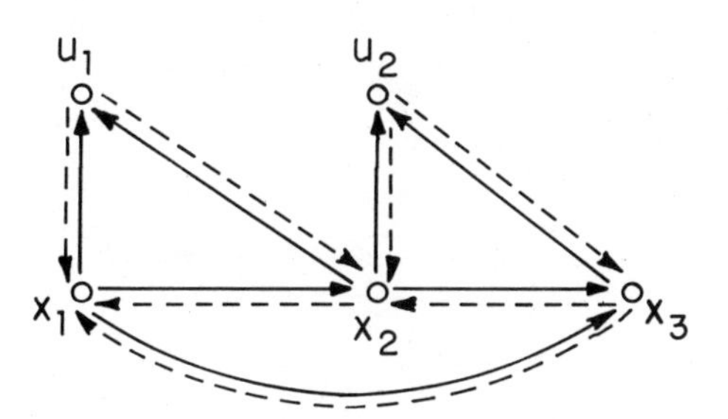

Figure 15

Figure 16 illustrates a portion of a model $\mathfrak{A}'$ for the $q = 4$ case corresponding to a model $\mathfrak{A}$ with only two distinct elements a, b. In addition to the old elements a, b, $\mathfrak{A}'$ has sixteen new elements $e^1_{aa}, \ldots, e^4_{bb}$. (To complete the model $\mathfrak{A}'$, we would also have to add additional new elements such as $e^1_{ae^1_{aa}}$; but in fact, all these elements need not be distinct.) The figure has an arrow between pairs of elements of which $P^{\mathfrak{A}'}$ holds. We assume that $P^{\mathfrak{A}}_1(b,a)$, $P^{\mathfrak{A}}_2(b,b)$, and $P^{\mathfrak{A}}_4(a,b)$ hold, so that $P^{\mathfrak{A}'}(e^1_{ba}, e^1_{ba})$, $P^{\mathfrak{A}'}(e^2_{bb}, e^2_{bb})$, and $P^{\mathfrak{A}'}(e^4_{ab}, e^4_{ab})$ hold (these are the three "loops" in the figure). Now the property we have chosen Figure 14 to satisfy is this: In Figure 16 (and in its completion, including all the "additional new" elements) if $a_1, \ldots, a_6$ are any six, not necessarily distinct, points, with arrows from a_1 to a_2 to a_3 to a_4 to a_1, from a_1 and a_2 to a_5, and from a_2 and a_3 to a_6, but with no arrows in the reverse directions, then $a_i = e^i_{a_5 a_6}$ for $i = 1, 2, 3, 4$.

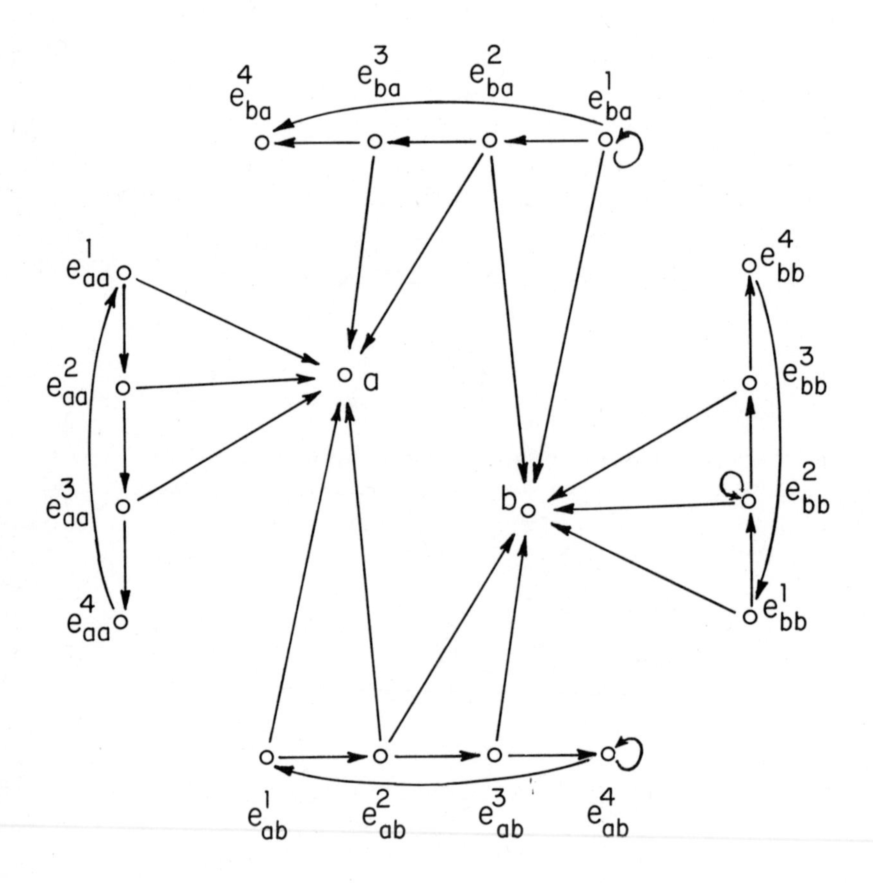

Figure 16

We now recapitulate and give a formal proof, which, however, is incomplete due to the omission of an exhaustive and general proof of the combinatorial property just asserted of Figure 14. This is deferred to Section IIB. 3.

$F = Q_1 v_1 \cdots Q_m v_m F^M$ is a schema whose only predicate letters are among the dyadic letters $P_1, \ldots, P_q$, with $q \geq 4$. P is a new dyadic letter. Let $u_1, u_2, x_1, \ldots, x_q$, and, for $1 \leq j, k \leq m$, $1 \leq i \leq q$, y^i_{jk}, be new variables. Let G^M_0 be the conjunction of atomic formulas obtained from the diagram in

Figure 14 by taking the conjunction of Pzz', when there is a solid arrow from z to z', and $\neg Pzz'$ when there is a dashed arrow from z to z'. Let $G_1 = F^M\{P_i v_j v_k / P y^i_{jk} y^i_{jk}\ (1 \leq j,k \leq m,\ 1 \leq i \leq q)\}$. Let $G_2 = \bigwedge_{1\leq j,k\leq m} G_0^M[u_1/v_1, u_2/v_2, x_1/y^1_{jk}, \ldots, x_q/y^q_{jk}]$. Finally, let

$$G_0 = \forall u_1 \forall u_2 \exists x_1 \cdots \exists x_q G_0^M; \qquad G = G_0 \wedge Q_1 v_1 \cdots Q_m v_m \forall y^1_{11} \cdots \forall y^q_{mm} (G_2 \to G_1).$$

Then F is satisfiable if and only if G is satisfiable.

Proof. We use the machinery of the Herbrand theory. Let $f_1, \ldots, f_q$ be the dyadic indicial correlates of $x_1, \ldots, x_q$. Then $D(G)$ is the domain generated from $\ddagger$, $f_1, \ldots, f_q$, and the function signs of F^*.

($F \rhd G$) Suppose F is satisfiable. By the Expansion Theorem $\mathcal{A} \models E(F)$ for some truth-assignment $\mathcal{A}$. Let $\gamma\colon D(G) \to D(F)$ be the following mapping:

$$\gamma(\ddagger) = \ddagger$$

$$\gamma(f_i(t_1 t_2)) = \ddagger \qquad (1 \leq i \leq q,\quad t_1, t_2 \in D(G))$$

$$\gamma(g(t_1 \cdots t_k)) = g(\gamma(t_1) \cdots \gamma(t_k)) \quad (g \text{ is a function sign of } F^*, \text{ and } t_1, \ldots, t_k \in D(G)).$$

Thus γ embeds $D(G)$ in $D(F)$ by identifying all x_i-terms with $\ddagger$. Then define a truth-assignment $\mathcal{B}$ on $E(G)$ as follows: $\mathcal{B} \models Pt_1t_2$ if and only if either

(a) $t_1 = t_2 = f_i(s_1 s_2)$ for some s_1, s_2, and $\mathcal{A} \models P_i\gamma(s_1)\gamma(s_2)$; or

(b) there are $s_1, s_2 \in D(G)$ such that $t_1 = f_i(s_1 s_2)$ and $t_2 = f_{i+1}(s_1 s_2)$ for some i, $1 \leq i < q$, or $t_1 = f_q(s_1 s_2)$ and $t_2 = f_1(s_1 s_2)$; or

(c) there are $s_1, s_2 \in D(G)$ such that $t_2 = s_1$ and $t_1 = f_1(s_1 s_2)$ or $f_2(s_1 s_2)$, or such that $t_2 = s_2$ and $s_1 = f_2(s_1 s_2)$ or $f_3(s_1 s_2)$.

Then $\mathcal{B} \models E(G)$. For let G' be an Herbrand instance of G. Then $\mathcal{B}$ verifies the Herbrand instance of G_1 occurring in G', by the relation of (b) and (c) to the construction of G_1; that is, if z_1 and z_2 are variables occurring in G_1, then there is a solid arrow from z_1 to z_2 in Figure 14 if and only if $\mathcal{B} \models Pt_1t_2$ by clause (b) or (c), where t_1, t_2 are the substituents in G' for the variables z_1, z_2 of G_0, respectively. Next, let $G_2' \to G_1'$ be the Herbrand instance of $G_2 \to G_1$ in G'. Then if $\mathcal{B} \models G_2'$, then by the relation of (b) and (c) to G_2 and the structure of Figure 14, $s^i_{jk} = f_i(t_j t_k)$ for $1 \leq j,k \leq m$, $1 \leq i \leq q$, where $t_1, \ldots, t_m$ are the substituents in G' for the variables $v_1, \ldots, v_m$ of G, and s^i_{jk} is the substituent for y^i_{jk}. Now let $v_{\ell_1}, \ldots, v_{\ell_p}$ be the y-variables among $v_1, \ldots, v_m$. Then there is an Herbrand instance of F, namely

$$F' = F^*\left[v_{\ell_1}/\gamma\left(t_{\ell_1}\right), \ldots, v_{\ell_p}/\gamma\left(t_{\ell_p}\right)\right]$$

to which G_1' bears the following intimate relationship: For each $w \in \overline{T}(F) = \overline{T}(G_1)$, if $\langle G_1' \rangle_w$ is the atomic formula $Pf_i(s_1 s_2)f_i(s_1 s_2)$, then $\langle F'_w \rangle = P_i \gamma(s_1)\gamma(s_2)$. But then $\mathscr{B} \models G_1'$ since $\mathscr{A} \models F'$, $T(F) = T(G_1)$, and, by (a) of the definition of $\mathscr{B}$, $\mathscr{B}(\langle G_1' \rangle_w) = \mathscr{A}(\langle F' \rangle_w)$ for each $w \in \overline{T}(F)$.

($G \triangleright F$) Now suppose G is satisfiable and let $\mathscr{B} \models E(G)$. A truth-assignment $\mathscr{A}$ verifying $E(F)$ can be constructed directly by letting $\mathscr{A} \models P_i t_1 t_2$ if and only if $\mathscr{B} \models Pf_i(t_1 t_2)f_i(t_1 t_2)$; this was the idea from the beginning. For if F' is an Herbrand instance of F, we can find an Herbrand instance $G_2' \to G_1'$ of $G_2 \to G_1$ such that

$$G_1' = F'\{P_i s_1 s_2 / Pf_i(s_1 s_2)f_i(s_1 s_2) \text{ (for all atomic subformulas } P_i s_1 s_2 \text{ of } F')\}$$

and such that

$$G_2' = \bigwedge_{1 \le j, k \le m} G_0^*[u_1/t_j, u_2/t_k] ,$$

where $t_1, \ldots, t_m$ are the substituents for $v_1, \ldots, v_m$ in F'. Since then $\mathscr{B} \models G_2'$ and $\mathscr{B} \models G_2' \to G_1'$, $\mathscr{B} \models G_1'$; and hence $\mathscr{A} \models F'$ by the definition of $\mathscr{A}$.

This completes the proof of the example. ∎

IIB. 2 Bigraphs

A bigraph is essentially a partial interpretation for a single predicate letter. If the predicate letter is dyadic, as in the example of the previous section, the partial interpretation can be pictured as a directed graph with arrows of two kinds, solid and dashed; the vertices are elements of the interpretation, the solid arrows are used for ordered pairs of elements of which the predicate is true, and the dashed arrows are used for pairs of which it is false. A pair not connected by an arrow of either type has an unspecified truth-value. Consistency requires that the same ordered pair not be connected by two arrows of different types (in the same direction). The purpose of much of the machinery introduced below is to facilitate the study of the combination and interlocking of bigraphs whose vertex sets overlap. This is the process through which a complete interpretation for a formula is constructed from partial interpretations, and the use of bigraphs makes it possible to demonstrate by explicit computation that these partial interpretations fit together in a logically consistent way.

Formally, a <u>bigraph</u> is a triple $\theta = \langle \theta^{(0)}, \theta^{(1)}, n \rangle$, where $n \geq 1$ and for some set V, $\theta^{(0)}$ and $\theta^{(1)}$ are disjoint subsets of the n-fold Cartesian product V^n. The <u>edges</u> of θ are the members of $\theta^{(0)} \cup \theta^{(1)}$; n is the <u>dimension</u> of θ; and the set of <u>vertices</u> of θ, $V(\theta)$, is the intersection of all sets V such that $\theta^{(0)} \cup \theta^{(1)} \subset V^n$. If η is also a bigraph of dimension n then

(a) η is a <u>subgraph</u> of θ, in symbols $\eta \subseteq \theta$, in case $\eta^{(0)} \subset \theta^{(0)}$ and $\eta^{(1)} \subset \theta^{(1)}$;

(b) η and θ are <u>disjoint</u> if they have no common edge;

(c) η and θ are <u>compatible</u> if $\theta^{(0)} \cap \eta^{(1)} = \theta^{(1)} \cap \eta^{(0)} = \emptyset$;

(d) θ is <u>complete</u> if $\theta^{(0)} \cup \theta^{(1)} = V(\theta)^n$;

(e) θ is a <u>completion</u> of η if $\eta \subset \theta$ and θ is complete;

(f) θ is the <u>minimal</u> completion of η if $\theta = \langle \eta^{(0)}, V(\eta)^n - \eta^{(0)}, n \rangle$;

(g) θ is <u>empty</u> if $\theta = \langle \emptyset, \emptyset, n \rangle$; in this case we write $\theta = \emptyset_n$;

(h) θ is <u>finite</u> if $V(\theta)$ is finite;

(i) $\theta = -\eta$ if $\theta = \langle \eta^{(1)}, \eta^{(0)}, n \rangle$;

(j) $1 \cdot \theta = \theta$;

(k) $0 \cdot \theta = \emptyset_n$;

(l) $-1 \cdot \theta = -\theta$;

(m) $\theta + \eta = \langle \theta^{(0)} \cup \eta^{(0)}, \theta^{(1)} \cup \eta^{(1)}, n \rangle$;

(n) $\theta - \eta = \theta + -\eta$;

(o) $\theta \cap \eta = \langle \theta^{(0)} \cap \eta^{(0)}, \theta^{(1)} \cap \eta^{(1)}, n \rangle$.

If C is any condition, then

$$\sum_C \theta = \left\langle \bigcup_C \theta^{(0)}, \bigcup_C \theta^{(1)}, n \right\rangle ,$$

the unions being over all bigraphs θ such that the condition holds, and the dimension being implicit in the condition. <u>Note that, in general,</u> $\theta + \eta$ <u>and</u> $\Sigma_C \theta$ <u>need not be bigraphs</u>; these sums are bigraphs only if the summands are (pairwise) compatible, and in this case the sums are said to be <u>proper</u>. A similar remark applies to $\theta - \eta$.

We are interested mainly in bigraphs of dimension 1 and 2. A bigraph of dimension 1 is essentially a partition of its vertex set. A bigraph of dimension 2 may be pictured as a directed graph with arcs of two kinds, for which we use solid and dashed arrows. Thus, in Figure 17,

$\theta_1 \subset \theta_5$ and $\theta_3 \subset \theta_4 \subset \theta_5$;

θ_1 and θ_3 are disjoint, but no other pair is disjoint;

$\theta_1, \theta_3, \theta_4, \theta_5$ are pairwise compatible; but θ_2 is not compatible with any of $\theta_1, \theta_3, \theta_4, \theta_5$;

θ_5 is complete;

θ_5 is a completion of each of $\theta_1, \theta_3, \theta_4$;

θ_5 is the minimal completion of θ_1, but not of θ_3 or θ_4 ;

$\theta_2 = -\theta_4$ and $\theta_4 = -\theta_2$;

$\theta_5 = \theta_1 + \theta_4$;

$\theta_5 = \theta_1 - \theta_2$.

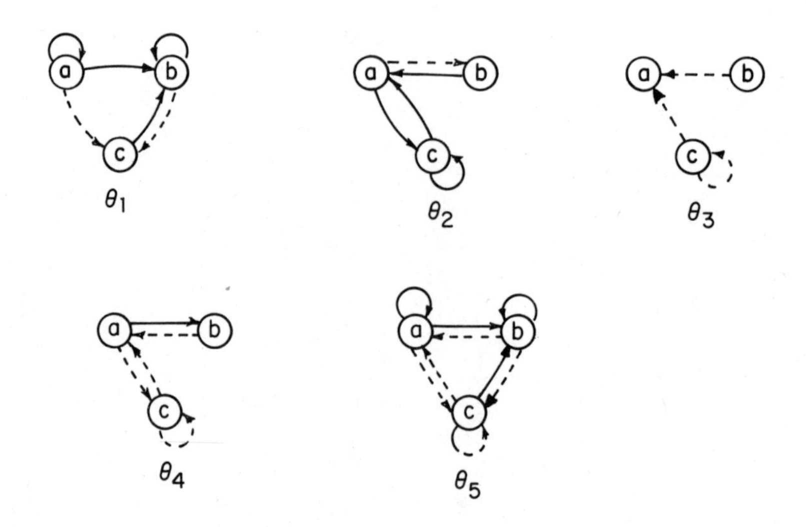

Figure 17

If η and θ are bigraphs of dimension n, and η is nonempty, then:

$$\chi_\theta(\eta) = \begin{cases} 1 & \text{if } \eta \subset \theta \\ -1 & \text{if } -\eta \subset \theta \\ 0 & \text{otherwise} \end{cases}$$

Note that χ_θ is well-defined, and that η need not be disjoint from θ in order for $\chi_\theta(\eta) = 0$. However, if θ is complete, $V(\eta) \subset V(\theta)$, and η has only one edge, then $\chi_\theta(\eta) = \pm 1$.

An interpretation for a formula with one n-place predicate letter is constructed by specifying a truth-value for each n-tuple of arguments. Sometimes, for example when the formula contains a biconditional, we wish to consider only those interpretations in which the n-tuples are clustered in certain ways: in each cluster the predicate letter must be true of all n-tuples, or false of all n-tuples. The next definitions deal with a corresponding situation for bigraphs.

If Ξ is a set of bigraphs of dimension n and θ is a bigraph of dimension n then a Ξ-extension of θ is any of the bigraphs

$$\theta + \sum_{\xi \in \Xi} e_\xi \cdot \xi$$

where $e_\xi = \pm 1$ and the sum is proper. (If θ is not disjoint from some member of Ξ, not every such sum will be proper and thus represent a Ξ-extension of θ. For example, if $\xi \in \Xi$ and θ is compatible with $-\xi$ but not with ξ, then e_ξ must be -1. And if θ is compatible with neither ξ nor $-\xi$, then there are no Ξ-extensions.) A Ξ-completion of θ is a completion of a Ξ-extension of θ, and a minimal Ξ-completion of θ is a minimal completion of a Ξ-extension of θ.

Again, let θ be a bigraph and let Ξ be a set of bigraphs. Then θ is Ξ-closed if and only if $\chi_\theta(\xi) = \pm 1$ whenever $\xi \in \Xi$ and ξ and θ are not disjoint. Thus every Ξ-extension of a bigraph if Ξ-closed. The Ξ-envelope of θ is the intersection of all Ξ-completions of θ. Since $\xi \cap -\xi = \emptyset_n$ for any ξ, the Ξ-envelope of a bigraph is Ξ-closed; it is the minimal (in the subgraph sense) Ξ-closed bigraph of which θ is a subgraph, and it is a subgraph of every Ξ-extension of θ. Any Ξ-closed bigraph is its own Ξ-envelope. If μ is a complete Ξ-closed bigraph, and η is the Ξ-envelope of a bigraph θ, then $\chi_\mu(\theta) = \chi_\mu(\eta)$.

Templates are devices for describing multiple bigraphs with similar structure. Formally, a p-template $(p \geq 1)$ is a bigraph τ such that $V(\tau) \subset \{0, \ldots, p-1\}$ and a template is a p-template for some p. If τ is a p-template and $a_0, \ldots, a_{p-1}$ are any (not necessarily distinct) elements, then

$$\tau(a_0, \ldots, a_{p-1}) = \langle \{\langle a_{i_1}, \ldots, a_{i_n} \rangle \mid \langle i_1, \ldots, i_n \rangle \in \tau^{(0)}\}, \{\langle a_{i_1}, \ldots, a_{i_n} \rangle \mid \langle i_1, \ldots, i_n \rangle \in \tau^{(1)}\}, n \rangle$$

where n is the dimension of τ. Once again, $\tau(a_0, \ldots, a_{p-1})$ need not be a bigraph; for example, if $\tau = \langle \{\langle 0, 1 \rangle\}, \{\langle 1, 0 \rangle\}, 2 \rangle$, then $\tau(0, 0)$ is not a bigraph. An expression $\tau(a_0, \ldots, a_{p-1})$ is proper if it denotes a bigraph, and an assertion of the form $\tau(a_0, \ldots, a_{p-1}) \subset \theta$ carries with it the assertion that $\tau(a_0, \ldots, a_{p-1})$ is proper.

An important template is

$$\omega(a_0, \ldots, a_{p-1}) = \langle \{\langle a_0, \ldots, a_{p-1} \rangle\}, \emptyset, p \rangle$$

(we use the same symbol ambiguously for all dimensions).

Let θ be a bigraph, $S \subset V(\theta)^q$, $a_0, \ldots, a_{p-1} \in V(\theta)$, and let τ be a $(p+q)$-template. Then $\tau(a_0, \ldots, a_{p-1}, \cdot, \ldots, \cdot)$ <u>extracts</u> S <u>from</u> θ if and only if

$$S = \{\langle a_p, \ldots, a_{p+q-1}\rangle \mid \tau(a_0, \ldots, a_{p+q-1}) \subset \theta\} \quad .$$

For example, if τ is the bigraph of Figure 14 (p. 77) for the $q = 4$ case, with the vertices numbered 0, 1, 2, 3, 4, 5 counterclockwise from the lower left, and θ is the bigraph of Figure 16 (p. 78) with dotted arrows supplied wherever there is no solid arrow, then $\tau(\cdot, \ldots, \cdot)$ extracts S from θ, where

$$S = \{\langle e^1_{a'b'}, e^2_{a'b'}, e^3_{a'b'}, e^4_{a'b'}, b', a'\rangle \mid a', b' \in \{a, b\}\} \quad .$$

The next definitions generalize the notion of bigraph so that interpretations for formulas with more than one predicate letter can be discussed. A <u>tagged bigraph</u> is a pair $\langle i, \theta\rangle$, where $i \geq 0$ and θ is a bigraph; i is the <u>tag</u> of $\langle i, \theta\rangle$. It is convenient to identify $\langle 0, \theta\rangle$ with θ itself. A <u>bistructure</u> is a set of tagged bigraphs not containing two tagged bigraphs of the same dimension with the same tag. If the bistructure has only one member we may identify it with that one member.

Most of the notions defined for bigraphs have natural generalizations for tagged bigraphs and bistructures. For example, a bistructure is <u>complete</u> if each tagged bigraph in it is complete and all share a common set of vertices. If $\langle i_1, \theta_1\rangle$ and $\langle i_2, \theta_2\rangle$ are tagged bigraphs, then $\langle i_1, \theta_1\rangle \subset \langle i_2, \theta_2\rangle$ if and only if $i_1 = i_2$ and $\theta_1 \subset \theta_2$; and $\langle i_1, \theta_1\rangle$ and $\langle i_2, \theta_2\rangle$ are <u>disjoint</u> if $i_1 \neq i_2$, or $i_1 = i_2$ and θ_1 and θ_2 are disjoint bigraphs. If η_1 and η_2 are bistructures, then $\eta_1 \subset \eta_2$ if and only if, for each $\langle i_1, \theta_1\rangle \in \eta_1$, there is an $\langle i_2, \theta_2\rangle \in \eta_2$ such that $\langle i_1, \theta_1\rangle \subset \langle i_2, \theta_2\rangle$ (no confusion with ordinary set inclusion will arise). <u>Templates</u> may now be bistructures as well as bigraphs; a template is simply a finite bistructure whose vertex set is a finite set of natural numbers. Then, for example, if $\tau = \{\langle 4, \omega(0)\rangle, \langle 0, \omega(1,0)\rangle\}$ then $\tau(a, b) = \{\langle 4, \omega(a)\rangle, \langle 0, \omega(a, b)\rangle\}$. We leave it to the reader to supply the definitions for similar notions when they arise.

We now come to the formal relation of bistructures to logical formulas. Corresponding to each tagged bigraph $\langle i, \theta\rangle$ there is a predicate letter $P^{\langle i, \theta\rangle}$, namely the ith predicate letter in lexical order whose degree is the dimension of θ. If η is a finite bistructure such that $V(\eta)$ is a set of terms, then we define a formula $\hat{\eta}$ corresponding to η as follows:

$$\hat{\eta} = P^{\langle i, \theta\rangle} t_1 \cdots t_n, \quad \text{if } \eta = \langle i, \theta\rangle \text{ and } \theta = \omega(t_1, \ldots, t_n)$$

$$\hat{\eta} = \neg P^{\langle i, \theta\rangle} t_1 \cdots t_n, \quad \text{if } \eta = \langle i, \theta\rangle \text{ and } \theta = -\omega(t_1, \ldots, t_n)$$

$\hat{\eta} = \wedge \hat{\theta}$, otherwise, the conjunction being over all tagged bigraphs $\theta \subset \eta$ with one edge, and the order being lexical.

We also define $\hat{\eta}$ in one case in which $V(\eta)$ is not a set of terms: if τ is a p-template, then $\hat{\tau} = \hat{\sigma}$, where $\sigma = \tau(y_0, \ldots, y_{p-1})$ and $y_0, y_1, \ldots$ is some fixed sequence of variables. Now, if $t_0, \ldots, t_{p-1}$ are terms, we write $\hat{\tau}(t_0, \ldots, t_{p-1})$ for the formula

$$\hat{\tau}[y_0/t_0, \ldots, y_{p-1}/t_{p-1}] \quad .$$

(Thus $\hat{\tau}(t_0, \ldots, t_{p-1})$ is a well-defined formula, even if $\tau(t_0, \ldots, t_{p-1})$ is not proper.) For example, let τ be the (q+2)-template illustrated in Figure 14, with the variables $x_1, \ldots, x_q, u_1, u_2$ replaced by the numbers $0, \ldots, q+1$, respectively. Then the formula G_0^M (p. 78) is $\hat{\tau}(x_1, \ldots, x_q, u_2, u_1)$, while G_2 is

$$\bigwedge_{1 \leq j, k \leq m} \hat{\tau}(y_{jk}^1, \ldots, y_{jk}^q, v_k, v_j) \quad .$$

For any finite nonempty bistructure η, let $\tilde{\eta}$ be a subgraph of η with one edge (chosen according to some rule). Note that the signed atomic formula $\hat{\tilde{\eta}}$ is a conjunct of $\hat{\eta}$, and $\hat{\tilde{\tau}}(t_0, \ldots, t_{p-1})$ is a conjunct of $\hat{\tau}(t_0, \ldots, t_{p-1})$.

If μ is a complete bistructure, F is a formula whose predicate letters are exactly those corresponding to the tagged bigraphs in μ, and γ is a mapping: $D(F) \to V(\mu)$, then $\mu(\gamma)$ is the truth-assignment such that

$$\mu(\gamma)\left(P^{\langle i, \theta \rangle} t_1 \cdots t_n\right) = \chi_{\langle i, \theta \rangle}(\omega(\gamma(t_1), \ldots, \gamma(t_n))) \quad ,$$

i.e., such that $\mu(\gamma) \models P^{\langle i, \theta \rangle} t_1 \cdots t_n$ if and only if $\omega(\gamma(t_1), \ldots, \gamma(t_n)) \subset \mu$, and $\mu(\gamma) \models \neg P^{\langle i, \theta \rangle} t_1 \cdots t_n$ if and only if $-\omega(\gamma(t_1), \ldots, \gamma(t_n)) \subset \mu$. The truth-assignment $\mu(\gamma)$ is called the truth-assingment <u>induced</u> by μ and γ.

Let us pause to note a few properties of induced truth-assingments. Recall that, formally, a truth-assignment is a mapping with range $\{-1, +1\}$. It need <u>not</u> be the case that

$$\mu(\gamma)(\hat{\tau}(t_0, \ldots, t_{p-1})) = \chi_\mu(\tau(\gamma(t_0), \ldots, \gamma(t_{p-1}))) \quad .$$

For example, if $\tau = \omega(0, 1) - \omega(1, 0)$, γ is the identity mapping, and $\chi_\mu(\omega(t_0, t_1)) = \chi_\mu(\omega(t_1, t_0)) = 1$, then $\mu(\gamma)(\hat{\tau}(t_0, t_1)) = \mu(\gamma)(Pt_0t_1 \wedge \neg Pt_1t_0) = -1$; but $\chi_\mu(\tau(\gamma(t_0), \gamma(t_1))) = \chi_\mu(\tau(t_0, t_1)) = 0$, since μ and $\tau(t_0, t_1)$ are incompatible, as are μ and $-\tau(t_0, t_1)$. But this situation cannot arise if τ has only one edge. In general,

$$\mu(\gamma)(\hat{\tilde{\tau}}(t_0, \ldots, t_{p-1})) = \chi_\mu(\tilde{\tau}(\gamma(t_0), \ldots, \gamma(t_{p-1})))$$

for any μ, γ and τ, as long as everything is defined and proper. However, the stronger equation (with τ instead of $\tilde{\tau}$) <u>is</u> true in certain important cases. Suppose that μ is complete and is Ξ-closed, and that $\tau(\gamma(t_0),\ldots,\gamma(t_{p-1}))\subset\tau^*$ for some bigraph $\tau^*\in\Xi$. Then either $\tau^*\subset\mu$ or $-\tau^*\subset\mu$, i.e., $\chi_\mu(\tau^*)=\pm 1$, and in this case $\mu(\gamma)(\hat{\tau}(t_0,\ldots,t_{p-1}))=\chi_\mu(\tau^*)=\chi_\mu(\tau(\gamma(t_0),\ldots,\gamma(t_{p-1})))$.

If $\gamma\colon D_1\to D_2$, D_1 is a set of terms, and η is a bistructure such that $V(\eta)\subset D_2$, then $\mathscr{L}_\gamma(\eta)$ is the set of all schemata F such that if F is satisfiable, then $\mu(\gamma)\models E(F)$ for some completion μ of η.

If $\mathscr{A}$ is a truth-assignment on an Herbrand expansion E(F) and s, t are terms in D(F), then s and t and $\mathscr{A}$-indistinguishable if and only if $\mathscr{A}(A)=\mathscr{A}(B)$ whenever A and B are atomic formulas $Ps_1\cdots s_n$, $Pt_1\cdots t_n$ and, for some i, $1\le i\le n$, $s_i=s$, $t_i=t$, and $s_j=t_j$ for $j\ne i$. As this is clearly an equivalence relation, we also say that a set of terms is $\mathscr{A}$-indistinguishable if the members are pairwise $\mathscr{A}$-indistinguishable. A set S of terms is $\mathscr{A}$-indistinguishable if and only if there is a bistructure μ and a mapping $\gamma\colon D(F)\to V(\mu)$ such that $\mathscr{A}=\mu(\gamma)$ and $\gamma(s)=\gamma(t)$ whenever $s,t\in S$; that is, if and only if the members of S cannot be distinguished by examination of $\mathscr{A}$.

Finally, a convention that will prove convenient is to denote the sequence $a_0,\ldots,a_{k-1}$, where k is fixed and the a_i are objects of any kind, by $\underset{\sim}{a}$. Here we have written the sequence without $\langle\ \rangle$ brackets, since we shall sometimes run several together: $\tau(\underset{\sim}{a},\underset{\sim}{b})$, for example, would be $\tau(a_0,\ldots,a_{k-1},b_0,\ldots,b_{k-1})$. We write $F[\underset{\sim}{v}/\underset{\sim}{t}]$, where F is a formula, $v_0,\ldots,v_k$ are distinct variables, and $t_0,\ldots,t_{k-1}$ are terms, for $F[v_0/t_0,\ldots,v_{k-1}/t_{k-1}]$. A distinct but closely related use of the wavy underline is when n is an integer: then by $\underset{\sim}{n}$ we mean the sequence $kn, kn+1,\ldots,kn+k-1$, and by $\underset{\sim}{n}_i$ the integer $kn+i$.

IIB. 3 Reprise of the Example

We now show how the example of Section IIB. 1 can be carried through using the concepts defined in the last section.

Theorem. There is an effective procedure that yields, for each schema F containing dyadic predicate letters only, a schema G with a single, dyadic predicate letter, such that G is satisfiable if and only if F is satisfiable.

Proof. Without loss of generality, assume F to be prenex, say $F=Q_1v_1\ldots Q_mv_mF^M$; and that the predicate letters of F are $P_1,\ldots,P_q$. We construct a suitable G in two stages: first we state some sufficient conditions for G to be constructible, then we show that these conditions can be met.

Let $x_1,\ldots,x_q$ be new variables, and let $f_1,\ldots,f_q$ be their dyadic indicial correlates. Let D_1 be the Herbrand domain of F, and let D_2 be the domain generated by ł, the indicial function signs of F^*, and $f_1,\ldots,f_q$. (D_2 will be the Herbrand domain of G.) Let ι be the identity mapping on D_2 (also on D_1, since $D_1\subset D_2$). Let $\gamma\colon D_2\to D_1$ be the following mapping:

$$\gamma(\ddagger) = \ddagger \; ;$$

$$\gamma(f_i(t_1t_2)) = \ddagger \quad (1 \le i \le q, \quad t_1, t_2 \in D_2) \; ;$$

$$\gamma(g(t_1 \dots t_k)) = g(\gamma(t_1) \dots \gamma(t_k)) \quad \text{(g is a function sign of } F^*, \text{ and } t_1, \dots, t_k \in D_2) \; .$$

Sufficient Conditions Lemma. Suppose there are $q+1$ templates of dimension 2, $\tau, \pi_1, \dots, \pi_q$ with the following properties:

Disjointness: The bigraphs $\tau(t_1, t_2, f_1(t_1t_2), \dots, f_q(t_1t_2))$ and $\pi_i(t_1, t_2, f_1(t_1t_2), \dots, f_q(t_1t_2))$, as t_1, t_2 range over D_2 and i over $\{1, \dots, q\}$, are pairwise disjoint.

Let

$$\Theta = \{\tau(t_1, t_2, f_1(t_1t_2), \dots, f_q(t_1t_2)) \mid t_1, t_2 \in D_2\}$$

$$\Phi = \{\pi_i(t_1, t_2, f_1(t_1t_2), \dots, f_q(t_1t_2)) \mid t_1, t_2 \in D_2, \; 1 \le i \le q\} \; .$$

By the Disjointness Condition,

$$\theta = \sum_{\alpha \in \Theta} \alpha$$

is a proper sum.

Extraction: $\tau(\cdot, \dots, \cdot)$ extracts

$$\{\langle t_1, t_2, f_1(t_1 \; t_2), \dots, f_q(t_1 \; t_2)\rangle \mid t_1, t_2 \in D_2\}$$

from any minimal Φ-completion of θ.

Schema: There is a schema G_0 such that the indicial function signs of G_0^* are $f_1, \dots, f_q$ and such that

(a) If μ is any minimal Φ-completion of θ then $\mu(\iota) \models E(G_0^*, D_2)$

(b) If $\mu(\iota) \models E(G_0^*, D_2)$, then μ is a completion of θ.

Then we can construct a schema G with a single, dyadic, predicate letter, such that G is satisfiable if and only if F is satisfiable.

Proof. We first note a few ellipses in the statement of this lemma characteristic of the way similar lemmata will be stated in the future. First, the lemma refers to the existence of the templates $\tau, \pi_1, \dots, \pi_q$, and the schema G_0; but what is really meant is that there is an effective procedure that supplies such templates and such a schema, given the schema F. Likewise, the phrase "we can construct" in the conclusion of the lemma refers to an effective procedure of which F is the input. Also, the Schema Condition does not precisely limit the class of μ under consideration; μ is to be, in each case, a complete bigraph of dimension 2 such that $V(\mu) = D_2$. Finally, it is implicitly assumed, but may be nontrivial to verify, that the expressions in the Disjointness Condition are proper; we return to this point in the proof of the Construction Lemma.

Now for the proof. Let y^i_{jk} $(1 \le i \le q,\ 1 \le j,k \le m)$ be new variables, and let

$$G = G_0 \wedge Q_1 v_1 \cdots Q_m v_m \forall y^1_{11} \cdots \forall y^q_{mm} (G_1 \to G_2),$$

where

$$G_1 = \bigwedge_{1 \le j,k \le m} \hat{\tau}(v_j, v_k, y^1_{jk}, \ldots, y^q_{jk})$$

and

$$G_2 = F^M\{P_i v_j v_k / \hat{\hat{\pi}}_i(v_j, v_k, y^1_{jk}, \ldots, y^q_{jk}), \quad (1 \le i \le q,\ 1 \le j,k \le m)\} .$$

<u>$(F \rhd G)$</u> If F is satisfiable, then there is a truth-assignment $\mathscr{A}$ verifying $E(F)$. Recall that $\mathscr{A}$ is a mapping with range $\{-1, 1\}$, and let μ be the minimal completion of

$$\theta + \sum_{\substack{t_1, t_2 \in D_2 \\ 1 \le i \le q}} \mathscr{A}(P_i \gamma(t_1)\gamma(t_2)) \cdot \pi_i(t_1, t_2, f_1(t_1 t_2), \ldots, f_q(t_1 t_2)) .$$

The sum is proper by the Disjointness Condition; that is, any combination of bigraphs φ and $-\varphi'$ for $\varphi, \varphi' \in \Phi$, may be properly added to θ (so long as the sum does not include both φ and $-\varphi$ for some single φ).

By Schema Condition (a), $\mu(\iota) \models E(G_0^*, D_2)$. Thus to show that $\mu(\iota) \models E(G)$ it suffices to show that $\mu(\iota)$ verifies each Herbrand instance $G_1' \to G_2'$ of $G_1 \to G_2$. If $\mu(\iota) \models G_1'$ then by the Extraction Condition the substituent for each y^i_{jk} in G_1' is $f_i(t_1 t_2)$, where t_1, t_2 are the substituents for v_j, v_k. Then let F' be the Herbrand instance of F in which, for $i = 1, \ldots, m$, if t is the substituent for v_i in G_2' then $\gamma(t)$ is the substituent for v_i in F'. (Such an F' exists because of the way γ is defined.) Then for each $w \in \overline{T}(F') \subset \overline{T}(G_2')$, if $\langle G_2 \rangle_w = \hat{\hat{\pi}}_i(v_j, v_k, y^1_{jk}, \ldots, y^q_{jk})$, then $\langle G_2' \rangle_w = \hat{\hat{\pi}}_i(t_1, t_2, f_1(t_1 t_2), \ldots, f_q(t_1 t_2))$ for some t_1, t_2 such that $\langle F' \rangle_w = P_i \gamma(t_1)\gamma(t_2)$. But then $\mu(\iota)(\langle G_2' \rangle_w) = \mu(\iota)(\hat{\hat{\pi}}_i(t_1, t_2, f_1(t_1 t_2), \ldots, f_q(t_1 t_2))) = \chi_\mu(\pi_i(t_1, t_2, f_1(t_1 t_2), \ldots, f_q(t_1 t_2))) = \mathscr{A}(P_i \gamma(t_1)\gamma(t_2)) = \mathscr{A}(\langle F' \rangle_w)$, by the definition of μ. Then since $\mathscr{A} \models F'$, $\mu(\iota) \models G_2'$.

<u>$(G \rhd F)$</u> Now suppose G is satisfiable. Then some truth-assignment verifies $E(G)$, so there is a complete bigraph μ such that $\mu(\iota) \models E(G)$. Then by Schema Condition (b), μ is a completion of θ. Define a truth-assignment $\mathscr{A}$ on $E(F)$ by

$$\mathscr{A}(P_i t_1 t_2) = \chi_\mu(\widetilde{\pi}_i(t_1, t_2, f_1(t_1 t_2), \ldots, f_q(t_1 t_2))) \quad ,$$

which is 1 or -1 in each case since $\widetilde{\pi}_i(-)$ has only one edge. Now let F' be any Herbrand instance of F; then there is an Herbrand instance $G_1' \to G_2'$ of $G_1 \to G_2$ such that the substituents for $v_1, \ldots, v_m$ are the same in $G_1' \to G_2'$ as in F', and such that the substituent for each y^i_{jk} is $f_i(t_1 t_2)$, where t_1, t_2 are the substituents for v_j, v_k. Since $\theta \subset \mu$, $\mu(\iota) \models G_1'$ by the definition of θ; hence $\mu(\iota) \models G_2'$. But since

$$G_2' = F'\{P_i t_1 t_2 / \hat{\widetilde{\pi}}_i(t_1, t_2, f_1(t_1 t_2), \ldots, f_q(t_1 t_2)) \text{ (for each atomic formula } P_i t_1 t_2 \text{ of } F')\} \quad ,$$

by the definition of $\mathscr{A}$, $\mathscr{A}(\langle F' \rangle_w) = \mu(\iota)(\langle G_2' \rangle_w)$ for each $w \in \overline{T}(F) \subset \overline{T}(G_2)$, so $\mathscr{A} \models F'$.

This completes the proof of the Sufficient Conditions Lemma. ■

<u>Construction Lemma</u>. Templates τ, $\pi_1, \ldots, \pi_q$ and a schema G_0 can be found as required by the Sufficient Conditions Lemma.

<u>Proof.</u> Let $\psi(a, b) = \omega(a, b) - \omega(b, a)$ for any a, b, and let

$$\begin{aligned} \tau(u_1, u_2, x_1, \ldots, x_q) = {} & \psi(x_1, x_2) + \psi(x_2, x_3) + \cdots + \psi(x_{q-1}, x_q) + \psi(x_q, x_1) \\ & + \psi(x_1, u_1) + \psi(x_2, u_1) \\ & + \psi(x_2, u_2) + \psi(x_3, u_2) \end{aligned}$$

as illustrated in Figure 14 (p. 77) and let

$$\pi_i(u_1, u_2, x_1, \ldots, x_q) = \omega(x_i, x_i) \quad .$$

<u>Disjointness Condition.</u> Note first that the expressions $\tau(t_1, t_2, f_1(t_1 t_2), \ldots, f_q(t_1 t_2))$ and $\pi_i(t_1, t_2, f_1(t_1 t_2), \ldots, f_q(t_1 t_2))$ are all proper. (This is not automatic; if the arrows between x_2 and u_2 were reversed (i.e., the above sum had $\psi(u_2, x_2)$ instead of $\psi(x_2, u_2)$), then $\tau(t, t, f_1(tt), \ldots, f_q(tt))$ would not be proper.) Each bigraph $\tau(t_1, t_2, f_1(t_1 t_2), \ldots, f_q(t_1 t_2))$ is disjoint from each $\pi_i(s_1, s_2, f_1(s_1 s_2), \ldots, f_q(s_1 s_2))$, since the latter are of the form $\omega(s, s)$ but none of the former has such an edge. Clearly, $\pi_i(s_1, s_2, f_1(s_1 s_2), \ldots, f_q(s_1 s_2))$ and $\pi_{i'}(s_1', s_2', f_1(s_1' s_2'), \ldots, f_q(s_1' s_2'))$ are disjoint if $i \neq i'$ or $s_1 \neq s_1'$ or $s_2 \neq s_2'$. Finally, the $\tau(t_1, t_2, f_1(t_1 t_2), \ldots, f_q(t_1 t_2))$ are disjoint from each other, since at least one vertex of each edge of $\tau(t_1, t_2, f_1(t_1 t_2), \ldots, f_q(t_1 t_2))$ is among $f_1(t_1 t_2), \ldots, f_q(t_1 t_2)$.

<u>Extraction.</u> Let μ be a minimal Φ-completion of θ and suppose that $\tau(t_1, t_2, s_1, \ldots, s_q) \subset \mu$ for some $t_1, t_2, s_1, \ldots, s_q \in D_2$. Note that $\psi(s_1', s_2') \subset \mu$, where $s_1', s_2' \in D_2$, only if either s_2' is an argument of s_1', or s_1' and s_2' are

the same except for their outermost function signs. The first possibility cannot hold if there also are $s'_3, \ldots, s'_q$ such that $\psi(s'_2, s'_3) + \cdots + \psi(s'_{q-1}, s'_q) + \psi(s'_q, s'_1) \subset \mu$, so it follows that $\langle s_1, \ldots, s_q \rangle$ is, for some t'_1, t'_2, a cyclic permutation of $\langle f_1(t'_1 t'_2), \ldots, f_q(t'_1 t'_2) \rangle$. To see that in fact $t'_1 = t_1$, $t'_2 = t_2$, and $s_i = f_i(t_1 t_2)$ for $i = 1, \ldots, q$, note that a "linked pair of triangles"

$$\psi(s'_1, t'_1) + \psi(s'_2, t'_1) + \psi(s'_1, s'_2) + \psi(s'_2, t'_2) + \psi(s'_3, t'_2) + \psi(s'_2, s'_3)$$

where $s'_1, s'_2, s'_3, t'_1, t'_2 \in D_2$, can be a subgraph of μ only if $s'_i = f_i(t'_1 t'_2)$ for $i = 1, 2, 3$; the result follows.

Schema. A schema G_0 as required is

$$\forall u_1 \forall u_2 \exists x_1 \cdots \exists x_q \hat{\tau}(u_1, u_2, x_1, \ldots, x_q) \quad .$$

This completes the proof of the Construction Lemma. ■

One closing remark. This example is simpler than those encountered in the rest of the chapter, in that Φ is a set of bigraphs with one edge each, so that "Φ-completion" is synonymous with "completion". In general the members Φ will have more than one edge, and so the notion of Φ-completion will play a more direct role than it does here.

Chapter IIC

THE PREFIX AND SIMILARITY CLASSES

This chapter considers the classical decision problem for first-order logic (see the Historical References, p. 136): to determine which quantifier prefixes yield solvable and unsolvable classes of schemata, and to refine these results on prefix by restricting, as much as possible, the number and degrees of the predicate letters a schema may have. The first section shows that, as a consequence of known decision procedures, the decision problem reduces naturally to nine special cases. Four theorems in the next four sections establish the unsolvability of these nine cases.

IIC.1 The Classification

A word over the two-symbol alphabet $\{\forall, \exists\}$ is called a <u>prefix type</u>; if $F = Q_1 v_1 Q_2 v_2 \ldots Q_n v_n F^M$ is a prenex schema, where $Q_1, \ldots, Q_n \in \{\forall, \exists\}$ and $v_1, \ldots, v_n$ are variables, then F is said to <u>have prefix type</u> $Q_1 \ldots Q_n$. If $\mathscr{P} \subset \{\forall, \exists\}^*$ is a set of prefix types, then $[\mathscr{P}]$ is the set of all prenex schemata whose prefix types are members of $\mathscr{P}$. To abbreviate the names of commonly used sets of prefix types we use regular expressions. Thus a word over $\{\forall, \exists\}$ stands for its singleton set, so that $[\forall\exists\forall]$ is $[\{\forall\exists\forall\}]$, i.e., the set of all prenex schemata $F = \forall v_1 \exists v_2 \forall v_3 F^M$. Also, $\forall^*$ or $\exists^*$ stands for a sequence of $\forall$'s or $\exists$'s of indeterminate length, so that, for example, $\forall^*\exists^* = \{w_1 w_2 \mid w_1 \in \{\forall\}^*, w_2 \in \{\exists\}^*\}$, and $[\forall^*\exists^*]$ is the set of all <u>Skolem</u> schemata, i.e., prenex schemata $F = \forall v_1 \ldots \forall v_m \exists v_{m+1} \ldots \exists v_n F^M$, where $n \geq m \geq 0$. Any class $[\mathscr{P}]$, where $\mathscr{P} \subset \{\forall, \exists\}^*$, is called a <u>prefix class</u>.

We carry this notation a bit further. If $Q_1, \ldots, Q_n$ are prefix types, then the word $Q_1 \wedge Q_2 \wedge \ldots \wedge Q_n$ is called an <u>extended prefix type</u>; it is the extended prefix type of any conjunction of prenex schemata $F_1 \wedge \ldots \wedge F_n$, where each F_i is of prefix type Q_i. Similarly, $\mathscr{P}_1 \wedge \ldots \wedge \mathscr{P}_n$ is the set of all extended prefix types $Q_1 \wedge \ldots \wedge Q_n$, with $Q_i \in P_i$ for each i, and $[\mathscr{P}_1 \wedge \ldots \wedge \mathscr{P}_n]$ is the class of all schemata whose extended prefix types are in $\mathscr{P}_1 \wedge \ldots \wedge \mathscr{P}_n$. These are the <u>extended prefix classes</u>. Thus $[\forall\exists \wedge \forall^*]$ is the class of all schemata of the form $\forall v_1 \exists v_2 M_1 \wedge \forall v_3 \ldots \forall v_n M_2$, where M_1 and M_2 are quantifier-free.

A <u>similarity type</u> is a mapping $\sigma : \mathbb{N} - \{0\} \to \mathbb{N}$; a schema F <u>has similarity type</u> σ provided that F has occurrences of exactly $\sigma(i)$ predicate letters of degree i for each i. (This is slightly at variance with the standard usage in model theory, where a similarity type is a set of predicate letters.) If σ is a mapping $\mathbb{N} - \{0\} \to \mathbb{N} \cup \{\infty\}$, then $[\sigma]$ is the class of all schemata F of some similarity type σ' such that $\sigma'(i) \le \sigma(i)$ for each i, with the symbol ∞ interpreted in the obvious way. Any such class $[\sigma]$ is called a similarity class. When possible, we abbreviate mappings $\sigma : \mathbb{N} - \{0\} \to \mathbb{N} \cup \{\infty\}$ by finite sequences. Thus $(\infty, 1)$ abbreviates the mapping σ such that $\sigma(1) = \infty$, $\sigma(2) = 1$, and $\sigma(i) = 0$ for all $i \ge 2$, so that $[(\infty, 1)]$ is the class of schemata with any number of monadic predicate letters, at most one dyadic predicate letter, and no k-place predicate letters for $k \ge 3$. We also use a combined notation when describing a class determined jointly by prefix and number of predicate letters: thus $[\forall\exists \wedge \forall^*(0, 1)]$ is the intersection of the $[\forall\exists \wedge \forall^*]$ extended prefix class with the $[(0, 1)]$ similarity class. Moreover, in later sections, since no confusion can arise, we omit the [ ] brackets, and refer directly, for example, to the $\forall\exists\forall$ or $(0, 1)$ or $\forall\exists \wedge \forall^*(0, 1)$ class.

When prefix and similarity classes are considered, five facts should be borne in mind.

(1) If $\mathscr{P}, \mathscr{P}' \subset \{\forall, \exists\}^*$ are sets of words such that for each word w in $\mathscr{P}$, there is a word w' in $\mathscr{P}'$ such that w may be obtained from w' by deleting some of the symbols, then the prefix class $[\mathscr{P}]$ is reducible to the prefix class $[\mathscr{P}']$.

This is a consequence of vacuous quantification. For example, the $[\forall\exists\forall]$ prefix class is reducible to the $[\forall\exists\exists\forall]$ prefix class, by transforming each schema $\forall y_1 \exists x_1 \forall y_2 F^M$ into $\forall y_1 \exists x_1 \exists x_2 \forall y_2 F^M$.

(2) If $\sigma, \sigma' : \mathbb{N} - \{0\} \to \mathbb{N} \cup \{\infty\}$ are such that $\Sigma_{i \ge n} \sigma(i) \le \Sigma_{i \ge n} \sigma'(i)$ for each $n \in \mathbb{N} - \{0\}$, then the similarity class $[\sigma]$ is reducible to the similarity class $[\sigma']$.

This is because, by replicating arguments, predicate letters of higher degree can play the role of predicate letters of lower degree.

(3) If $\mathscr{P} \subset \{\forall, \exists\}^*$ is finite, and $\sigma : \mathbb{N} \to \{\forall, \exists\}^*$ is finite, and $\sigma : \mathbb{N} \to \{0\} \to \mathbb{N} \cup \{\infty\}$ is such that $\sigma(n)$ is finite $(\ne \infty)$ for each n, then the intersection $[\mathscr{P}(\sigma)]$ of prefix class $[\mathscr{P}]$ with similarity class $[\sigma]$ is solvable.

This is because $[\mathscr{P}(\sigma)]$ contains only finitely many inequivalent schemata. That is, simply by substituting for the variables and predicate letters of a schema in $[\mathscr{P}(\sigma)]$ the earliest (in the lexical order) variables and predicate letters of the appropriate degree, and then making truth-functional simplifications, each schema in $[\mathscr{P}(\sigma)]$ may be reduced to one of finitely many forms. Of course, there can be no method for solving these "essentially finite" $[\mathscr{P}(\sigma)]$ classes that is uniform in $\mathscr{P}$ and σ, but every such class is trivially solvable.

(4) The similarity class $[(\infty)]$, i.e., the class of schemata with monadic predicate letters only, is solvable.

(5) The prefix classes $[\exists^*\forall^*]$, $[\exists^*\forall\exists^*]$, and $[\exists^*\forall\forall\exists^*]$ are solvable.

The conjunction of (1)-(5) implies that the classification by prefix and similarity type can be completely settled, if the following nine classes can be shown to be unsolvable:

$[\forall\exists\forall(\infty, 1)]$
$[\forall\forall\forall\exists(\infty, 1)]$
$[\forall^*\exists(0, 1)]$
$[\forall\exists\forall^*(0, 1)]$
$[\forall\forall\forall\exists^*(0, 1)]$
$[\exists^*\forall\forall\forall\exists(0, 1)]$
$[\exists^*\forall\exists\forall(0, 1)]$
$[\forall\exists^*\forall(0, 1)]$
$[\forall\exists\forall\exists^*(0, 1)]$

For by (5), if $[\mathscr{P}]$ is an unsolvable prefix class then $\mathscr{P}$ must contain a prefix type of the form $\ldots\forall\ldots\exists\ldots\forall\ldots$ or $\ldots\forall\ldots\forall\ldots\forall\ldots\exists\ldots$, and by (4) any unsolvable similarity class $[\sigma]$ must have $\sigma(k) > 0$ for at least one $k \geq 2$. Thus by (3), $[\forall\exists\forall(\infty, 1)]$ and $[\forall\forall\forall\exists(\infty, 1)]$ are the minimal classes if the prefix is to be bounded in length. If the prefix is not bounded, then the $\forall\exists\forall$ and $\forall\forall\forall\exists$ prefix types may be expanded by adding universals (yielding $\forall^*\exists$ from $\forall\forall\forall\exists$ and $\forall\exists\forall^*$ from $\forall\exists\forall$; $\forall^*\exists\forall$ from $\forall\exists\forall$ would be redundant, by (1)) or by adding existentials (yielding $\forall\forall\forall\exists^*$ and $\exists^*\forall\forall\forall\exists$ from $\forall\forall\forall\exists$, and $\exists^*\forall\exists\forall$, $\forall\exists^*\forall$, and $\forall\exists\forall\exists^*$ from $\forall\exists\forall$).

In fact all these classes are unsolvable. In the next section we prove the unsolvability of three special classes $\mathscr{V}_0$, $\mathscr{V}_1$, $\mathscr{V}_2$ from which the unsolvability of the nine classes listed follows by the reductions illustrated in Figure 18.

Because we are concerned for the rest of this chapter with formulas having monadic and dyadic predicate letters only, we make a notational convention. P is the first dyadic predicate letter in the lexical order, and $R_0, R_1, \ldots$ are the monadic predicate letters in lexical order. Thus $\hat{\omega}(x, y)$ is the atomic formula Pxy, and $\hat{\rho}(x)$, where $\rho = \langle i, \omega(0)\rangle$, is $R_i x$.

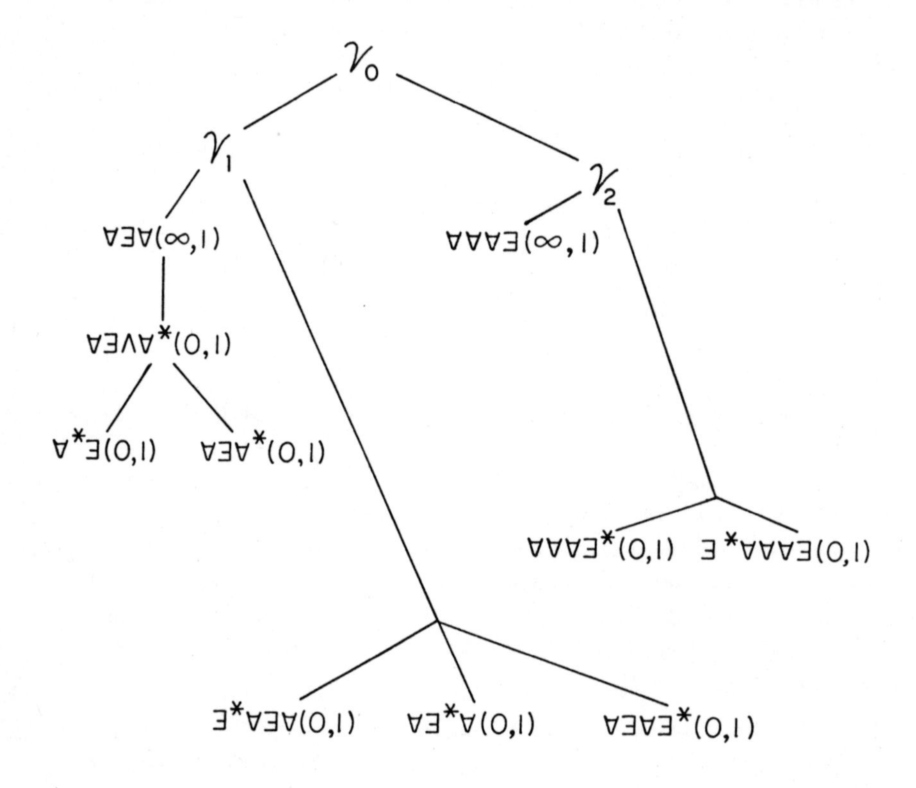

Figure 18

IIC.2 The $(\infty, 1)$ Similarity Class

All the results presented here on prefix and similarity classes follow from the unsolvability of the class $\mathcal{V}_0$ defined below. After reducing the tiling problem for $\Lambda_2^{(2)}$-systems (p. 43) to the decision problem for $\mathcal{V}_0$, we reduce $\mathcal{V}_0$ to subclasses of $\forall\exists\forall(\infty, 1)$ and $\forall\forall\forall\exists(\infty, 1)$. The other sections of this chapter reduce these subclasses to prefix subclasses of the $(0, 1)$ similarity class.

Unsolvability of $\mathcal{V}_0$. Let f be a monadic function sign, let D_1 be the domain generated by f and the constant $\mathbf{1}$, and let $\delta: D_1 \to \mathbb{N}$ be defined by $\delta(f^n(\mathbf{1})) = n$ for each n. Let α_0 be the bigraph $-\sum_{n=0}^{\infty} \omega(n, n)$. Let $\mathcal{V}_0$ be the class of all closed, rectified, prenex formulas $F = \forall y_1 \forall y_2 F^M$ such that F^M contains no function sign but f and only the predicate letters P (dyadic) and $R_0, R_1, \ldots$ (monadic), and such that

(a) the only atomic formulas of F are Py_1y_2, $Pf(y_1)f(y_2)$, and R_iy_1, R_iy_2, $R_if(y_1)$ $(i \geq 0)$; and

(b) $F \in \mathscr{L}_\delta(\alpha_0)$.

Then $\mathscr{V}_0$ is unsolvable.

(The thrust of (b) is that, <u>if</u> F is a satisfiable member of $\mathscr{V}_0$, <u>then</u> there is a truth-assignment $\mathscr{A}$ that verifies $E(F)$ but falsifies Ptt for each $t \in D(F)$. The membership problem for $\mathscr{V}_0$ need not be solvable.)

<u>Proof.</u> We prove the unsolvability of $\mathscr{V}_0$ by reducing to it the second version of the linear sampling problem, including the 0-1 restriction. In fact, in order to accommodate restriction (b) we must strengthen the Linear Sampling Theorem (Second Version) slightly. We need to show:

> The tiling problem is unsolvable for $\Lambda_2^{(2)}$-systems $\mathscr{S}$ satisfying the 0-1 restriction and with the further property that if $\mathscr{S}$ accepts a tiling, then $\mathscr{S}$ accepts a tiling τ such that $\tau(0, 3) = 0$ (i.e., such that cell 0 of the third tape contains a 0, rather than a 1).

This is easy to argue from the details of the proof in Chapter IE, but a direct argument is also easy. Suppose $\mathscr{S} = (T, L, G)$; we construct a $\Lambda_2^{(2)}$-system $\mathscr{S}' = (T', L', G')$ with $T' = T \cup \overline{T}$, where $\overline{t}$ is a new tile for each $t \in T$ and $\overline{T}$ is the set of all these copies $\overline{t}$ of tiles in T. The tilings accepted by $\mathscr{S}'$ are those accepted by $\mathscr{S}$, plus the "complements" obtained by replacing each tile t on one of the first two tapes by its copy $\overline{t}$, and replacing 0 by 1 and 1 by 0 on the third tape. Thus either the original or the complementary tiling will have the property that $\tau(0, 3) = 0$. Formally, the construction of $\mathscr{S}'$ is straightforward: $\mathscr{S}' = (T', L', G')$, where

$$L' = L \cup \{\langle \overline{s}, \overline{t} \rangle \mid \langle s, t \rangle \in L\}$$

$$G' = G \cup \{\langle \overline{s}, \overline{t}, 1-e \rangle \mid \langle s, t, e \rangle \in G\} .$$

Now let $\mathscr{S} = (T, L, G)$ be any $\Lambda_2^{(2)}$-system satisfying the stated conditions. For convenience let $T = \{0, \ldots, q-1\}$. We construct a formula $F \in \mathscr{V}_0$ such that F is satisfiable if and only if $\mathscr{S}$ accepts some tiling. In addition to the dyadic predicate letter P, F contains the monadic predicate letters $R_0, \ldots, R_{2q-1}$. A tiling τ accepted by $\mathscr{S}$ is to be associated with a truth-assignment $\mathscr{A}$ verifying $E(F)$ in the following way: for $i = 0, \ldots, q-1$, $\mathscr{A} \models R_if^n(\text{ł})$ if and only if $\tau(n, 1) = i$; for $i = q, \ldots, 2q-1$, $\mathscr{A} \models R_if^n(\text{ł})$ if and only if $\tau(n, 2) = i-q$; and $\mathscr{A} \models Pf^m(\text{ł})f^n(\text{ł})$ if and only if $\tau(n-m, 3) = 1$. Let $F = \forall y_1 \forall y_2(M_1 \wedge M_2 \wedge M_3 \wedge M_4)$, where $M_1, \ldots, M_4$ are as follows:

$$M_1 : Py_1y_2 \leftrightarrow Pf(y_1)f(y_2)$$

$$M_2 : \left(\mathop{\nabla}_{i=0}^{q-1} R_i y_1 \right) \wedge \left(\mathop{\nabla}_{i=q}^{2q-1} R_i y_1 \right)$$

where ∇ denotes "exclusive or";

$$M_3 : \left(\bigvee_{\langle i,j \rangle \in L} (R_i y_1 \wedge R_j f(y_1)) \right) \wedge \left(\bigvee_{\langle i,j \rangle \in L} (R_{q+i} y_1 \wedge R_{q+j} f(y_1)) \right)$$

$$M_4 : \bigvee_{\langle i,j,e \rangle \in G} (R_i y_1 \wedge R_{q+j} y_2 \wedge H_e)$$

where H_e is Py_1y_2 if $e = 1$, $\neg Py_1y_2$ if $e = 0$.

M_1 states that the truth-value of an atomic formula $Pf^m(\ddagger)f^n(\ddagger)$ appearing in $E(F)$ depends only on the difference between m and n. M_2 states that each cell of each of the first two tapes contains exactly one symbol. M_3 imposes the local condition on the first two tapes; M_4 imposes the global condition.

Note first that F satisfies the syntactic restrictions given under (a) and is therefore a member of $\mathscr{V}_0$. Second, if τ is a tiling accepted by $\mathscr{S}$ then the truth-assignment $\mathscr{A}$ described above does indeed verify $E(F)$. For example, in the Herbrand instance $F^*[y_1/f^m(\ddagger), \; y_2/f^n(\ddagger)]$, the verified disjunct of the Herbrand instance of M_4 is that in which $i = \tau(m,1)$, $j = \tau(n,2)$, and $e = \tau(n-m,3)$. Moreover, if $\tau(0,3) = 0$ then $\mathscr{A} \models \neg Pf^m(\ddagger)f^m(\ddagger)$ for each m.

To prove that $\mathscr{S}$ accepts a tiling if F is satisfiable, let $\mathscr{A} \models E(F)$ and note that because $\mathscr{A}$ verifies every Herbrand instance of M_1, $\mathscr{A}(Pf^m(\ddagger)f^n(\ddagger)) = \mathscr{A}(Pf^{m+p}(\ddagger)f^{n+p}(\ddagger))$ for all $m, n, p \in \mathbb{N}$. Thus a mapping $\tau_3 : \mathbb{Z} \to \{0,1\}$ can be defined by:

$$\tau_3(n-m) = \begin{cases} 1, & \text{if } \mathscr{A} \models Pf^m(\ddagger)f^n(\ddagger) \\ 0, & \text{if } \mathscr{A} \models \neg Pf^m(\ddagger)f^n(\ddagger) \end{cases}$$

for all $m, n \in \mathbb{N}$. Moreover, because $\mathscr{A}$ verifies every Herbrand instance of M_2, mappings τ_1 and $\tau_2 : \mathbb{N} \to T = \{0, \ldots, q-1\}$ can be defined by

$$\tau_1(n) = i \quad \text{if and only if} \quad \mathscr{A} \models R_i f^n(\ddagger)$$

$$\tau_2(n) = i \quad \text{if and only if} \quad \mathscr{A} \models R_{q+i} f^n(\ddagger) \, .$$

Now define $\tau : \mathbb{N} \times \{1,2\} \cup \mathbb{Z} \times \{3\} \to T$ by $\tau(m,i) = \tau_i(m)$. Because $\mathscr{A}$ verifies every Herbrand instance of M_3 or M_4, τ fulfills the conditions for acceptability by $\mathscr{S}$ in the domain where it is defined. That is, $\langle \tau(m,i), \tau(m+1,i) \rangle \in L$ for

$i = 1, 2, 3$ and for all m such that $\tau(m,i)$ and $\tau(m+1,i)$ are defined, and $\langle \tau(m,1), \tau(n,2), \tau(n-m,3)\rangle \in G$ for all m, n such that $\tau(m,1)$ and $\tau(n,2)$ are defined. Moreover, for each $p \in \mathbb{N}$ the mapping $\tau^{(p)}$ defined by

$$\tau^{(p)}(n,i) = \begin{cases} \tau(n+p, i) & (i = 1 \text{ or } 2) \\ \tau(n,i) & (i = 3) \end{cases}$$

fulfills the conditions for acceptability by $\mathscr{S}$ in the domain $\{-p, -p+1, \ldots\} \times \{1,2\} \cup \mathbb{Z} \times \{3\}$. But then the Infinity Lemma implies that there is a tiling of $3\mathbb{Z}$ accepted by $\mathscr{S}$.

Thus $\mathscr{S}$ accepts a tiling if and only if F is satisfiable; and if $\mathscr{S}$ accepts a tiling τ with $\tau(0,3) = 0$ then there is a truth-assignment $\mathscr{A}$ verifying $E(F)$ such that $\mathscr{A} \models \neg Ptt$ for each t. But $\mathscr{S}$ was selected so that if $\mathscr{S}$ accepts any tiling, then $\mathscr{S}$ accepts one with this property; so if any truth-assignment verifies $E(F)$, then some such truth-assignment does.

This completes the proof of the unsolvability of $\mathscr{V}_0$. ∎

The unsolvability of $\mathscr{V}_0$ implies the unsolvability of $\forall\exists\forall(\infty, 1)$ and $\forall\forall\forall\exists(\infty, 1)$ via the next theorem. Since we shall need the unsolvability of subclasses of these classes in order to obtain further reductions later on, the theorem refers to these subclasses $\mathscr{V}_1$ and $\mathscr{V}_2$.

<u>Bounded Prefix Theorem</u>. Let f be the monadic indicial correlate of the variable x, and let D_1 be the domain generated by f and $\ddagger$. Let $\delta: D_1 \to \mathbb{N}$ be defined by $\delta(f^n(\ddagger)) = n$ for all n, and let

$$\alpha_0 = -\sum_{n \in \mathbb{N}} \omega(n,n) \quad .$$

Let g be the 3-place indicial correlate of x, and let D_2 be the domain generated by $\ddagger$ and g. Let

$$\alpha_1 = -\sum_{n=0}^{\infty} (\omega(n+1, n) + \omega(n,n) + \omega(n, n+1))$$

$$\alpha_2 = -\sum_{t \in D_2} \omega(t,t) \quad .$$

Then the following are unsolvable classes:

(a) The class $\mathscr{V}_1$ of all schemata $F = \forall y_1 \exists x \forall y_2 F^M$ in $\forall\exists\forall(\infty, 1) \cap \mathscr{L}_\delta(\alpha_1)$ whose only atomic formulas are among Py_1y_2, Py_2x, R_iy_1, R_iy_2, and R_ix $(i \in \mathbb{N})$;

(b) $\forall\exists \wedge \forall\forall\forall(\infty, 1) \cap \mathscr{L}_\delta(\alpha_0)$;

(c) $\mathscr{V}_2 = \forall\forall\forall\exists(\infty, 1) \cap \mathscr{L}_\iota(\alpha_2)$, where ι is the identity mapping on D_2.

Proof. As (c) is a consequence of (b), we concentrate on the proofs of (a) and (b). Call two atomic formulas directly related if they appear in the same Herbrand instance of F. Then for F in the class $\mathscr{V}_0$, the atomic formulas Pst and $Pf(s)f(t)$ are directly related for any $s, t \in D_1$, as are $Pf(s)f(t)$ and $Pf^2(s)f^2(t)$, as are $Pf^2(s)f^2(t)$ and $Pf^3(s)f^3(t)$, etc. On the other hand, for F in $\mathscr{V}_1$ the following pairs are directly related, for any s, t: Pst and $Ptf(s)$; $Ptf(s)$ and $Pf(s)f(t)$; $Pf(s)f(t)$ and $Pf^2(t)f(s)$; $Pf^2(t)f(s)$ and $Pf^2(s)f^2(t)$; etc. Thus some pairs that are directly related for $F \in \mathscr{V}_0$ are "more distantly" related for $F \in \mathscr{V}_1$. In order to capture in a formula G in $\mathscr{V}_1$ the direct relationships of a formula F in $\mathscr{V}_0$, we imagine the Herbrand domain of G to be that for F, with every term subdivided into three successive pieces. (Of course, this is only conceptual; in fact $D(F) = D(G) = D_1$.) New monadic predicate letters will be introduced in G to describe which "third" of a term of D(F) a term of D(G) represents. (Actually, for the unsolvability of $\mathscr{V}_1$ it would suffice to divide terms in half, but for the unsolvability of $\forall\exists \wedge \forall\forall\forall(\infty, 1)$ we must use the single dyadic predicate letter P not only to represent an arbitrary binary relation on D_1, but also to represent the relation of each term t to its "successor" f(t); this requires a finer subdivision.)

To be precise, let F be any formula in $\mathscr{V}_0$, and without loss of generality let $R_0, \ldots, R_{q-1}$ be the monadic predicate letters of F. (Recall that these are the first q monadic predicate letters in lexical order.)

Let

$$\underset{\sim}{n}_i = 3n + i, \qquad \text{for each } n \in \mathbb{N} \text{ and } i = 0, 1, 2 \ .$$

Thus $\delta^{-1}(\underset{\sim}{n}_0)$, $\delta^{-1}(\underset{\sim}{n}_1)$, $\delta^{-1}(\underset{\sim}{n}_2)$ will represent in D(G) the three thirds of $\delta^{-1}(n) \in D(F)$. Let $\oplus$ denote addition mod 3, and for $i = 0, 1, 2$ let θ_i be the bigraph of dimension 1 defined as follows:

$$\theta_i = \sum_{n \in \mathbb{N}} (\omega(\underset{\sim}{n}_i) - \omega(\underset{\sim}{n}_{i \oplus 1}) - \omega(\underset{\sim}{n}_{i \oplus 2})) \qquad (0 \le i \le 2) \ .$$

Thus θ_i is true for numbers congruent to i (mod 3). Then $\theta_0, \theta_1, \theta_2$ combine to form a bistructure:

$$\theta = \{\langle q, \theta_0\rangle, \langle q+1, \theta_1\rangle, \langle q+2, \theta_2\rangle\}$$

(recall that q is the number of monadic predicate letters of F; the tags q, q+1, q+2 will correspond to three new monadic predicate letters of G). Next let π_{mn} be the bigraphs of dimension 2 defined as follows:

$$\pi_{m0} = \left(\sum_{j=0}^{2} \omega(\underset{\sim}{m}_j, \underset{\sim}{0}_j)\right) + \left(\sum_{j=0}^{1} \omega(\underset{\sim}{0}_j, \underset{\sim}{m}_{j+1})\right) \qquad (m \in \mathbb{N})$$

$$\pi_{mn} = \left(\sum_{j=0}^{2} \omega(\underset{\sim}{m}_j, \underset{\sim}{n}_j)\right) + \left(\sum_{j=0}^{1} \omega(\underset{\sim}{n}_j, \underset{\sim}{m}_{j+1})\right) + \omega(\underset{\sim}{n-1}_2, \underset{\sim}{m}_0) \qquad (m \in \mathbb{N},\ n>0).$$

Each π_{mn} will correspond in the construction of a truth-assignment on E(G) to a single atomic formula $Pf^m(\mathfrak{t})f^n(\mathfrak{t})$ appearing in E(F). Also let

$$\zeta = \sum_{n \in \mathbb{N}} \omega(n+1, n) ;$$

ζ will be used to represent the relation of a term t to its successor f(t). Note that ζ is disjoint from each of the π_{mn}.

By analogy with the π_{mn}, we need bigraphs of dimension 1 for the monadic atomic formulas of E(F):

$$\rho_n = \sum_{j=0}^{2} \omega(\underset{\sim}{n}_j) \quad .$$

Finally, let Φ be the collection of the π_{mn} and q tagged bigraphs of dimension 1 formed from the ρ_n:

$$\Phi = \{\pi_{mn} \mid m, n \in \mathbb{N}\} \cup \{\langle i, \rho_n \rangle \mid n \in \mathbb{N},\quad 0 \leq i \leq q-1\} \quad .$$

We now construct schemata $G \in \mathscr{V}_1$ and $H \in \forall\exists \wedge \forall\forall\forall(\infty, 1) \cap \mathscr{L}_{\delta}(\alpha_0)$ such that F is satisfiable if and only if G is satisfiable, if and only if H is satisfiable. The construction is in two stages. We first state some conditions on subformulas of G and H, and prove that if subformulas meeting these conditions can be found then the construction of G and H can be completed. We then construct the subformulas.

<u>Sufficient Conditions Lemma</u>. For $i = 0, 1, 2$ let $\delta_i(f^n(\mathfrak{t})) = n+i$ for each $n \in \mathbb{N}$, so that $\delta_0 = \delta$. Suppose there are schemata $G_0 \in \forall\exists\forall(\infty, 1)$ and $G_1 \in \forall\exists \wedge \forall\forall\forall(\infty, 1)$ such that the following conditions are satisfied:

(1)(a) If μ is any Φ-completion of θ then $\mu(\delta) \models E(G_0)$.

(b) If $\mathscr{A} \models E(G_0)$, then $\mathscr{A} = \mu(\delta_i)$ for some i $(0 \leq i \leq 2)$, where μ is a Φ-completion of θ.

(2)(a) If μ is any minimal Φ-completion of $\theta + \zeta$, then $\mu(\delta) \models E(G_1)$.

(b) If $\mathscr{A} \models E(G_1)$ then $\mathscr{A} = \mu(\delta_i)$ for some i $(0 \le i \le 2)$, where μ is a Φ-completion of $\theta + \zeta$.

(Note that $\theta + \zeta$ is proper.) Then we can construct schemata $G \in \mathscr{V}_1$ and $H \in \forall\exists\wedge\forall\forall\forall(\infty, 1) \cap \mathscr{L}_\delta(\alpha_0)$ such that either F, G, and H are all satisfiable or none is.

Proof. By a change of variables if necessary, let $G_0 = \forall y_1 \exists x \forall y_2 G_0^M$ and let $G_1 = \forall y \exists x G_{11} \wedge \forall y_1 \forall y_2 \forall y_3 G_{12}$. Then let

$$G = \forall y_1 \exists x \forall y_2 (G_0^M \wedge G_2) \quad ,$$

where

$$G_2 = R_{q+2} y_1 \wedge R_{q+2} y_2 \rightarrow G_3$$

and

$$G_3 = F^M \{ Pf(y_1) f(y_2)/Py_2x, \; R_i f(y_1)/R_i x \quad (i = 0, \ldots, q-1) \}$$

Also let

$$H = \forall y \exists x (Pxy \wedge G_{11}) \wedge \forall y_1 \forall y_2 \forall y_3 (G_{12} \wedge G_4) \quad ,$$

where

$$G_4 = (R_{q+2} y_1 \wedge R_{q+2} y_2 \wedge R_q y_3 \wedge P y_3 y_1) \rightarrow G_3[x/y_3] .$$

(F $\triangleright$ G, H) First suppose that F is satisfiable; we show that G and H are satisfiable, $G \in \mathscr{L}_\delta(\alpha_1)$, and $H \in \mathscr{L}_\delta(\alpha_0)$. Since $F \in \mathscr{L}_\delta(\alpha_0)$, there is a λ such that $\lambda(\delta) \models E(F)$ and $\alpha_0 \subset \lambda$. Define μ to be the minimal completion of

$$\eta = \theta + \sum_{i=0}^{q-1} \sum_{n=0}^{\infty} \chi_\lambda(\langle i, \omega(n)\rangle) \cdot \langle i, \rho_n \rangle + \sum_{m=0}^{\infty} \sum_{n=0}^{\infty} \chi_\lambda(\omega(m,n)) \cdot \pi_{mn} \, .$$

This sum is proper since θ, the $\langle i, \rho_n \rangle$, and the π_{mn} are pairwise disjoint. (The truth-assignment $\mu(\delta)$ verifies an atomic formula $R_j \delta^{-1}(\underset{\sim}{n}_i)$, where $0 \le j \le q-1$, $0 \le i \le 2$, and $n \in \mathbb{N}$, if and only if $\lambda(\delta) \models R_j \delta^{-1}(n)$; verifies $R_{q+j} \delta^{-1}(\underset{\sim}{n}_i)$, where $j = 0, 1$, or 2, if and only if $i = j$; and verifies $P\delta^{-1}(\underset{\sim}{m}_i)\delta^{-1}(\underset{\sim}{n}_j)$ if and only if either $j = i$ or $i+1$ and $\lambda(\delta) \models P\delta^{-1}(m)\delta^{-1}(n)$, or $i = 2$, $j = 0$, $m > 0$, and $\lambda(\delta) \models P\delta^{-1}(m-1)\delta^{-1}(n)$.) We claim that $\mu(\delta) \models E(G)$. Clearly $\mu(\delta) \models E(G_0)$ by (1a). Now consider the Herbrand instance of G_2 occurring in $G^*[y_1/\delta^{-1}(m), y_2/\delta^{-1}(n)]$. If the antecedent is verified by $\mu(\delta)$ then $\langle q+2, \omega(m)\rangle \subset \mu$ and $\langle q+2, \omega(n)\rangle \subset \mu$. Since $\langle q+2, \theta_2\rangle \subset \theta \subset \mu$ and θ_2 is complete it follows that $m \equiv n \equiv 2 \pmod 3$, so that $m = \underset{\sim}{m}'_2$, $n = \underset{\sim}{n}'_2$ for some m', n'. Let G'_3 be the Herbrand instance of G_3 occurring in this Herbrand instance of G, that is, $G'_3 = G_3[y_1/\delta^{-1}(m), y_2/\delta^{-1}(n),$

$x/\delta^{-1}(m+1)] = G_3[y_1/\delta^{-1}(\underset{\sim}{m}'_2), y_2/\delta^{-1}(\underset{\sim}{n}'_2), x/\delta^{-1}(\underset{\sim}{m'+1_0})]$. Also, let $F' = F^*[y_1/\delta^{-1}(m'), y_2/\delta^{-1}(n')]$; we claim that $\mu(\delta)(G'_3) = \lambda(\delta)(F')$, which will suffice to show that $\mu(\delta) \models E(G)$ since $F' \in E(F)$.

So let $w \in \overline{T}(F)$. If $\langle F \rangle_w = Py_1y_2$, then $\langle F' \rangle_w = P\delta^{-1}(m')\delta^{-1}(n')$ and $\langle G'_3 \rangle_w = P\delta^{-1}(\underset{\sim}{m}'_2)\delta^{-1}(\underset{\sim}{n}'_2)$. But then

$$\begin{aligned}\mu(\delta)(\langle G'_3 \rangle_w) &= \chi_\mu(\omega(\underset{\sim}{m}'_2, \underset{\sim}{n}'_2)) \\ &= \chi_\mu(\pi_{m'n'}) \quad \text{since } \mu \text{ is a } \Phi\text{-completion and } \omega(\underset{\sim}{m}'_2, \underset{\sim}{n}'_2) \subset \pi_{m'n'} \\ &= \chi_\lambda(\omega(m', n')) \\ &= \lambda(\delta)(\langle F' \rangle_w) \quad .\end{aligned}$$

A similar argument applies when $\langle F \rangle_w$ is R_iy_1 or R_iy_2. If $\langle F \rangle_w$ is $Pf(y_1)f(y_2)$ then $\langle F' \rangle_w = P\delta^{-1}(m'+1)\delta^{-1}(n'+1)$ and $\langle G'_3 \rangle_w = P\delta^{-1}(n)\delta^{-1}(m+1) = P\delta^{-1}(\underset{\sim}{n}'_2)\delta^{-1}(\underset{\sim}{m'+1_0})$. Then

$$\begin{aligned}\mu(\delta)(\langle G'_3 \rangle_w) &= \chi_\mu(\omega(\underset{\sim}{n}'_2, \underset{\sim}{m'+1_0})) \\ &= \chi_\mu(\pi_{m'+1,\, n'+1}) \\ &= \lambda(\delta)(\langle F' \rangle_w)\end{aligned}$$

and a similar argument applies when $\langle F \rangle_w$ is $R_if(y_1)$.

Next we show that $\alpha_1 \subset \mu$. That $-\omega(n+1, n) \subset \mu$ for each n follows from the fact that $\omega(n+1, n)$ is disjoint from all the $\pi_{mn'}$ $(m, n' \in \mathbb{N})$ and the fact that μ is a minimal completion of η. That $-\omega(n, n) - \omega(n, n+1) \subset \mu$ follows from the definition of π_{nn} and the fact that $\alpha_0 \subset \lambda$ so that $\chi_\lambda(\omega(n, n)) = -1$, and hence $-\pi_{nn} \subset \mu$, for each $n \in \mathbb{N}$.

Now define ν to be the minimal completion of $\eta + \zeta$; this sum is proper since ζ is disjoint from all the π_{mn} $(m, n \in \mathbb{N})$. We show that $\nu(\delta) \models E(H)$ and $\alpha_0 \subset \nu$. Since $\zeta \subset \nu$, $\nu(\delta) \models Pf(t)t$ for each $t \in D(H)$, and by 2(a) $\nu(\delta) \models E(G_1)$. Consider the Herbrand instance of G_4 occurring in $H^*[y_1/\delta^{-1}(m), y_2/\delta^{-1}(n), y_3/\delta^{-1}(p)]$. If the antecedent is verified then $m \equiv n \equiv 2 \pmod 3$ and $p \equiv 0 \pmod 3$, and also $\omega(p, m) \subset \nu$. But $\omega(p, m) \subset \nu$, with $p \equiv m+1 \pmod 3$, only if $p = m+1$, so in this case the consequent is exactly the formula G'_3 considered above and the same argument shows that $\nu(\delta) \models G'_3$. Also, $\alpha_0 \subset \nu$ by the same argument as before (though it is not true that $\alpha_1 \subset \nu$, since $\zeta \subset \nu$).

<u>$(G, H \rhd F)$</u> Now we show that if G is satisfiable, then so is F; the argument for H is similar and is omitted. Let $\mathcal{A} \models E(G)$; then $\mathcal{A} \models E(G_0)$ and by (1b)

$\mathscr{A} = \mu(\delta_i)$, where $0 \le i \le 2$ and μ is a Φ-completion of θ. Define λ to be

$$\sum_{\substack{0 \le j \le q-1 \\ n \in \mathbb{N}}} \chi_\mu(\langle j, \rho_n \rangle) \cdot \langle j, \omega(n) \rangle + \sum_{m, n \in \mathbb{N}} \chi_\mu(\pi_{mn}) \cdot \omega(m, n) \quad .$$

Note that λ is complete since $\chi_\mu(\langle j, \rho_n \rangle) \ne 0$ and $\chi_\mu(\pi_{mn}) \ne 0$ because μ is a Φ-completion of θ. Now let $F' = F^*[y_1/\delta^{-1}(m), y_2/\delta^{-1}(n)]$; we need to show that $\lambda(\delta) \models F'$. Note that $\delta_i^{-1}(p)$ is defined for all $p \ge 2$, and consider $G^*[y_1/\delta_i^{-1}(\underset{\sim}{m}_2), \delta_i^{-1}(\underset{\sim}{n}_2)]$; since it is a member of $E(G)$ it is verified by $\mu(\delta_i)$. In particular, consider the Herbrand instance of G_2 occurring in this instance of G^*,

$$R_{q+2}\delta_i^{-1}(\underset{\sim}{m}_2) \wedge R_{q+2}\delta_i^{-1}(\underset{\sim}{n}_2) \rightarrow G_3' \quad .$$

The antecedent is verified by $\mu(\delta_i)$ since

$$\begin{aligned} \mu(\delta_i)(R_{q+2}\delta_i^{-1}(\underset{\sim}{m}_2)) &= \chi_\mu(\langle q+2, \omega(\underset{\sim}{m}_2) \rangle) \\ &= \chi_{\theta_2}(\omega(\underset{\sim}{m}_2)) \\ &= 1 \quad , \end{aligned}$$

and similarly for $\underset{\sim}{n}_2$, so that $\mu(\delta_i) \models G_3'$. Thus it remains only to show that $\lambda(\delta)(\langle F' \rangle_w) = \mu(\delta_i)(\langle G_3' \rangle_w)$ for each $w \in \overline{T}(F)$; this argument is essentially identical to that above.

This completes the proof of the Sufficient Conditions Lemma. ∎

To complete the proof we need to construct G_0 and G_1 as described by the Sufficient Conditions Lemma.

<u>Construction Lemma</u>. Schemata G_0 and G_1 as described can be constructed.

<u>Proof.</u> Write E_0, E_1, E_2 for R_q, R_{q+1}, R_{q+2}, and define G_0 to be $\forall y_1 \exists x \forall y_2 (G_5 \wedge G_6 \wedge G_7)$, where

$$G_5 = \bigvee_{i=0}^{2} (E_i y_1 \wedge E_{i \oplus 1} x \wedge \neg E_{i \oplus 1} y_1 \wedge \neg E_{i \oplus 2} y_1) \quad ;$$

$$G_6 = (\vee(E_i y_1 \wedge E_j y_2)) \rightarrow (P y_1 y_2 \leftrightarrow P y_2 x) \quad ,$$

the disjunction being over all i, j $(0 \le i, j \le 2)$ such that $j = i \oplus 1$ or $i = j \ne 2$;

$$G_7 = (E_0 y_1 \vee E_1 y_1) \rightarrow \bigwedge_{i=0}^{r-1} (R_i y_1 \leftrightarrow R_i x) \quad .$$

Clearly, $\mu(\delta) \models E(\forall y_1 \exists x \forall y_2 G_5)$ if $\theta \subset \mu$, and if $\mathscr{A} \models E(\forall y_1 \exists x \forall y_2 G_5)$ then $\mathscr{A} = \mu(\delta_i)$ for some i and some μ that is a completion of θ. And from G_6 and G_7 it follows that

(i) If $\theta \subset \mu$ and $\chi_\mu(\alpha) = \chi_\mu(\beta)$ for all α, β such that $\alpha, \beta \subset \pi_{mn}$ for some m, n or $\alpha, \beta \subset \langle i, \rho_n \rangle$ for some i, n, then $\mu(\delta) \models E(G_0)$;

(ii) if $\mu(\delta_i) \models E(G_0)$ then $\chi_\mu(\alpha) = \chi_\mu(\beta)$ for all α, β as in (i).

Thus G_6 and G_7 force certain directly related atomic formulas, in the sense of the earlier discussion, to take the same truth-value; this effect propagates to force more distantly related pairs to have the same truth-value. Thus (i) and (ii) show that G_0 meets Sufficient Condition (1).

Now for (2). Let G_1 be $\forall y_1 \exists x G_8 \wedge \forall y_1 \forall y_2 \forall y_3 G_9$ where G_8 and G_9 are as follows:

$$G_8: \qquad Pxy_1 \wedge G_5$$

$$G_9: \left(Py_3y_1 \wedge \bigvee_{i=0}^{2} (E_i y_1 \wedge E_{i \oplus 1} y_3) \right) \rightarrow (G_6 \wedge G_7)[x/y_3]$$

Now if μ is any minimal Φ-completion of $\theta + \zeta$, then $\mu(\delta)$ verifies each Herbrand instance of G_8, and if $\mu(\delta)$ verifies the antecedent of an Herbrand instance of G_9, say,

$$\mu(\delta) \models P\delta^{-1}(p)\delta^{-1}(m) \wedge \bigvee_{i=0}^{2} (E_i \delta^{-1}(m) \wedge E_{i \oplus 1} \delta^{-1}(p))$$

then $\omega(m) \subset \theta_i$, $\omega(p) \subset \theta_{i \oplus 1}$, and $\omega(p, m) \subset \mu$. But by the definition of π_{mn}, $\omega(p, m)$ is then not a subgraph of any member of Φ; hence $\omega(p, m) \subset \theta + \zeta$ and $p = m+1$. Hence the Herbrand instance of $G_6 \wedge G_7$ in the consequent is the same as in an Herbrand instance of G_0 and is verified for the same reason.

Conversely, if $\mathscr{A} \models E(G_1)$ then $\mathscr{A} = \mu(\delta_i)$ for some i and some μ that is a completion of $\theta + \zeta$. To show that μ is a Φ-completion, it suffices to show that each Herbrand instance of $G_6 \wedge G_7$ occurs as the consequent of some Herbrand instance of G_9 whose antecedent is verified by $\mu(\delta_i)$. This is easily done; instead of $G_6 \wedge G_7[y_1/\delta^{-1}(m), y_2/\delta^{-1}(n), x/\delta^{-1}(m+1)]$, consider $G_9[y_1/\delta^{-1}(m), y_2/\delta^{-1}(n), y_3/\delta^{-1}(m+1)]$.

This completes the proof of the Construction Lemma and the proof of the Bounded Prefix Theorem. ∎

IIC. 3 The Unbounded-Universals Classes

In this section we prove the unsolvability of the $\forall\exists\wedge\forall^*(0,1)$ class, which implies the unsolvability of $\forall^*\exists(0,1)$ and $\forall\exists\forall^*(0,1)$. The next two sections deal with subclasses of the $(0,1)$ similarity class in which the number of existential quantifiers is unbounded, rather than the number of universals, as here.

Unbounded-Universals Theorem. $\forall\exists\wedge\forall^*(0,1)$ is unsolvable.

Proof. As in the proofs of the last section, let f be the monadic indicial correlate of the variable x, and let D_1 be the domain generated by f and $\dagger$. For each i and n in $\mathbb{N}$, let $\delta_i(f^n(\dagger)) = n+i$, and let $\delta = \delta_0$.

Now let F be any schema in $\forall\exists\forall(\infty,1)$; we construct a schema $G\in\forall\exists\wedge\forall^*(0,1)$ that is satisfiable if and only if F is satisfiable. As in the proof of the Bounded Prefix Theorem there are two stages to the construction. First we define conditions on bigraphs and a schema and prove that these conditions suffice for the construction of an appropriate G; we then show that objects meeting these conditions can be found.

Without loss of generality, assume that F contains occurrences of the dyadic predicate letter P and the r monadic predicate letters $R_0,\ldots,R_{r-1}$.

Sufficient Conditions Lemma. Let $q>0$; for each $n\in\mathbb{N}$ let $\underset{\sim}{n}$ be the sequence $qn, qn+1,\ldots,qn+q-1$, and for $0\le i\le q-1$ let $\underset{\sim}{n}_i = qn+i$. Let τ and $\rho_0,\ldots,\rho_{r-1}$ be q-templates, let σ be a 2-template, and let π be a 2q-template, all of dimension 2. Suppose the following conditions are satisfied.

Disjointness. The bigraphs

$$\tau(\underset{\sim}{m}),\quad \sigma(\underset{\sim}{m}_{q-1}, \underset{\sim}{m+1}_0),$$

$$\rho_i(\underset{\sim}{m}),\quad \pi(\underset{\sim}{m},\underset{\sim}{n}),$$

as m and n range over $\mathbb{N}$ and i over $0,\ldots,r-1$, are pairwise disjoint.

Let

$$\Theta = \{\tau(\underset{\sim}{m}), \sigma(\underset{\sim}{m}_{q-1}, \underset{\sim}{m+1}_0) \mid m\in\mathbb{N}\}$$

$$\Phi = \{\rho_i(\underset{\sim}{m}), \pi(\underset{\sim}{m},\underset{\sim}{n}) \mid 0\le i\le r-1,\ m,n\in\mathbb{N}\},$$

and let

$$\theta = \sum_{\eta\in\Theta} \eta ,$$

which is a proper sum by the Disjointness Condition.

Extraction. The 3q-template

$$\psi = \tau(\underset{\sim}{0}) + \tau(\underset{\sim}{1}) + \tau(\underset{\sim}{2}) + \sigma(\underset{\sim}{0}_{q-1}, \underset{\sim}{1}_0)$$

extracts $\{\langle \underset{\sim}{m}, \underset{\sim}{m+1}, \underset{\sim}{n}\rangle \mid m, n \in \mathbb{N}\}$ from any minimal Φ-completion of θ.

Schema. There is a schema $G_0 \in \forall\exists\wedge\forall^*(0,1)$ such that

(a) if μ is any minimal Φ-completion of θ then $\mu(\delta) \models E(G_0)$;

(b) if $\mathscr{A} \models E(G_0)$ then $\mathscr{A} = \mu(\delta_i)$ for some i $(0 \le i \le q-1)$, where μ is a completion of θ.

Then we can construct a schema $G \in \forall\exists\wedge\forall^*(0,1)$ such that F is satisfiable if and only if G is satisfiable.

Proof. Intuitively, the bigraphs $\rho_i(\underset{\sim}{m})$ and $\pi(\underset{\sim}{m}, \underset{\sim}{n})$ are the correlates for G of the atomic formulas $R_i f^m(\ddagger)$ and $Pf^m(\ddagger)f^n(\ddagger)$ in $E(F)$. That is, every element of $D(F)$ is subdivided into a sequence of q elements of $D(G)$, and relations among the q elements are used to represent the r monadic predicate letters of F. The template τ represents the relation in which the q elements of any such sequence stand to each other and will be used in the construction of G for distinguishing such sequences from other q-tuples of elements that do not all arise from a single element of $D(F)$. Also, σ represents the relation of the last element in the sequence for $f^m(\ddagger)$ to the first in the sequence for $f^{m+1}(\ddagger)$. Thus the Disjointness Condition guarantees that even when all the relations in Θ are asserted, any combination of bigraphs φ, $-\varphi'$, where $\varphi, \varphi' \in \Phi$, may be consistently added.

Let $F = \forall x_1 \exists x_2 \forall x_3 F^M$. For $i = 1, 2, 3$ and for $j = 0, \ldots, q-1$ let y_{ij} be a new variable and let $\underset{\sim}{y}_i$ be the sequence $y_{i0}, \ldots, y_{i,q-1}$. Let

$$G = G_0 \wedge \forall \underset{\sim}{y}_1 \forall \underset{\sim}{y}_2 \forall \underset{\sim}{y}_3 (\hat{\psi}(\underset{\sim}{y}_1, \underset{\sim}{y}_2, \underset{\sim}{y}_3) \to G_1)$$

where

$$G_1 = F^M \Big\{ Px_i x_j / \hat{\hat{\pi}}(\underset{\sim}{y}_i, \underset{\sim}{y}_j) \ (1 \le i, j \le 3), \ \ R_i x_j / \hat{\hat{\rho}}_i(\underset{\sim}{y}_j) \ (0 \le i \le r-1, \ 1 \le j \le 3) \Big\}$$

and where we have written $\forall \underset{\sim}{y}_i$ for $\forall y_{i0} \ldots \forall y_{i,q-1}$. Thus G_1 is formed from the matrix of F by replacing atomic formulas by signed atomic formulas.

($F \triangleright G$) Suppose F is satisfiable, and let $\mathscr{A} \models E(F)$. Let μ be the minimal completion of

$$\theta + \sum_{m=0}^{\infty} \sum_{i=0}^{r-1} \mathscr{A}(R_i f^m(\ddagger)) \cdot \rho_i(\underset{\sim}{m}) + \sum_{m=0}^{\infty} \sum_{n=0}^{\infty} \mathscr{A}(Pf^m(\ddagger)f^n(\ddagger)) \cdot \pi(\underset{\sim}{m}, \underset{\sim}{n}) \quad .$$

This is a proper sum by the Disjointness Condition. We claim that $\mu(\delta) \models E(G)$. Clearly $\mu(\delta) \models E(G_0)$ by Schema Condition (a). Now let $t_{ij} \in D_1$ for $i = 1, 2, 3$, $j = 0, \ldots, q-1$; to show that $\mu(\delta) \models E(G)$ it will suffice to show that

$$\mu(\delta) \models \hat{\psi}(\underset{\sim}{t_1}, \underset{\sim}{t_2}, \underset{\sim}{t_3}) \rightarrow G_1'$$

where

$$G_1' = G_1[\underset{\sim}{y_1}/\underset{\sim}{t_1}, \underset{\sim}{y_2}/\underset{\sim}{t_2}, \underset{\sim}{y_3}/\underset{\sim}{t_3}] \quad ,$$

and where we have written $\underset{\sim}{y_i}/\underset{\sim}{t_i}$ for $y_{i0}/t_{i0}, \ldots, y_{i,q-1}/t_{i,q-1}$. If $\mu(\delta) \models \hat{\psi}(\underset{\sim}{t_1}, \underset{\sim}{t_2}, \underset{\sim}{t_3})$ then by the Extraction Condition there are $n_1, n_2, n_3 \in \mathbb{N}$ such that $\delta(t_{ij}) = \underset{\sim}{n_i}_j$ for each i, j and such that $n_2 = n_1 + 1$. We complete the proof that $\mu(\delta) \models E(G)$ by showing that $\mu(\delta)(G_1') = \mathscr{A}(F')$, where

$$F' = F^M[x_1/\delta^{-1}(n_1), x_2/\delta^{-1}(n_2), x_3/\delta^{-1}(n_3)] \in E(F) \quad .$$

Let $w \in \overline{T}(F)$. If $\langle F \rangle_w = Px_ix_j$, then $\langle G_1 \rangle_w = \hat{\hat{\pi}}(\underset{\sim}{y_i}, \underset{\sim}{y_j})$ and $\langle G_1' \rangle_w = \hat{\hat{\pi}}(\underset{\sim}{t_i}, \underset{\sim}{t_j})$, so

$$\begin{aligned} \mu(\delta)(\langle G_1' \rangle_w) &= \chi_\mu(\tilde{\pi}(\underset{\sim}{n_i}, \underset{\sim}{n_j})) = \chi_\mu(\pi(\underset{\sim}{n_i}, \underset{\sim}{n_j})) \\ &= \mathscr{A}(P\delta^{-1}(n_i)\delta^{-1}(n_j)) \\ &= \mathscr{A}(\langle F' \rangle_w) \quad . \end{aligned}$$

The argument in case $\langle F \rangle_w = R_ix_j$ is similar.

(G ▷ F) If G is satisfiable, then let $\mathscr{A} \models E(G)$; then by Schema Condition (b), $\mathscr{A} = \mu(\delta_k)$ for some $k = 0, \ldots, q-1$ and some completion μ of θ. Let

$$\lambda = \sum_{n=0}^{\infty} \sum_{j=0}^{r-1} \chi_\mu(\tilde{\rho}_j(\underset{\sim}{n+1})) \cdot \langle j, \omega(n) \rangle + \sum_{n=0}^{\infty} \sum_{m=0}^{\infty} \chi_\mu(\tilde{\pi}(\underset{\sim}{m+1}, \underset{\sim}{n+1})) \cdot \omega(m, n) \quad .$$

The sum is obviously proper. Note that $\chi_\mu(\tilde{\rho}_j(\underset{\sim}{n+1})) = \pm 1$ and $\chi_\mu(\tilde{\pi}(\underset{\sim}{m+1}, \underset{\sim}{n+1})) = \pm 1$ since $\tilde{\rho}_j(\underset{\sim}{n+1})$, $\tilde{\pi}(\underset{\sim}{m+1}, \underset{\sim}{n+1})$ each has one edge and μ is complete.

Now let

$$F' = F^M[x_1/\delta^{-1}(n_1), x_2/\delta^{-1}(n_2), x_3/\delta^{-1}(n_3)] \in E(F) \quad ,$$

so that $n_2 = n_1 + 1$. Note that $\delta_k^{-1}(m)$ is defined for all $m \geq k$, and let

$$t_{ij} = \underbrace{n_i + 1}_{j} \qquad (i = 1, 2, 3, \quad j = 0, \ldots, q-1) \quad .$$

Finally, let $G_1' = G_1[\underset{\sim}{y_1}/\underset{\sim}{t_1}, \underset{\sim}{y_2}/\underset{\sim}{t_2}, \underset{\sim}{y_3}/\underset{\sim}{t_3}]$; then clearly $\lambda(\delta)(F') = \mu(\delta_k)(G_1')$. But $\mu(\delta_k) \models G_1'$ since

$$\hat{\psi}(\underset{\sim}{t_1}, \underset{\sim}{t_2}, \underset{\sim}{t_3}) \rightarrow G_1'$$

is an Herbrand instance of a conjunct of G^M, and

$$\begin{aligned}\mu(\delta_k)(\hat{\psi}(\underset{\sim}{t}_1, \underset{\sim}{t}_2, \underset{\sim}{t}_3)) &= \chi_\mu(\psi(\underset{\sim}{n_1+1}, \underset{\sim}{n_2+1}, \underset{\sim}{n_3+1})) \\ &= \chi_\theta(\psi(\underset{\sim}{n_1+1}, \underset{\sim}{n_2+2}, \underset{\sim}{n_3+3})) \\ &= \chi_\theta(\tau(\underset{\sim}{n_1+1}) + \tau(\underset{\sim}{n_1+2}) + \tau(\underset{\sim}{n_3+1}) + \sigma(\underset{\sim}{n_1+1}_{q-1}, \underset{\sim}{n_1+2}_0)) \\ &= 1 \quad ,\end{aligned}$$

by the definition of θ.

This completes the proof of the Sufficient Conditions Lemma. ∎

Construction Lemma. Let $q = r+5$, and let τ, σ, ρ_i, and π be defined as follows:

$$\tau = \omega(0,0) + \sum_{i=0}^{q-2} (\omega(i,i+1) - \omega(i+1,i) - \omega(i+1,i+1))$$

$$\sigma = \omega(0,1) - \omega(1,0)$$

$$\pi = \omega(0,q+2) + \omega(q+2,0)$$

$$\rho_i = \omega(0,i+3) + \omega(i+3,0) \quad , \qquad (i = 0,\dots,r-1) \quad .$$

(These templates are illustrated in Figure 19.) Then the Sufficient Conditions are satisfied.

Proof. The Disjointness Condition follows from the facts that

$$\pm\,\omega(k,\ell) \subset \tau(\underset{\sim}{m}) + \sigma(\underset{\sim}{m}_{q-1}, \underset{\sim}{m+1}_0)$$

only if

$$\ell - k \equiv 0, 1, \text{ or } q-1 \pmod q \quad ;$$

$$\pm\,\omega(k,\ell) \subset \pi(\underset{\sim}{m}, \underset{\sim}{n})$$

only if

$$k \equiv 0 \text{ and } \ell \equiv 2 \pmod q \quad , \quad \text{or } k \equiv 2 \text{ and } \ell \equiv 0 \pmod q \; ;$$

and

$$\pm\,\omega(k,\ell) \subset \rho_i(\underset{\sim}{m})$$

only if

$$k \equiv 0 \text{ and } \ell \equiv i+3 \pmod q \quad , \quad \text{or } k \equiv i+3 \text{ and } \ell \equiv 0 \pmod q ,$$

where, $3 \le i+3 \le q-2$. Now

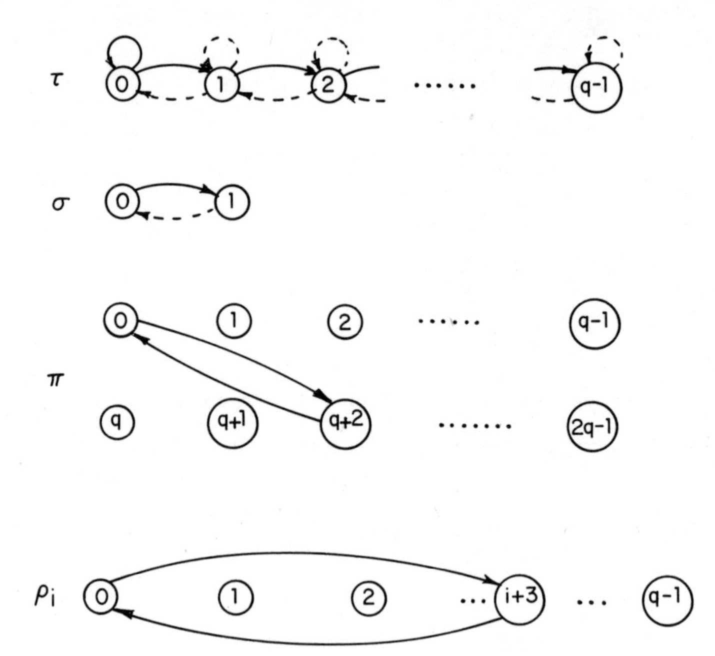

Figure 19

$$\theta = \sum_{m=0}^{\infty} (\tau(\underset{\sim}{m}) + \sigma(\underset{\sim}{m}_{q-1}, \underset{\sim}{m+1}_0))$$

$$= \sum_{m=0}^{\infty} \left(\omega(\underset{\sim}{m}_0, \underset{\sim}{m}_0) - \sum_{i=1}^{q-1} \omega(\underset{\sim}{m}_i, \underset{\sim}{m}_i)\right) + \sum_{n=0}^{\infty} (\omega(n, n+1) - \omega(n+1, n)) \quad ,$$

and for each $\varphi \in \Phi$, $\chi_\varphi(\omega(m,n)) = \chi_\varphi(\omega(n,m))$ for all m, n, and $\chi_\varphi(\omega(m,m)) = 0$ for all m. Then the Extraction Condition follows from the facts that if μ is a minimal Φ-completion of θ then

(a) $\omega(m,m) \subset \mu$ if and only if $m \equiv 0 \pmod q$,

and

(b) $\omega(m,n) - \omega(n,m) \subset \mu$ if and only if $n = m+1$.

Finally, a schema G_0 as required by the Schema Condition is

$$\forall y \exists x (Pyx \wedge \neg Pxy)) \wedge \forall y_0 \cdots \forall y_{q-1}\left(\left(\bigwedge_{i=0}^{q-2} (Py_iy_{i+1} \wedge \neg Py_{i+1}y_i)\right) \rightarrow \left(\mathop{\nabla}_{i=0}^{q-1} Py_iy_i\right)\right)$$

where we have used ∇ to indicate "exclusive or". Both parts (a) and (b) may be checked without difficulty.

This completes the proof of the Construction Lemma and the proof of the Unbounded-Universals Theorem. ■

IIC. 4 The Unbounded-Existentials Classes (I)

In this section we prove the unsolvability of $\exists^*\forall\forall\forall\exists(0,1)$ and $\forall\forall\forall\exists^*(0,1)$ from the unsolvability of the class $\mathscr{V}_2 \subset \forall\forall\forall\exists(\infty,1)$ defined in the Bounded Prefix Theorem. The intuitive idea is to replace the monadic atomic formulas of a schema in $\mathscr{V}_2$ by formulas involving new x-variables $x_1,\ldots,x_q$, but since these x-variables may be governed by y-variables, the x_i terms need not be constants. Thus the constructed formula G must have a subformula guaranteeing the $\mathscr{A}$-indistinguishability of all x_i-terms, for any truth-assignment $\mathscr{A}$ verifying E(G). (The case in which the new x-variables are not governed by y-variables could be treated more simply by itself, but our proof handles both cases at the same time.) Another complication in this proof is that not every truth-assignment verifying F can be transformed into one verifying G; we make use of the fact that if F is satisfiable at all, then F has a verifying truth-assignment that falsifies Ptt for every $t \in D(F)$.

Unbounded-Existentials Theorem (I). $\exists^*\forall\forall\forall\exists(0,1)$ and $\forall\forall\forall\exists^*(0,1)$ are unsolvable.

Proof. As defined in the Bounded Prefix Theorem, let g be the triadic indicial correlate of the variable x, let D_2 be the domain generated by g and ł, and let $\alpha_2 = -\Sigma_{t\in D_2}\,\omega(t,t)$, so that $\mathscr{V}_2 = \forall\forall\forall\exists(\infty,1) \cap \mathscr{L}_\iota(\alpha_2)$, where ι is the identity mapping on D_2. Let F be a schema in $\mathscr{V}_2$; without loss of generality let $F = \forall y_1 \forall y_2 \forall y_3 \exists x F^M$, where F^M has occurrences of the dyadic predicate letter P and the monadic predicate letters $R_0,\ldots,R_{q-1}$ only. We now fix once and for all one of the prefix classes $\exists^*\forall\forall\forall\exists$ or $\forall\forall\forall\exists^*$, call it $\mathscr{P}$; the proof works equally well for either. Let $x_0,\ldots,x_q$ be new variables, and let $f_0,\ldots,f_q$ be their 0-place indicial correlates in case $\mathscr{P}$ is $\exists^*\forall\forall\forall\exists$ and their triadic indicial correlates in case $\mathscr{P}$ is $\forall\forall\forall\exists^*$. In either case, let D_3 be the domain generated by ł, g, and $f_0,\ldots,f_q$. Also let $a_0,\ldots,a_q$ be new constants; a_i will represent the common value of all x_i-terms, provided all x_i-terms are indistinguishable. Then let $D_3^- = D_3 - \{ł\} - \{t \mid t \text{ is an } x_i\text{-term for some } i = 0,\ldots,q\}$; thus D_3^- consists of the x-terms of D_3 (which, however, may contain occurrences of the f_i). Also, let $D_3^* = D_3^- \cup \{a_0,\ldots,a_q\} \cup \{ł\}$, so that D_3^* is the result of replacing the x_i-terms in D_3 by the new constants. Thus D_3 will be the Herbrand domain of the schema constructed, D_3^- will be the part of it that corresponds, intuitively, to "old" elements, i.e., members of $D_2 = D(F)$, and D_3^* will be the result of augmenting D_3^- with ł and the "ideal" elements $a_0,\ldots,a_q$ with which the x_i-terms are to be identified. Define a mapping $\gamma\colon D_3 \to D_3^*$ as follows:

$\gamma(t) = a_i$ if t is an x_i-term $(i = 0, \ldots, q)$,

$\gamma(t) = t$ otherwise .

Like the earlier proofs, this one first proves certain conditions to suffice for the construction of an appropriate G, and then discharges these conditions.

Sufficient Conditions Lemma. Let θ be a bigraph and let τ, ρ, π be 2-templates, all of dimension two. Suppose that the following conditions are satisfied:

Disjointness:

(a) The bigraphs $\rho(a_i, t)$, $\pi(s,t)$, as s, t range over D_3^- and i over $1, \ldots, q$, are pairwise disjoint.

(b) Each $\rho(a_i, t)$ $(i = 1, \ldots, q,\ t \in D_3^-)$ is disjoint from θ.

(c) If $\pi(s,t)$ $(s, t \in D_3^-)$ is not disjoint from θ, then $s = t$ and $\chi_\theta(\omega(t,t)) = -1$.

Let

$$\Phi = \{\rho(a_i, t),\ \pi(s,t) \mid i = 1, \ldots, q,\ s, t \in D_3^-\} .$$

Extraction: $\tau(a_0, \cdot)$ extracts D_3^- from any minimal Φ-completion of θ.

Subgraph: $\tau(a_0, t) \subset \theta$ for each $t \in D_3^-$.

Schema: There is a schema $G_0 \in \mathscr{P}(0,1)$ such that

(a) if μ is any minimal Φ-completion of θ then $\mu(\gamma) \models E(G_0)$;

(b) if $\mathscr{A} \models E(G_0)$, then $\mathscr{A} = \mu(\gamma)$, where μ is a completion of θ.

Then we can construct a schema $G \in \mathscr{P}(0,1)$ such that F is satisfiable if and only if G is satisfiable.

Proof. The bigraphs $\rho(a_i, t)$ and $\pi(s,t)$ are to be the correlates of the atomic formulas $R_i t$ and Pst, respectively, except that s and t need not be terms in $D(F)$, and so will have to be mapped into $D(F)$ (by the mapping β defined below) for the association to be made precise. Then the Disjointness Condition will provide the needed flexibility in the construction of truth-assignments on $E(G)$ from truth-assignments on $E(F)$. The Extraction and Subgraph conditions essentially state that $\tau(a_0, \cdot)$ can be used to distinguish "old" elements from "new", i.e., members of D_3^- from members of $D_3^* - D_3^-$. And implicit in the Schema condition is the assertion that the indistinguishability of x_i-terms can be guaranteed.

Let

$$G = Q\left(G_0^M \wedge \left(\left(\bigwedge_{i=1}^{3} \hat{\tau}(x_0, y_i)\right) \rightarrow G_1\right)\right)$$

where

$$G_1 = F^M \Big\{ Pv_1v_2/\hat{\hat{\pi}}(v_1, v_2)\,(v_1, v_2 \in \{y_1, y_2, y_3, x\}),\ R_i v/\hat{\hat{\rho}}(x_{i+1}, v)$$

$$(i = 0, \ldots, q-1, \quad v \in \{y_1, y_2, y_3, x\}) \Big\}$$

and where Q is one of the prefixes $\exists x_0 \cdots \exists x_q \forall y_1 \forall y_2 \forall y_3 \exists x$ or $\forall y_1 \forall y_2 \forall y_3 \exists x \exists x_0 \ldots \exists x_q$, as appropriate.

(<u>$F \rhd G$</u>) If F is satisfiable then let $\mathscr{A} \models E(F)$, and assume without loss of generality that $\mathscr{A} = \lambda(\iota)$ for some completion λ of α_2 (recall that $F \in \mathscr{L}_\iota(\alpha_2)$). Define a mapping $\beta: D_3^* \to D_2$ as follows:

$$\beta(t) = \text{ł} \qquad , \qquad \text{if } t \text{ is not an x-term}$$

$$= g(\beta(t_1)\beta(t_2)\beta(t_3)) , \qquad \text{if } t = g(t_1t_2t_3) \quad .$$

Let μ be the minimal completion of

$$\theta + \sum_{i=0}^{q-1} \sum_{t \in D_3^-} \chi_\lambda(\langle i, \beta(t)\rangle) \cdot \rho(a_{i+1}, t) + \sum_{s,t \in D_3^-} \chi_\lambda(\omega(\beta(s), \beta(t))) \cdot \pi(s,t) .$$

That this sum is proper follows from the Disjointness Conditions. In particular, if $\varphi \in \Phi$ is not disjoint from θ then by (c) $\varphi = \pi(t,t)$ for some t and $\chi_\theta(\omega(t,t)) = -1$; but $\chi_\lambda(\omega(\beta(t), \beta(t))) = -1$ since $\alpha_2 \subset \lambda$.

We claim that $\mu(\gamma) \models E(G)$. Clearly, $\mu(\gamma) \models E(G_0)$ by Schema Condition (a). Consider the second conjunct of a member of E(G), say

$$\bigwedge_{i=1}^{3} \hat{\tau}(s, t_i) \to G_1' \quad .$$

Then s is an x_0-term. Suppose that

$$\mu(\gamma) \models \bigwedge_{i=1}^{3} \hat{\tau}(s, t_i) \quad ;$$

then $\tau(a_0, \gamma(t_i)) \subset \mu$ for $i = 1, 2, 3$ and by the Extraction Condition $\gamma(t_i) \in D_3^-$ for $i = 1, 2, 3$. We claim that $\mu(\gamma)(G_1') = \mathscr{A}(F')$, where

$$F' = F^*[y_1/\beta(t_1),\ y_2/\beta(t_2),\ y_3/\beta(t_3)] \in E(F) \quad ,$$

so that $\mu(\gamma) \models G_1'$. Let $w \in \overline{T}(F)$. If $\langle F \rangle_w = Pv_1v_2$ for some v_1, v_2 then $\langle G_1' \rangle_w = \hat{\hat{\pi}}(s_1, s_2)$ for some s_1, s_2 such that $\langle F' \rangle_w = P\beta(s_1)\beta(s_2)$; and if $\langle F \rangle_w = R_i v$ for some v then $\langle G_1' \rangle_w = \hat{\hat{\rho}}(s,t)$ for some s, t such that $\gamma(s) = a_{i+1}$ and $\langle F' \rangle_w = R_i\beta(t)$.

$(\underline{G \rhd F})$ If G is satisfiable then let $\mathscr{A} \models E(G)$; by Schema Condition (b), $\mathscr{A} = \mu(\gamma)$ for some completion μ of θ. Define a mapping $\beta': D_2 \to D_3$ by

$$\beta'(\ddagger) = g(\ddagger\ddagger\ddagger)$$

$$\beta'(g(t_1 t_2 t_3)) = g(\beta'(t_1)\beta'(t_2)\beta'(t_3)) \quad .$$

Now let

$$\lambda = \sum_{s,t \in D_2} \chi_\mu(\tilde{\pi}(\beta'(s), \beta'(t))) \cdot \omega(s,t) + \sum_{i=0}^{q-1} \sum_{t \in D_2} \chi_\mu(\tilde{\rho}(a_{i+1}, \beta'(t))) \cdot \langle i, \omega(t) \rangle$$

which is proper and complete. We claim that $\lambda(\iota) \models E(F)$. For let $F' = F^*[y_1/t_1, y_2/t_2, y_3/t_3] \in E(F)$, and let G_1' be the Herbrand instance of G_1 in the Herbrand instance of G,

$$G' = G^*[y_1/\beta'(t_1), y_2/\beta'(t_2), y_3/\beta'(t_3)] \quad .$$

First note that $\mu(\gamma) \models G_1'$, since

$$\left(\bigwedge_{i=1}^{3} \hat{\tau}(s, \beta'(t_i)) \right) \to G_1'$$

is a conjunct of G' for some x_0-term s, and $\mu(\gamma)$ verifies the antecedent since by the Subgraph Condition

$$\mu(\gamma)(\hat{\tau}(s, \beta'(t_i))) = \chi_\mu(\tau(\gamma(s), \gamma(\beta'(t_i))))$$

$$= \chi_\mu(\tau(a_0, t_i')) = 1 \quad ,$$

where t_i' is an x-term. Thus to complete the proof of the lemma it suffices to show that $\lambda(\iota)(F') = \mathscr{A}(G_1')$. So let $w \in \overline{T}(F)$. If $\langle F' \rangle_w = Pst$ then $\langle G_1' \rangle_w = \hat{\tilde{\pi}}(\beta'(s), \beta'(t))$, so that

$$\mathscr{A}(\langle G_1' \rangle_w) = \chi_\mu(\tilde{\pi}(\beta'(s), \beta'(t))) = \chi_\lambda(\omega(s,t))$$

$$= \lambda(\iota)(Pst) = \lambda(\iota)(\langle F' \rangle_w) \quad .$$

And if $\langle F' \rangle_w = R_i t$ then $\langle G_1' \rangle_w = \hat{\tilde{\rho}}(s_{i+1}, \beta'(t))$, where s_{i+1} is some x_{i+1}-term, so

$$\mathscr{A}(\langle G_1' \rangle_w) = \chi_\mu(\tilde{\rho}(a_{i+1}, \beta'(t))) = \chi_\lambda(\langle i, \omega(t) \rangle)$$

$$= \lambda(\iota)(R_i t) = \lambda(\iota)(\langle F' \rangle_w) \quad .$$

This completes the proof of the Sufficient Conditions Lemma. ■

The next lemma reduces the problem of meeting the Sufficient Conditions to that of constructing certain templates.

Reduction Lemma. The Subgraph and Schema Conditions are implied by the hypothesis that there is a 1-template σ_0 and a 2-template σ such that

$$(\theta) \qquad \theta = \sigma_0(a_0) + \sum_{i=0}^{q-1} \sigma(a_i, a_{i+1}) + \sum_{t \in \bar{D}_3} \tau(a_0, t) \quad ,$$

this being a proper sum;

(σ) if μ is any minimal Φ-completion of θ then $\sigma_0(\cdot)$ extracts a_0 from μ, and for $i = 0, \ldots, q-1$, $\sigma(a_i, \cdot)$ extracts a_{i+1} from μ.

Proof. Clearly (θ) implies the Subgraph Condition. Now let $G_0 = Q(M_1 \wedge \ldots \wedge M_5)$, where Q is one of the prefixes $\exists x_0 \ldots \exists x_q \forall y_1 \forall y_2 \forall y_3 \exists x$ or $\forall y_1 \forall y_2 \forall y_3 \exists x \exists x_0 \ldots \exists x_q$ as appropriate, and $M_1, \ldots, M_5$ are as follows:

M_1: $\hat{\sigma}_0(x_0)$

M_2: $\bigwedge_{i=0}^{q-1} \hat{\sigma}(x_i, x_{i+1})$

M_3: $(\hat{\sigma}_0(y_1) \wedge \hat{\sigma}_0(y_2)) \rightarrow (Py_1y_3 \leftrightarrow Py_2y_3) \wedge (Py_3y_1 \leftrightarrow Py_3y_2)$

M_4: $\bigwedge_{i=0}^{q-1} \Big((\hat{\sigma}(x_i, y_1) \wedge \hat{\sigma}(x_i, y_2)) \rightarrow (Py_1y_3 \leftrightarrow Py_2y_3) \wedge (Py_3y_1 \leftrightarrow Py_3y_2)\Big)$

M_5: $\hat{\tau}(x_0, x)$

First let μ be a minimal Φ-completion of θ; we must show that $\mu(\gamma) \models E(G_0)$. That $\mu(\gamma)$ verifies each Herbrand instance of M_1, M_2, or M_5 follows immediately from the definition of γ and from Condition (θ). That $\mu(\gamma)$ verifies an Herbrand instance of M_3 or M_4 follows from (σ) and the $\mu(\gamma)$-indistinguishability of x_i-terms for each $i = 0, \ldots, q$. For example, if

$$(\hat{\sigma}_0(t_1) \wedge \hat{\sigma}_0(t_2)) \rightarrow (Pt_1t_3 \leftrightarrow Pt_2t_3) \wedge (Pt_3t_1 \leftrightarrow Pt_3t_2)$$

is an Herbrand instance of M_3, and $\mu(\gamma)$ verifies the antecedent, then $\sigma_0(\gamma(t_1)) + \sigma_0(\gamma(t_2)) \subset \mu$ and by Condition (σ), $\gamma(t_1) = \gamma(t_2) = a_0$, whence

$$\mu(\gamma)(Pt_1t_3 \leftrightarrow Pt_2t_3) = 1$$

since

$$\chi_\mu(\omega(\gamma(t_1), \gamma(t_3))) = \chi_\mu(\omega(\gamma(t_2), \gamma(t_3)))$$

$$= \chi_\mu(\omega(a_0, \gamma(t_3))) \quad .$$

Now let $\mathscr{A} \models E(G_0)$; we need to show that $\mathscr{A} = \mu(\gamma)$, where μ is a completion of θ. Define mappings γ_i $(i = 0, \ldots, q)$ and bigraphs η_i $(i = 0, \ldots, q)$ as follows:

$$\gamma_i(t) = a_j \;, \qquad \text{if } t \text{ is an } x_j\text{-term} \quad (j = 0, \ldots, i)$$

$$\gamma_i(t) = t \;, \qquad \text{otherwise}$$

$$\eta_i(t) = \sigma_0(a_0) + \sum_{j=0}^{i-1} \sigma(a_j, a_{j+1}) \quad .$$

Claim. For $i = 0, \ldots, q$, $\mathscr{A} = \mu_i(\gamma_i)$ for some completion μ_i of η_i.

Before proving the claim, note that the claim establishes that $\mathscr{A} = \mu(\gamma)$ for some completion μ of θ. For $\gamma = \gamma_q$, and by Condition (θ), $\theta = \eta_q + \Sigma_{t \in \bar{D}_3} \tau(a_0, t)$; but $\tau(a_0, t) \subset \mu_q$, for each x-term $t \in D_3$, because of M_5, so it suffices to let $\mu = \mu_q$.

The claim is proved by induction on i. For $i = 0$ it is a consequence of Herbrand instances of M_1 and M_3; thus if t_0, t_0' are any x_0-terms and t is any term in D_3 then $\mathscr{A} \models \hat{\sigma}_0(t_0) \wedge \hat{\sigma}_0(t_0')$ by Herbrand instances of M_1, and hence $\mathscr{A}(Pt_0t) = \mathscr{A}(Pt_0't)$ and $\mathscr{A}(Ptt_0) = \mathscr{A}(Ptt_0')$ by Herbrand instances of M_3. Thus $\mathscr{A}$ may be written as $\mu_0(\gamma_0)$, where $V(\mu_0) = D_3 - \{t \mid t \text{ is an } x_0\text{-term}\} \cup \{a_0\}$. Given the claim for i, $0 \le i < q$, the claim for $i+1$ follows in a similar way from the $(i+1)$ st conjuncts of M_2 and of M_4.

This completes the proof of the Reduction Lemma. ∎

Thus to complete the proof of the theorem it is necessary to find templates satisfying the revised Sufficient Conditions.

Construction Lemma. The following templates satisfy the Disjointness and Extraction Conditions, and also Conditions (θ) and (σ):

$$\sigma_0 = \omega(0,0)$$

$$\sigma = \omega(0,1) - \omega(1,0)$$

$$\tau = \omega(0,1) + \omega(1,0) - \omega(1,1)$$

$$\rho = \omega(0,1) + \omega(1,0)$$

$$\pi = \omega(0,1).$$

(These templates are illustrated in Figure 20.)

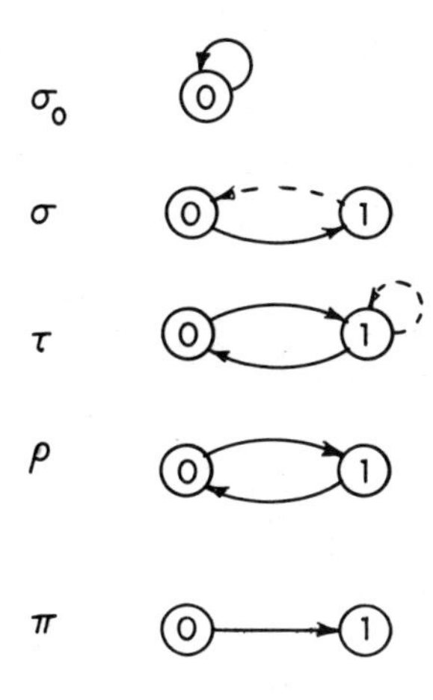

Figure 20

Proof. The sum in Condition (θ) is

$$\theta = \omega(a_0, a_0) + \sum_{i=0}^{q-1} \omega(a_i, a_{i+1}) + \sum_{t \in D_3^-} (\omega(a_0, t) + \omega(t, a_0) - \omega(t, t)) \quad ,$$

which is proper since $a_i \notin D_3^-$ for any i. Also, the members of $\Phi = \{\omega(a_i, t) + \omega(t, a_i), \omega(s, t) \mid i = 1, \ldots, q, \; s, t \in D_3^-\}$ are pairwise disjoint, and the bigraphs $\rho(a_i, t) = \omega(a_i, t) + \omega(t, a_i)$ are disjoint from θ, so Disjointness Conditions (a) and (b) are satisfied. The only bigraphs $\pi(s, t)$ $(s, t \in D_3^-)$ not disjoint from θ are those with $s = t$; in this case $\pi(s, t) = \omega(t, t)$ and $\tau(a_0, t) = \omega(a_0, t) + \omega(t, a_0) - \omega(t, t)$, so that Disjointness Condition (c) is satisfied.

Finally, let μ be a minimal Φ-completion of θ. To check the Extraction Condition, let $\tau(a_0, d) \subset \mu$; then $d \neq a_0$, since $\tau(a_0, a_0)$ is not proper; $d \neq a_1$, since $-\omega(a_1, a_0) \subset \sigma(a_0, a_1) \subset \mu$ but $\omega(d, a_0) \subset \tau(a_0, d)$; and d is not $\ddagger$ or a_i $(2 \leq i \leq q)$ since by the minimality of μ

$$-\omega(a_0, \ddagger) - \sum_{i=2}^{q} \omega(a_0, a_i) \subset \mu \quad ,$$

but $\omega(a_0, d) \subset \tau(a_0, d)$. To check Condition ($\sigma$), first note that $\omega(d, d) \subset \mu$ only if $d = a_0$; the possibility that $d = \ddagger$ or $d = a_i$ for some $i > 0$ is ruled out by minimality, and the possibility that $d \in D_3^-$ is ruled out since in this case $-\omega(d, d) \subset \tau(a_0, d) \subset \theta$. Finally, for $i = 0, \ldots, q-1$, if $\sigma(a_i, d) = \omega(a_i, d) - \omega(d, a_i) \subset \mu$, then $d = a_{i+1}$; for $-\omega(a_i, \ddagger) \subset \mu$ by minimality; $-\omega(a_i, a_j) \subset \mu$ by minimality, if $j \neq i, i+1$; $\chi_\mu(\omega(a_i, d)) = \chi_\mu(\omega(d, a_i))$, if $d \in D_3^-$, since $\omega(a_0, d) + \omega(d, a_0) \subset \tau(a_0, d)$ and $\omega(a_i, d) + \omega(d, a_i) \subset \rho(a_i, d)$ $(i = 1, \ldots, q)$.

This completes the proof of the Construction Lemma and the proof of the Unbounded-Existentials Theorem (I). ∎

IIC. 5 The Unbounded-Existentials Classes (II)

We now come to the most difficult cases, those arising from the $\forall\exists\forall$ prefix by the addition of existential quantifiers. As with the cases of the last section, the basic strategy is to reduce schemata in the $(\infty, 1)$ similarity class by using new x-variables x_i instead of monadic predicate letters; and the chief difficulty is to guarantee the indistinguishability of x_i-terms, in case the x_i are governed by y-variables. However, it is also necessary to "subdivide" the terms arising from the $\forall\exists\forall$ prefix as was done in the proof of the Unbounded-Universals Theorem.

Unbounded-Existentials Theorem (II). $\exists^*\forall\exists\forall(0,1)$, $\forall\exists^*\forall(0,1)$, and $\forall\exists\forall\exists^*(0,1)$ are unsolvable.

Proof. As in the Bounded Prefix Theorem let f be the monadic indicial correlate of the variable x, let D_1 be the domain generated by f and $\dagger$, and define $\delta: D_1 \to \mathbb{N}$ by $\delta(f^n(\dagger)) = n$. Then let

$$\alpha_1 = -\sum_{n=0}^{\infty} (\omega(n+1, n)+\omega(n,n)+\omega(n,n+1))$$

so that $\mathscr{V}_1$ is the set of all schemata $F = \forall y_1 \exists x \forall y_2 F^M$ in $\forall\exists\forall(\infty,1) \cap \mathscr{L}_\delta(\alpha_1)$ whose only atomic formulas are among Py_1y_2, Py_2x, R_iy_1, R_iy_2, and $R_i x$ $(i \in \mathbb{N})$. Let F be a schema in $\mathscr{V}_1$ with monadic predicate letters $R_0, \ldots, R_{r-1}$; we propose to construct a formula G in one of the three classes named in the statement of the theorem such that F is satisfiable if and only if G is satisfiable. We fix our attention on one of these three prefix classes $\exists^*\forall\exists\forall$, $\forall\exists^*\forall$, $\forall\exists\forall\exists^*$; call it $\mathscr{P}$. G will have new variables $x_0, \ldots, x_{2q-1}$, for some $q \in \mathbb{N}$ depending on r. Let Q be one of the quantifier prefixes $\exists z \exists x_0 \cdots \exists x_{2q-1} \forall y_1 \exists x \forall y_2$, $\forall y_1 \exists z \exists x_0 \cdots \exists x_{2q-1} \exists x \forall y_2$, or $\forall y_1 \exists x \forall y_2 \exists z \exists x_0 \cdots \exists x_{2q-1}$, as appropriate. Let D_4 be the domain generated by $\dagger$ and the appropriate indicial correlates of the variables $x, z, x_0, \ldots, x_{2q-1}$; thus D_4 is to be the Herbrand domain of G. Let $a_0, \ldots, a_{2q-1}$ be new constants; let

$$D_4^* = D_4 - \{f^n(t) \mid t \text{ is a z-term and } n \geq 0\} - \{t \mid t \text{ is an } x_i\text{-term, } 0 \leq i \leq 2q-1\} \cup \mathbb{N} \cup \{a_0, \ldots, a_{2q-1}\} .$$

All x_i-terms are to be $\mathscr{A}$-indistinguishable, for any truth-assignment $\mathscr{A}$ verifying $E(G)$; the new constant a_i is to represent the common value of all x_i-terms. The variable z stands for the number zero, and any term $f^n(t)$, where t is a z-term, is a representative in D_4 for the number n. Thus in D_4^* terms are replaced by the elements they represent. The mapping $\gamma: D_4 \to D_4^*$ makes the correspondence precise:

$$\gamma(f^n(t)) = n \quad , \qquad \text{if } t \text{ is a z-term and } n \geq 0 \;;$$

$$\gamma(t) = a_i \quad , \qquad \text{if } t \text{ is an } x_i\text{-term} \;;$$

$$\gamma(t) = t \quad , \qquad \text{otherwise} \;.$$

Clearly γ is onto; let γ^{-1} be any inverse of γ such that $\gamma^{-1}(n+1) = f(\gamma^{-1}(n))$, $(n \in \mathbb{N})$. For $d \in D_4^*$, $n \in \mathbb{N}$ let

$$d \oplus n = \gamma(f^n(\gamma^{-1}(d))) \quad ,$$

so that $\gamma^{-1}(d \oplus 1) = f(\gamma^{-1}(d))$, $d \oplus (n+m) = (d \oplus n) \oplus m$, etc. Also, for $d \in D_4 - D_4^*$, $n \in \mathbb{N}$ define

$$d \oplus n = f^n(d) \quad .$$

Finally, for $n, i \in \mathbb{N}$ let $\underset{\sim}{n}_i = qn+i$; thus $\underset{\sim}{n}_q = \underset{\sim}{n+1}_0$.

We first establish some sufficient conditions for the construction of G. The next two lemmata, called Reduction Lemmata, weaken these conditions. The last lemma, called the Construction Lemma, shows that templates meeting the revised conditions can be found.

Sufficient Conditions Lemma. Let σ^*, θ_0, θ_1, τ_n^*, ρ_{in}^*, π_{mn}^* $(m, n \in \mathbb{N};$ $i = 0, \ldots, r-1)$ be bigraphs of dimension 2. Let τ and ρ_i $(i = 0, \ldots, r-1)$ be $(q+1)$-templates and let π be a 2-template. Let $\Phi = \{\rho_{in}^*, \pi_{mn}^* \mid m, n \in \mathbb{N};\ i = 0, \ldots, r-1\}$. Suppose the following conditions are satisfied.

Disjointness:

(a) The bigraphs ρ_{in}^*, π_{mn}^*, as m, n range over $\mathbb{N}$ and i over $\{0, \ldots, r-1\}$, are pairwise disjoint.

(b) Each ρ_{in}^* is disjoint from θ_1.

(c) If π_{mn}^* and θ_1 are not disjoint, say, $\chi_{\theta_1}(\eta) \cdot \chi_{\pi_{mn}^*}(\eta) = e \neq 0$ for some η with one edge, then $e = \chi_{\alpha_1}(\omega(m, n))$.

(θ_0) $$\theta_0 = \sigma^* + \sum_{n=0}^{\infty} \tau_n^* \;, \qquad \text{the sum being proper.}$$

Extraction: For $i = 0, \ldots, q$, $\tau(a_{q-i}, \ldots, a_{2q-i-1}, \cdot)$ extracts $\{\underset{\sim}{n}_i \mid n \in \mathbb{N}\}$ from any minimal Φ-completion of θ_1.

Subgraph:

(τ) $\quad \tau(a_0, \ldots, a_{q-1}, \underset{\sim}{n}_0) \subset \tau^*_n \qquad (n \in \mathbb{N})$

(ρ) $\quad \rho_i(a_{q-j}, \ldots, a_{2q-j-1}, \underset{\sim}{n-1}_j) \subset \rho^*_{in} \quad (i = 0, \ldots, r-1;\ j = 0, \ldots, q;\ n \in \mathbb{N};\ \underset{\sim}{n-1}_j \geq 0)$

(π) $\quad \pi(\underset{\sim}{m}_0, \underset{\sim}{n-1}_j) \subset \pi^*_{mn} \quad (j = 1, \ldots, q;\ m, n \in \mathbb{N};\ \underset{\sim}{n-1}_j \geq 0)$.

Schema: There is a schema $G_0 \in \mathscr{P}(0, 1)$ such that

(a) if μ is any minimal Φ-completion of θ_1 then $\mu(\gamma) \models E(G_0)$;

(b) if $\mathscr{A} \models E(G_0)$, then $\mathscr{A} = \mu(\gamma)$ for some Φ-completion of μ of θ_0.

Then we can construct a schema $G \in \mathscr{P}(0, 1)$ such that F is satisfiable if and only if G is satisfiable.

Proof. Let

$$G = Q(G_0 \wedge (\hat{\tau}(x_0, \ldots, x_{q-1}, y_1) \wedge \hat{\tau}(x_0, \ldots, x_{q-1}, y_2) \rightarrow G_1))$$

where

$$G_1 = F^M \Big\{ R_i y_j / \hat{\hat{\rho}}_i(x_0, \ldots, x_{q-1}, y_j) \ (i = 0, \ldots, r-1,\ j = 1, 2),$$

$$R_i x / \hat{\hat{\rho}}_i(x_q, \ldots, x_{2q-1}, y_1) \ (i = 0, \ldots, r-1),$$

$$P v_1 v_2 / \hat{\hat{\pi}}(v_1, v_2) \ (v_1, v_2 \in \{y_1, x, y_2\}) \Big\} .$$

($F \triangleright G$) Suppose that F is satisfiable, and let $\mathscr{A} \models E(F)$; without loss of generality, assume that $\mathscr{A} = \lambda(\delta)$ for some completion λ of α_1. Let μ be the minimal completion of

$$\theta_1 + \sum_{i=0}^{r-1} \sum_{m=0}^{\infty} \chi_\lambda(\langle i, \omega(m) \rangle) \cdot \rho^*_{im} + \sum_{m=0}^{\infty} \sum_{n=0}^{\infty} \chi_\lambda(\omega(m, n)) \cdot \pi^*_{mn} \quad .$$

This sum is proper by the Disjointness Conditions. For by (a) and (b), the summands are disjoint, except that θ_1 need not be disjoint from the π^*_{mn}. But if $\chi_{\theta_1}(\eta) = e_1 \neq 0$ and $\chi_{\pi^*_{mn}}(\eta) = e_2 \neq 0$, then by (c) $e_1 e_2 = \chi_{\alpha_1}(\omega(m, n))$. Since $\alpha_1 \subset \lambda$ and $e_2 \in \{-1, 1\}$, $e_1 = \chi_\lambda(\omega(m, n)) \cdot e_2$ or $\chi_{\theta_1}(\eta) = \chi_\lambda(\omega(m, n)) \cdot \chi_{\pi^*_{mn}}(\eta)$, which suffices to show that $\theta_1 + \chi_\lambda(\omega(m, n)) \cdot \pi^*_{mn}$ is a proper sum.

We claim that $\mu(\gamma) \models E(G)$. Clearly, $\mu(\gamma) \models E(G_0)$ by Schema Condition (a). Now let

$$\hat{\tau}(t_0, \ldots, t_{q-1}, s_1) \wedge \hat{\tau}(t_0, \ldots, t_{q-1}, s_2) \rightarrow G_1'$$

be an Herbrand instance of the second conjunct of G^M. Then each t_j is an x_j-term and by the Extraction Condition the antecedent is verified only if $\gamma(s_1) \in \mathbb{N}$, $\gamma(s_2) \in \mathbb{N}$, and $\gamma(s_1) \equiv \gamma(s_2) \equiv 0 \pmod{q}$. So let $\gamma(s_1) = \underset{\sim}{m}_0$, $\gamma(s_2) = \underset{\sim}{n}_0$; we claim that $\mu(\gamma)(G_1') = \mathscr{A}(F')$, where $F' = F^*[y_1/\delta^{-1}(m), y_2/\delta^{-1}(n)]$. For let $w \in \overline{T}(F)$. If $\langle F \rangle_w = R_i y_1$ for some $i = 0, \ldots, r-1$, then

$$\begin{aligned}
\mu(\gamma)(\langle G_1' \rangle_w) &= \mu(\gamma)(\widehat{\widetilde{\rho}}_i(t_0, \ldots, t_{q-1}, s_1)) \\
&= \chi_\mu(\widetilde{\rho}_i(a_0, \ldots, a_{q-1}, \gamma(s_1))) \\
&= \chi_\lambda(\langle i, \omega(m) \rangle) \qquad \text{(by Subgraph Condition } (\rho)) \\
&= \mathscr{A}(\langle F' \rangle_w) \quad .
\end{aligned}$$

The case in which $\langle F \rangle_w = R_i y_2$ is similar.

If $\langle F \rangle_w = R_i x$ then $\langle G_1' \rangle_w = \widehat{\rho}_i(t_q, \ldots, t_{2q-1}, s_1)$, where each t_i is an x_i-term, so that

$$\begin{aligned}
\mu(\gamma)(\langle G_1' \rangle_w) &= \chi_\mu(\widetilde{\rho}_i(a_q, \ldots, a_{2q-1}, \underset{\sim}{m}_0)) \\
&= \chi_\mu(\rho^*_{i,m+1}) \qquad \text{(by Subgraph Condition } (\rho)) \\
&= \chi_\lambda(\langle i, \omega(m+1) \rangle) \\
&= \mathscr{A}(\langle F' \rangle_w) \quad .
\end{aligned}$$

If $\langle F \rangle_w = P y_1 y_2$ then

$$\begin{aligned}
\mu(\gamma)(\langle G_1' \rangle) &= \chi_\mu(\widetilde{\pi}(\underset{\sim}{m}_0, \underset{\sim}{n}_0)) = \chi_\mu(\pi^*_{m,n}) \\
&= \mathscr{A}(P\delta^{-1}(m)\delta^{-1}(n)) \\
&= \mathscr{A}(\langle F' \rangle_w) \quad .
\end{aligned}$$

Finally, if $\langle F \rangle_w = P y_2 x$, then

$$\begin{aligned}
\mu(\gamma)(\langle G_1' \rangle_w) &= \mu(\gamma)(\widehat{\widetilde{\pi}}(s_2, f(s_1))) = \chi_\mu(\widetilde{\pi}(\underset{\sim}{n}_0, \underset{\sim}{m}_1)) \\
&= \chi_\mu(\pi^*_{n,m+1}) \qquad \text{(by Subgraph Condition } (\pi)) \\
&= \mathscr{A}(P\delta^{-1}(n)\delta^{-1}(m+1)) \\
&= \mathscr{A}(\langle F' \rangle_w) \quad .
\end{aligned}$$

Since $F \in \mathscr{V}_1$, these are the only possible atomic subformulas.

(<u>$G \rhd F$</u>) Now suppose that G is satisfiable, and let $\mathscr{A} \models E(G)$. Then by Schema Condition (b), $\mathscr{A} = \mu(\gamma)$ for some Φ-completion μ of θ_0. Define λ by

$$\lambda = \sum_{i=0}^{r-1} \sum_{n=0}^{\infty} \chi_\mu(\tilde{\rho}_i(a_0, \ldots, a_{q-1}, \underset{\sim}{n}_0)) \cdot \langle i, \omega(n) \rangle + \sum_{m=0}^{\infty} \sum_{n=0}^{\infty} \chi_\mu(\tilde{\pi}(\underset{\sim}{m}_0, \underset{\sim}{n}_0)) \cdot \omega(m, n) \quad ,$$

which is proper and complete. We claim that $\lambda(\delta) \models E(F)$. Let $F' = F^*[y_1/\delta^{-1}(m), y_2/\delta^{-1}(n)]$, and let G_1' be the instance of G_1 in the Herbrand instance of G,

$$G' = G^*[y_1/\gamma^{-1}(\underset{\sim}{m}_0), y_2/\gamma^{-1}(\underset{\sim}{n}_0)] \quad .$$

Now $\mu(\gamma) \models G_1'$ since

$$\hat{\tau}(t_0, \ldots, t_{q-1}, \gamma^{-1}(\underset{\sim}{m}_0)) \wedge \hat{\tau}(t_0, \ldots, t_{q-1}, \gamma^{-1}(\underset{\sim}{n}_0)) \rightarrow G_1$$

is a conjunct of $G' \in E(G)$ for some $t_0, \ldots, t_{q-1}$ such that each t_i is an x_i-term, and by Schema Condition (b), Subgraph Condition (τ), and Condition (θ_0),

$$\mu(\gamma)(\hat{\tau}(t_0, \ldots, t_{q-1}, \gamma^{-1}(\underset{\sim}{m}_0))) = \chi_\mu(\tau(a_0, \ldots, a_{q-1}, \underset{\sim}{m}_0)) = 1 \quad ,$$

and similarly for n. Thus it remains to show that $\mu(\gamma)(\langle G_1' \rangle_w) = \lambda(\delta)(\langle F' \rangle_w)$ for each $w \in \overline{T}(F)$.

If $\langle F \rangle_w = Py_1y_2$ then

$$\begin{aligned} \mu(\gamma)(\langle G_1' \rangle_w) &= \mu(\gamma)(\hat{\tilde{\pi}}(\gamma^{-1}(\underset{\sim}{m}_0), \gamma^{-1}(\underset{\sim}{n}_0))) \\ &= \chi_\mu(\tilde{\pi}(\underset{\sim}{m}_0, \underset{\sim}{n}_0)) = \chi_\lambda(\omega(m, n)) \\ &= \lambda(\delta)(\langle F' \rangle_w) \quad . \end{aligned}$$

The calculation in case $\langle F \rangle_w = R_i y_j$ is similar. If $\langle F \rangle_w = Pxy_2$ then

$$\begin{aligned} \mu(\gamma)(\langle G_1' \rangle_w) &= \mu(\gamma)(\hat{\tilde{\pi}}(\gamma^{-1}(\underset{\sim}{n}_0), f(\gamma^{-1}(\underset{\sim}{m}_0)))) \\ &= \chi_\mu(\tilde{\pi}(\underset{\sim}{n}_0, \underset{\sim}{m}_1)) \\ &= \chi_\mu(\pi^*_{n, m+1}) && \text{(by Subgraph Condition } (\pi) \text{ since } \mu \text{ is a } \Phi\text{-completion)} \\ &= \chi_\mu(\tilde{\pi}(\underset{\sim}{n}_0, \underset{\sim}{m+1}_0)) && \text{(by Subgraph Condition } (\pi)) \\ &= \chi_\lambda(\omega(n, m+1)) = \lambda(\delta)(\langle F' \rangle_w) \quad . \end{aligned}$$

Finally, if $\langle F \rangle = R_i x$ then

$$\begin{aligned}\mu(\gamma)(\langle G_1' \rangle_w) &= \chi_\mu(\tilde{\rho}_i(a_q, \ldots, a_{2q-1}, \underset{\sim}{m}_0)) \\ &= \chi_\mu(\rho^*_{i,m+1}) \qquad \text{(by Subgraph Condition } (\rho)) \\ &= \chi_\mu(\tilde{\rho}_i(a_0, \ldots, a_{q-1}, \underset{\sim}{m+1}_0)) \\ &= \chi_\lambda(\langle i, \omega(m+1)\rangle) \\ &= \lambda(\delta)(\langle F' \rangle_w) \quad .\end{aligned}$$

This completes the proof of the Sufficient Conditions Lemma. ∎

The first weakening of the Sufficient Conditions is that the bigraphs τ^*_n, ρ^*_{in}, and π^*_{mn} are generated from "smaller" bigraphs by the process of forming Ξ-envelopes, where Ξ is a set of "constituent" bigraphs. Under this assumption, Schema Condition (b) can be replaced by one referring to Ξ-completions instead of Φ-completions.

First Reduction Lemma. The five Sufficient Conditions (Disjointness, Extraction, (θ_0), Subgraph, and Schema, p. 117) are implied by the Disjointness, Extraction, and (θ_0) Conditions, and the existence of a set Ξ of bigraphs satisfying the following Envelope and Schema$_2$ Conditions:

Envelope: π and each ρ_i $(i = 0, \ldots, r-1)$ has one edge; and

(τ) τ^*_n is the Ξ-envelope of

$$\tau(a_{q-j}, \ldots, a_{2q-j-1}, \underset{\sim}{n-1}_j) \qquad (j = 0, \ldots, q, \quad n \in \mathbb{N}, \quad \underset{\sim}{n-1}_j \geq 0)$$

(ρ) ρ^*_{in} is the Ξ-envelope of

$$\rho_i(a_{q-j}, \ldots, a_{2q-j-1}, \underset{\sim}{n-1}_j) \quad (i = 0, \ldots, r-1, j = 0, \ldots, q, \quad n \in \mathbb{N}, \quad \underset{\sim}{n-1}_j \geq 0)$$

(π) π^*_{mn} is the Ξ-envelope of

$$\sum_{\substack{1 \leq j \leq q \\ \underset{\sim}{n-1}_j \geq 0}} \pi(\underset{\sim}{m}_0, \underset{\sim}{n-1}_j) \quad ,$$

which is a proper sum.

Schema$_2$: There is a schema G_2 such that

(a) if μ is any minimal Φ-completion of θ_1 then $\mu(\gamma) \models E(G_2)$;

(b) if $\mathscr{A} \models E(G_2)$, then $\mathscr{A} = \mu(\gamma)$ for some Ξ-completion μ of σ^*.

Proof. It is easy to see that the Envelope Condition implies the Subgraph Condition. To see that the Schema Condition is implied by the Disjointness, Extraction, (θ_0), Envelope, and Schema$_2$ Conditions, let the schema G_0 of the Schema Condition be $Q(G_2^M \wedge M_1 \wedge M_2 \wedge M_3)$, where Q is the appropriate prefix as before, and M_1, M_2, M_3 are:

M_1: $\hat{\tau}(x_0, \ldots, x_{q-1}, z)$

M_2: $\hat{\tau}(x_0, \ldots, x_{q-1}, y_1) \rightarrow \hat{\tau}(x_q, \ldots, x_{2q-1}, y_1)$

M_3: $\bigwedge_{i=1}^{q} (\hat{\tau}(x_0, \ldots, x_{q-1}, y_2) \wedge \hat{\tau}(x_{q-i}, \ldots, x_{2q-i-1}, y_1) \rightarrow (\hat{\pi}(y_2, y_1) \leftrightarrow \hat{\pi}(y_2, x)))$

(a) First let μ be a minimal Φ-completion of θ_1; we need to show that $\mu(\gamma) \models E(G_0)$. Since $\mu(\gamma) \models E(G_2)$ by Schema$_2$ Condition (a), it suffices to show that $\mu(\gamma)$ verifies each Herbrand instance of $M_1 \wedge M_2 \wedge M_3$. If $\hat{\tau}(t_0, \ldots, t_{q-1}, t)$ is an Herbrand instance of M_1 then each t_i is an x_i-term and t is a z-term; then $\mu(\gamma) \models \hat{\tau}(t_0, \ldots, t_{q-1}, t)$ since $\tau(a_0, \ldots, a_{q-1}, 0) \subset \mu$ by Condition (θ_0). If $\hat{\tau}(t_0, \ldots, t_{q-1}, t) \rightarrow \hat{\tau}(t_q, \ldots, t_{2q-1}, t)$ is an Herbrand instance of M_2 then each t_i is an x_i-term so $\mu(\gamma)$ verifies the antecedent only if $\tau(a_0, \ldots, a_{q-1}, \gamma(t)) \subset \mu$, and by the Extraction Condition this implies that $\gamma(t) = \underset{\sim}{n}_q$ for some $n \in \mathbb{N}$. But then the consequent is verified since $\gamma(t) = \underset{\sim}{n+1}_0$.

Next consider an Herbrand instance of a conjunct of M_3,

$$\hat{\tau}(t_0, \ldots, t_{q-1}, s_2) \wedge \hat{\tau}(t_{q-i}, \ldots, t_{2q-i-1}, s_1) \rightarrow (\hat{\pi}(s_2, s_1) \leftrightarrow \hat{\pi}(s_2, f(s_1))) \quad .$$

Suppose the antecedent is verified; then

$$\tau(a_0, \ldots, a_{q-1}, \gamma(s_2)) \subset \mu$$

and

$$\tau(a_{q-i}, \ldots, a_{2q-i-1}, \gamma(s_1)) \subset \mu \quad ,$$

so by the Extraction Condition $\gamma(s_1) = \underset{\sim}{m}_i$, $\gamma(s_2) = \underset{\sim}{n}_0$ for some $m, n \in \mathbb{N}$. Then the consequent will be verified if

$$\mu(\gamma)(\hat{\pi}(\gamma^{-1}(\underset{\sim}{n}_0), \gamma^{-1}(\underset{\sim}{m}_i))) = \mu(\gamma)(\hat{\pi}(\gamma^{-1}(\underset{\sim}{n}_0), \gamma^{-1}(\underset{\sim}{m}_{i+1}))) \quad .$$

But μ is a Φ-completion, so by Envelope Condition (π)

$$\chi_\mu(\pi(\underset{\sim}{n}_0, \underset{\sim}{m}_i)) = \chi_\mu(\pi(\underset{\sim}{n}_0, \underset{\sim}{m}_{i+1}))$$

$$= \chi_\mu(\pi^*_{n,m+1}) \in \{-1, 1\} \quad .$$

(b) Now let $\mathscr{A} \models E(G_0)$, so that by Schema$_2$ Condition (b) $\mathscr{A} = \mu(\gamma)$ for some Ξ-completion μ of σ^*. We first show that $\tau^*_n \subset \mu$ for each $n \in \mathbb{N}$, so that by Condition (θ_0) μ is a completion of θ_0. For $n = 0$, consider any Herbrand instance $\hat{\tau}(t_0, \ldots, t_{q-1}, t)$ of M_1; since $\mu(\gamma) \models \hat{\tau}(t_0, \ldots, t_{q-1}, t)$, $\tau(a_0, \ldots, a_{q-1}, 0) \subset \mu$ and hence $\tau^*_0 \subset \mu$ since by Envelope Condition (τ), τ^*_0 is the Ξ-envelope of $\tau(a_0, \ldots, a_{q-1}, 0)$. Now we proceed by induction, assuming that $\tau^*_n \subset \mu$. There is an instance of M_2

$$\hat{\tau}(t_0, \ldots, t_{q-1}, \gamma^{-1}(\underset{\sim}{n}_0)) \rightarrow \hat{\tau}(t_q, \ldots, t_{2q-1}, \gamma^{-1}(\underset{\sim}{n}_0))$$

and by the induction hypothesis the antecedent is verified since $\tau(a_0, \ldots, a_{q-1}, \underset{\sim}{n}_0) \subset \tau^*_n$. Hence, $\tau(a_q, \ldots, a_{2q-1}, \underset{\sim}{n}_0) \subset \mu$; but by Envelope Condition (τ) again τ^*_{n+1} is the Ξ-envelope of $\tau(a_q, \ldots, a_{2q-1}, \underset{\sim}{n}_0)$.

Next we must show that for each $\varphi \in \Phi$ there is an $e \in \{-1, 1\}$ such that $e \cdot \varphi \subset \mu$. For $\varphi = \rho^*_{i,n}$ or $\pi^*_{m,0}$ $(i = 0, \ldots, r-1,\ m, n \in \mathbb{N})$ this follows immediately from the fact that by Envelope Condition (ρ) or (π), φ is the Ξ-envelope of a bigraph with a single edge. Now let $\varphi = \pi^*_{m,n+1}$. Then for $i = 1, \ldots, q-1$ there are terms $t_0, \ldots, t_{2q-1}$ such that each t_j is an x_j-term and

$$\mu(\gamma) \models \hat{\tau}(t_0, \ldots, t_{q-1}, \gamma^{-1}(\underset{\sim}{m}_0)) \wedge \hat{\tau}(t_{q-i}, \ldots, t_{2q-i-1}, \gamma^{-1}(\underset{\sim}{n}_i))$$

$$\rightarrow (\hat{\pi}(\gamma^{-1}(\underset{\sim}{m}_0), \gamma^{-1}(\underset{\sim}{n}_i)) \leftrightarrow \hat{\pi}(\gamma^{-1}(\underset{\sim}{m}_0), f(\gamma^{-1}(\underset{\sim}{n}_i)))) \quad .$$

The antecedent is verified by Envelope Condition (τ), since $\tau^*_m, \tau^*_{n+1} \subset \mu$, so $\chi_\mu(\hat{\pi}(\underset{\sim}{m}_0, \underset{\sim}{n}_i)) = \chi_\mu(\hat{\pi}(\underset{\sim}{m}_0, \underset{\sim}{n}_{i+1}))$. But π has only one edge, so by induction on i

$$\chi_\mu\left(\sum_{i=1}^{q} \pi(\underset{\sim}{m}_0, \underset{\sim}{n}_i)\right) \in \{-1, 1\} \quad .$$

By Envelope Condition (π), $\pi^*_{m,n+1}$ is the Ξ-envelope of this sum.

This completes the proof of the First Reduction Lemma. ■

Before stating and proving the second weakening of the Sufficient Conditions we partially define the set Ξ referred to in the Envelope and Schema$_2$ Conditions. Ξ is specified in terms of certain bigraphs ξ_{ie} which are completely defined here, and certain templates σ_i to be defined later. Just as the First Reduction Lemma replaced conditions on the ρ^*_{in}, π^*_{mn}, and τ^*_n by conditions on Ξ, the Second Reduction Lemma replaces conditions on σ^* and Ξ by conditions on the newly introduced templates σ_i. The Construction Lemma then shows that these conditions can be met.

Let ω_0 and ω_1 be 2-templates as follows: $\omega_0 = \omega(0,1)$, $\omega_1 = \omega(1,0)$. For $i = 0,\ldots,2q-1$, $e = 0,1$, and for each $t \in D_4 \cup D_4^*$ let

$$\xi_{ie}(t) = \omega_e(t \oplus i+1,\ t \oplus i) + \sum_{j=0}^{i} \omega_e(t \oplus j,\ a_{i-j}) \quad .$$

The bigraphs $\xi_{ie}(t)$ are related by a property that is used repeatedly below. Note that $\xi_{i+1,e}(t) = \xi_{ie}(t \oplus 1) + \omega_e(t, a_{i+1})$. Hence

$$\xi_{ie}(t \oplus j) \subset \xi_{i+j,e}(t)$$

whenever $i+j \leq 2q-1$.

For each $i = 0,\ldots,2q-1$, σ_i is an $(i+1)$-template to be specified later (p. 128). Let

$$\Xi = \{\sigma_0(a_0)\} \cup \{\xi_{ie}(t) \mid i = 0,\ldots,2q-1,\ e = 0,1,\ t \in D_4^* - \{\ddagger\},\ \text{and } i \neq 0 \text{ or } t \neq a_0\} \ .$$

<u>Second Reduction Lemma</u>. Condition Schema$_2$ is implied by the Disjointness, Extraction, (θ_0), and Envelope Conditions, and the following three conditions:

(σ) For $i = 0,\ldots,2q-1$, $\sigma_i(a_0,\ldots,a_{i-1},\cdot)$ extracts a_i from any minimal Φ-completion of θ_1;

(σ^*) σ^* is the Ξ-envelope of

$$\sum_{i=0}^{2q-1} \sigma_i(a_0,\ldots,a_i) \quad ,$$

which is a proper sum;

(θ_1)
$$\theta_1 = \theta_0 - \sum_{|m-n| \leq 1} \pi^*_{mn} \quad ,$$

which is a proper sum.

<u>Proof</u>. Let

$$G_1 = Q\left(\bigwedge_{i=0}^{2q-1} N_i\right) \quad ,$$

where $N_0 = N_{01} \wedge N_{02} \wedge N_{03}$ and for $i = 1,\ldots,2q-1$, $N_i = N_{i1} \wedge N_{i2}$, and the N_{ij} are as follows:

N_{01}: $\hat{\sigma}_0(x_0) \wedge \neg\, \hat{\sigma}_0(z) \wedge \neg\, \hat{\sigma}_0(x) \wedge \bigwedge_{i=1}^{2q-1} \neg\, \hat{\sigma}_0(x_i)$;

N_{02}: $(\hat{\sigma}_0(y_1) \wedge \hat{\sigma}_0(y_2)) \to (Py_1y_2 \leftrightarrow Py_1y_1 \leftrightarrow Py_2y_1)$;

N_{03}: $(\neg\,\hat{\sigma}_0(y_1) \wedge \hat{\sigma}_0(y_2)) \to ((Py_1y_2 \leftrightarrow Pxy_1) \wedge (Py_2y_1 \leftrightarrow Py_1x))$;

for $i = 1, \ldots, 2q-1$,

N_{i1}: $\hat{\sigma}_i(x_0, \ldots, x_i)$;

N_{i2}: $\hat{\sigma}_i(x_0, \ldots, x_{i-1}, y_2) \to ((Py_1y_2 \leftrightarrow Pxx_{i-1}) \wedge (Py_2y_1 \leftrightarrow Px_{i-1}x))$.

(a) First suppose that μ is a minimal Φ-completion of θ_1; we need to show that $\mu(\gamma) \models E(G_1)$. Clearly, $\mu(\gamma) \models \hat{\sigma}_0(t)$ for any x_0-term t by Condition (σ). Also, if t is not an x_0-term, then $\gamma(t) \neq a_0$ and by Condition (σ) $\chi_\mu(\sigma_0(\gamma(t))) \neq 1$. Then since σ_0 can have only one edge and μ is complete, $-\sigma_0(\gamma(t)) \subset \mu$ and $\mu(\gamma) \models \neg\,\hat{\sigma}_0(t)$. Thus each Herbrand instance of N_{01} is verified by $\mu(\gamma)$. Each Herbrand instance of N_{i1} $(i > 0)$ is verified by $\mu(\gamma)$ since $\sigma_i(a_0, \ldots, a_i) \subset \mu$ by Condition (σ). Next, consider an Herbrand instance of N_{02},

$$(\hat{\sigma}_0(t_1) \wedge \hat{\sigma}_0(t_2)) \to (Pt_1t_2 \leftrightarrow Pt_1t_1 \leftrightarrow Pt_2t_1) \quad .$$

If the antecedent is verified by $\mu(\gamma)$ then by Condition (σ) $\gamma(t_1) = \gamma(t_2) = a_0$ so the consequent is verified as well.

Before continuing we observe that μ is Ξ-closed. For by Conditions (θ_0) and (θ_1), μ is a minimal Φ-completion of

$$\sigma^* + \sum_{n=0}^{\infty} \tau_n^* - \sum_{|m-n| \leq 1} \pi_{mn}^* \quad ,$$

and by the Envelope Conditions and Condition (σ^*), σ^*, each τ_n^*, and each member of Φ is Ξ-closed.

Next consider an Herbrand instance of N_{03},

$$(\neg\,\hat{\sigma}_0(t_1) \wedge \hat{\sigma}_0(t_2)) \to ((Pt_1t_2 \leftrightarrow Pf(t_1)t_1) \wedge (P\,t_2t_1 \leftrightarrow Pt_1f(t_1))) \quad .$$

If the antecedent is verified, then $\gamma(t_1) \neq a_0$, $\gamma(t_2) = a_0$, so the consequent will be verified provided that

$$\chi_\mu(\omega_e(t, a_0)) = \chi_\mu(\omega_e(t \oplus 1, t))$$

for $e = 0, 1$ and $t \neq a_0$. But each of these bigraphs is a subgraph of $\xi_{0e}(t)$, which is a member of Ξ, so the equation follows from our observation (p. 124). Finally, consider an Herbrand instance of N_{i2} $(i > 0)$,

$$\hat{\sigma}_i(t_0, \ldots, t_{i-1}, t) \rightarrow ((Pst \leftrightarrow Pf(s)t_{i-1}) \wedge (Pts \leftrightarrow Pt_{i-1}f(s))) \quad .$$

If $\mu(\gamma)$ verifies the antecedent, then by Condition (σ) $\gamma(t) = a_i$, so the consequent will be verified provided that, for any $s \in D_4$, $e = 0, 1$,

$$\chi_\mu(\omega_e(\gamma(s), a_i)) = \chi_\mu(\omega_e(\gamma(s) \oplus 1, a_{i-1})) \quad .$$

But both these bigraphs are subgraphs of $\xi_{ie}(\gamma(s))$, which is a member of Ξ since $i > 0$; so the equation follows once again.

(b) Now let $\mathscr{A} \models E(G_1)$; we must show that $\mathscr{A} = \mu(\gamma)$ for some Ξ-completion μ of σ^*. For $i = 0, \ldots, 2q-1$ define a mapping

$$\gamma_i : D_4 \rightarrow D_4^* - \{a_{i+1}, \ldots, a_{2q-1}\} \cup \{t \mid t \text{ is an } x_j\text{-term for } i+1 \leq j \leq 2q-1\}$$

as follows:

$$\gamma_i(f^n(t)) = n \quad , \qquad \text{if } t \text{ is a z-term}$$

$$\gamma_i(t) = a_j \quad , \qquad \text{if } t \text{ is an } x_j\text{-term and } 0 \leq j \leq i$$

$$\gamma_i(t) = t \quad , \qquad \text{otherwise.}$$

Thus $\gamma_{2q-1} = \gamma$. We prove by induction on i that for $i = 0, \ldots, 2q-1$, $\mathscr{A} = \mu_i(\gamma_i)$ for some μ_i such that

(i) $$\sum_{j=0}^{i} \sigma_i(a_0, \ldots, a_i) \subset \mu_i$$

and

(ii) for $j = 0, \ldots, i$, for each $t \in D_4 - \{\ddagger\}$ such that $j \neq 0$ or t is not an x_0-term, and for $e = 0, 1$, $\chi_{\mu_i}(\xi_{je}(\gamma_i(t))) \in \{-1, 1\}$.

For $i = 2q-1$, this will complete the proof.

$i=0$: We first show that all x_0-terms are $\mathscr{A}$-indistinguishable, that is, if s and t are x_0-terms and u is any term, then

$$\mathscr{A} \models (Psu \leftrightarrow Ptu) \wedge (Pus \leftrightarrow Put) \quad .$$

Clearly, $\mathscr{A} \models \hat{\sigma}_0(s) \wedge \hat{\sigma}_0(t)$. If also $\mathscr{A} \models \hat{\sigma}_0(u)$, then from two Herbrand instances of N_{02},

$$\mathscr{A} \models Pus \leftrightarrow Puu \leftrightarrow Put \leftrightarrow Psu \leftrightarrow Ptu \quad .$$

On the other hand, if $\mathscr{A} \models \neg\hat{\sigma}_0(u)$, then from two Herbrand instances of N_{03},

$$\mathscr{A} \models (Pus \leftrightarrow Pf(u)u \leftrightarrow Put) \wedge (Psu \leftrightarrow Puf(u) \leftrightarrow Ptu) \quad .$$

Hence $\mathscr{A} = \mu_0(\gamma_0)$ for some μ_0, and clearly $\sigma_0(a_0) \subset \mu_0$ from N_{01}. And if t is neither ŧ nor an x_0-term, then

$$\xi_{0e}(\gamma_0(t)) = \omega_e(\gamma_0(t) \oplus 1, \gamma_0(t)) + \omega_e(\gamma_0(t), a_0) \quad .$$

By N_{01}, $\mu_0(\gamma_0) \models \neg\hat{\sigma}_0(t)$, so by N_{03} there is an x_0-term t_0 such that

$$\mu_0(\gamma_0) \models (Ptt_0 \leftrightarrow Pf(t)t) \wedge (Pt_0t \leftrightarrow Ptf(t))$$

whence

$$\chi_{\mu_0}(\omega_e(\gamma_0(t), a_0)) = \chi_{\mu_0}(\omega_e(\gamma_0(t) \oplus 1, \gamma_0(t)))$$

for $e = 0, 1$, so $\chi_{\mu_0}(\xi_{0e}(\gamma_0(t))) \in \{-1, 1\}$.

$i > 0$: Let s and t be x_i-terms; by N_{i1} there are terms $s_0, \ldots, s_{i-1}$, $t_0, \ldots, t_{i-1}$, where s_j and t_j are x_j-terms for $0 \leq j < i$, such that

$$\mathscr{A} \models \hat{\sigma}_i(s_0, \ldots, s_{i-1}, s) \wedge \hat{\sigma}_i(t_0, \ldots, t_{i-1}, t) \quad .$$

By the induction hypothesis $\mathscr{A} = \mu_{i-1}(\gamma_{i-1})$ for some μ_{i-1}, so

$$\mathscr{A} \models \hat{\sigma}_i(r_0, \ldots, r_{i-1}, s) \wedge \hat{\sigma}_i(r_0, \ldots, r_{i-1}, t)$$

for any $r_0, \ldots, r_{i-1}$ such that each r_j is an x_j-term. Now let u be any term in D_4; then by N_{i2} there are terms $s'_0, \ldots, s'_{i-1}$, $t'_0, \ldots, t'_{i-1}$, each s'_j and each t'_j an x_j-term, such that

$$\mathscr{A} \models \hat{\sigma}_i(s'_0, \ldots, s'_{i-1}, s) \rightarrow ((Pus \leftrightarrow Pf(u)s'_{i-1}) \wedge (Psu \leftrightarrow Ps'_{i-1}f(u)))$$

and

$$\mathscr{A} \models \hat{\sigma}_i(t'_0, \ldots, t'_{i-1}, t) \rightarrow ((Put \leftrightarrow Pf(u)t'_{i-1}) \wedge (Ptu \leftrightarrow Pt'_{i-1}f(u))) \quad .$$

The antecedents are verified, and by the induction hypothesis again s'_{i-1} and t'_{i-1} are $\mathscr{A}$-indistinguishable, so $\mathscr{A} \models (Pus \leftrightarrow Put) \wedge (Psu \leftrightarrow Ptu)$. Hence s and t are $\mathscr{A}$-indistinguishable and $\mathscr{A} = \mu_i(\gamma_i)$ for some μ_i. Now (i) follows immediately from N_{i1}. Also (ii) follows. For, by N_{i2},

$$\chi_{\mu_i}(\omega_e(\gamma_i(t), a_i)) = \chi_{\mu_i}(\omega_e(\gamma_i(t) \oplus 1, a_{i-1}))$$

for any $t \in D_2$ and for $e = 0, 1,$ and as observed just after the definition of ξ_{ie} (p. 124)

$$\xi_{ie}(\gamma_i(t)) = \xi_{i-1,e}(\gamma_i(t) \oplus 1) + \omega_e(\gamma_i(t), a_i) \quad .$$

But by the induction hypothesis,

$$\chi_{\mu_i}(\xi_{i-1,e}(\gamma_i(t) \oplus 1)) \in \{-1, 1\} \quad ;$$

this completes the proof of the Second Reduction Lemma. ■

Now let the templates σ_i, ω, ρ_i, π be defined as follows. Let

$$q = 2r + 4 \quad ,$$

$$\pi = \omega(0, 1) \quad ,$$

$$\rho_i = \omega(2i + 5, q) \qquad (i = 0, \ldots, r-1) \quad ,$$

$$\tau = \sum_{j=0}^{3} \sum_{e=0}^{1} c_{je}\omega_e(q, j) \quad ,$$

$$\sigma_0 = \omega(0, 0) \quad ,$$

$$\sigma_i = \sum_{j=0}^{i-1} \sum_{e=0}^{1} d_{ije}\omega_e(i, j) \qquad (i = 1, \ldots, 2q-1) \quad ,$$

where

$$c_{00} = -1 \qquad c_{01} = -1$$

$$c_{10} = 1 \qquad c_{11} = 1$$

$$c_{20} = -1 \qquad c_{21} = -1$$

$$c_{30} = -1 \qquad c_{31} = 1$$

$$d_{100} = 1 \qquad d_{101} = -1$$

$$d_{200} = -1 \qquad d_{201} = 1$$

$$d_{210} = -1 \qquad d_{211} = 1$$

$$d_{300} = 1 \qquad d_{301} = 1$$

$$d_{310} = -1 \qquad d_{311} = -1$$

$$d_{320} = 1 \qquad d_{321} = -1$$

and for $i = 4, \ldots, 2q-1$,

$$d_{i00} = 1 \qquad d_{i01} = 1$$

$$d_{ij0} = -1 \qquad d_{ij1} = -1 \qquad (j = 1, \ldots, i-2)$$

$$d_{i,i-1,0} = 1 \qquad d_{i,i-1,1} = 1 \quad .$$

These templates are illustrated in Figure 21.

Construction Lemma. The Disjointness, Extraction, (θ_0), Envelope, (σ), (σ^*), and (θ_1) Conditions are satisfied (pp. 117, 121, 124).

Proof. We need to find the Ξ-envelopes of various bigraphs and to show that they are disjoint from various others. A major tool is the following:

Sublemma 1. For $k = 1,2$, let $e_k \in \{0,1\}$, $i_k \in \{0, \ldots, 2q-1\}$, $t_k \in D_4^* - \{\ddagger\}$ be such that $i_k \neq 0$ or $t_k \neq a_0$, and assume that $e_1 \neq e_2$ or $i_1 \neq i_2$ or $t_1 \neq t_2$. Then $\xi_{i_1 e_1}(t_1) \cap \xi_{i_2 e_2}(t_2)$ is nonempty if and only if

(a) $e_1 = e_2$, $t_1 \oplus i_1 = t_2 \oplus i_2$, in which case

$$\xi_{i_k e_k}(t_k) \subset \xi_{i_\ell e_\ell}(t_\ell) \quad ,$$

where $\ell = 1$ if $i_1 > i_2$, 2 if $i_2 > i_1$, and $k = 3 - \ell$; or

(b) $e_2 = 1 - e_1$, $t_2 = a_{i_1}$, $t_1 = a_{i_2}$, in which case the intersection is $\omega_e(a_{i_2}, a_{i_1}) = \omega_e(t_1, t_2)$; or

(c) $e_2 = e_1$, $t_1 = t_2 = a_j$ for some j, $1 \leq j \leq 2q-2$, $i_1 = 0$, and $i_2 = j+1$, in which case the intersection is $\omega_e(a_j \oplus 1, a_j)$; or

(d) a case symmetric to (c).

Proof of Sublemma 1. There are six cases to consider:

(i) $\langle t_1 \oplus i_1 + 1,\ t_1 \oplus i_1 \rangle = \langle t_2 \oplus i_2 + 1,\ t_2 \oplus i_2 \rangle$

(ii) $\langle t_1 \oplus i_1 + 1,\ t_1 \oplus i_1 \rangle = \langle t_2 \oplus i_2,\ t_2 \oplus i_2 + 1 \rangle$

(iii) $\langle t_1 \oplus i_1 + 1,\ t_1 \oplus i_1 \rangle = \langle t_2 \oplus j_2,\ a_{i_2 - j_2} \rangle$ for some $j_2 \leq i_2$

(iv) $\langle t_1 \oplus i_1 + 1,\ t_1 \oplus i_1 \rangle = \langle a_{i_2 - j_2},\ t_2 \oplus j_2 \rangle$ for some $j_2 \leq i_2$

(v) $\langle t_1 \oplus j_1,\ a_{i_1 - j_1} \rangle = \langle t_2 \oplus j_2,\ a_{i_2 - j_2} \rangle$ for some $j_1 \leq i_1$, $j_2 \leq i_2$

(vi) $\langle t_1 \oplus j_1,\ a_{i_1 - j_1} \rangle = \langle a_{i_2 - j_2},\ t_2 \oplus j_2 \rangle$ for some $j_1 \leq i_1$, $j_2 \leq i_2$.

Either (i) or (v) yields (a); (ii) and (iv) are syntactically impossible; (iii) yields (c); and (vi) yields (b). This completes the proof of Sublemma 1. ∎

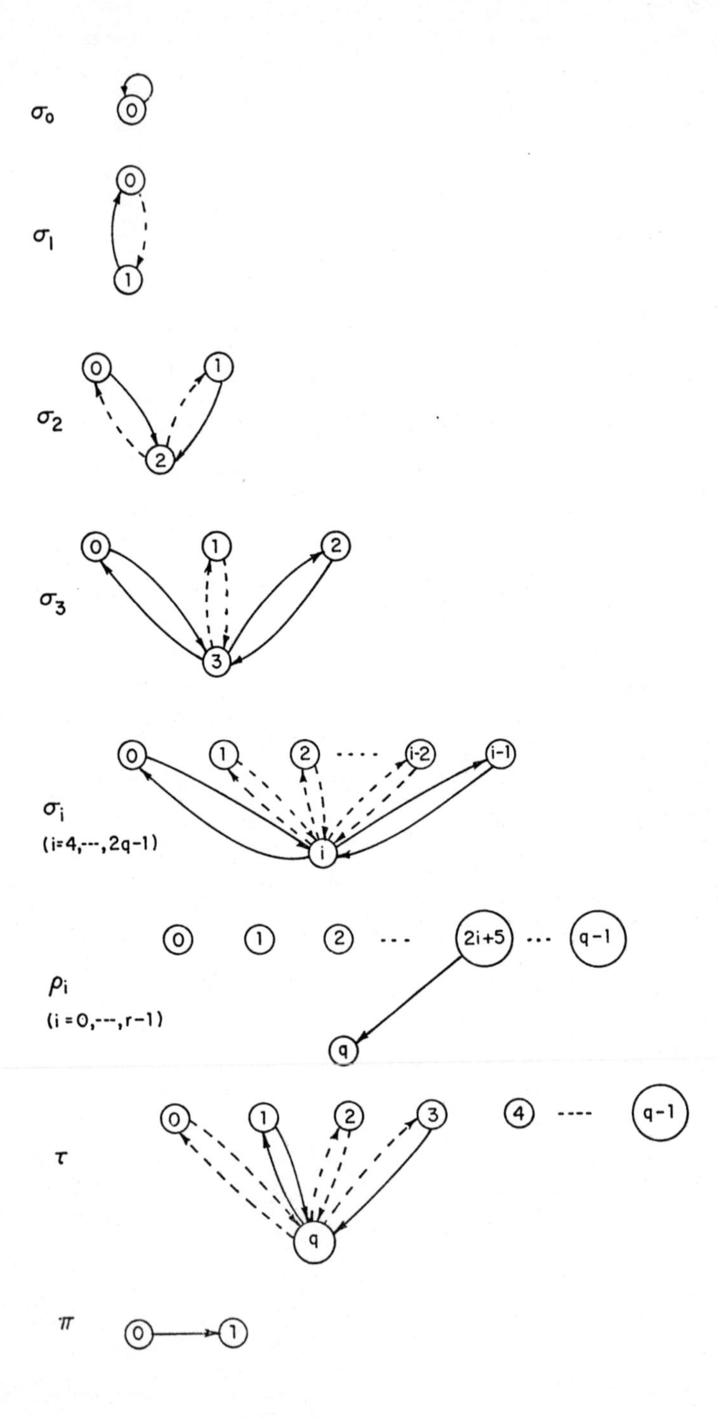

σ_0
0
σ_1
0
1
σ_2
0
1
2
σ_3
0
1
2
3
σ_i
(i=4,---,2q-1)
0
1
2
i-2
i-1
i
ρ_i
(i = 0,---,r-1)
0
1
2
2i+5
q-1
q
τ
0
1
2
3
4
q-1
q
π
0
1

Figure 21

Returning to the proof of the Construction Lemma, we first show that the sum in Condition (σ^*),

$$\sigma' = \sum_{i=0}^{2q-1} \sigma_i(a_0, \ldots, a_i) \quad ,$$

is proper and that its Ξ-envelope σ^* is, in fact,

$$\sigma'' = \omega(a_0, a_0) + \sum_{i=1}^{2q-1} \sum_{j=0}^{i-1} \sum_{e=0}^{1} d_{ije}(\xi_{je}(a_i) + \xi_{i,1-e}(a_j)) \quad .$$

We first observe that if σ'' is a proper sum, then so is σ' and $\sigma' \subset \sigma'' \subset \sigma^*$. For

$$\sigma_i(a_0, \ldots, a_i) = \sum_{j=0}^{i-1} \sum_{e=0}^{1} d_{ije}\omega_e(a_i, a_j) \quad ,$$

and $\omega_e(a_i, a_j)$ is a subgraph of both $\xi_{je}(a_i)$ and $\xi_{i,1-e}(a_j)$. Thus it remains only to show that σ'' is proper and is Ξ-closed; it will then follow that $\sigma'' = \sigma^*$. To do this it suffices to prove a second sublemma. For any i, j, e, $0 \le j < i \le 2q-1$, $e = 0, 1$, define $d_{jie} = d_{i,j,1-e}$, so that

$$\sigma'' = \omega(a_0, a_0) + \sum_{\substack{i,j=1 \\ i \ne j}}^{2q-1} \sum_{e=0}^{1} d_{ije} \cdot \xi_{je}(a_i) \quad .$$

Sublemma 2. For any i, j, e, $0 \le i, j \le 2q-1$, $i \ne j$, and $e = 0, 1$, and for any t, j', e', $t \in D_4^* - \{\ddagger\}$, $0 \le j' \le 2q-1$, $e' = 0, 1$, if $\xi_{je}(a_i) \cap \xi_{j'e'}(t) \ne \phi_2$ then either $t \ne a_{i'}$ for all i', and $\xi_{j'e'}(t) \subset \xi_{je}(a_i)$, or $t = a_{i'}$ for some i', and $d_{i'j'e'} = d_{ije}$.

Proof of Sublemma 2. We proceed according to the cases of Sublemma 1. If (a) applies then $e' = e$ and $t \oplus j' = a_i \oplus j$; but then $j > j'$ and $t = a_i \oplus (j-j')$, so $\xi_{j'e'}(t) \subset \xi_{je}(a_i)$. In case (b) we have $t = a_{i'}$, where $i' = j$, $e' = 1-e$, and $j' = i$; then $d_{i'j'e'} = d_{j,i,1-e} = d_{ije}$. In case (c) we have $e' = e$, $t = a_{i'}$, where $i' = i$ and $j = 0$, $j' = i+1$; we must show that $d_{i0e} = d_{i,i+1,e}$, i.e., $d_{i0e} = d_{i+1,i,1-e}$. Referring to the definition (p. 128) we have that

$$d_{100} = d_{211} = 1 \; , \qquad d_{101} = d_{210} = -1 \; ;$$

$$d_{200} = d_{321} = -1 \; , \qquad d_{201} = d_{320} = 1 \; ;$$

and for $i = 3, \ldots, 2q-2$,

$$d_{i00} = d_{i+1,i,1} = 1 \; , \qquad d_{i01} = d_{i+1,i,0} = 1 \quad .$$

We omit the rest of the proof, since (d) is symmetric to (c). This completes the proof of Sublemma 2. ■

This completes the proof of Condition (σ^*). ■

Next we show that, as required by Envelope Condition (τ), for each $n \geq 0$ all the bigraphs $\tau(a_{q-j}, \ldots, a_{2q-j}, \underset{\sim}{n-1}_j)$ ($j = 0, \ldots, q$, $\underset{\sim}{n-1}_j \geq 0$) have the same Ξ-envelope, which is

$$\sum_{i=0}^{3} \sum_{e=0}^{1} c_{ie} \xi_{2q-1,e}(\underset{\sim}{n-2}_{i+1}) \qquad \text{for} \quad n \geq 2 \quad ,$$

$$\sum_{i=0}^{3} \sum_{e=0}^{1} c_{ie} \xi_{qn+i}(0) \qquad \text{for} \quad n \leq 1 \quad .$$

We treat just the $n \geq 2$ case; the other case is similar. Call the sum τ'_n temporarily. First, note that if the sum is proper, then $\tau(a_{q-j}, \ldots, a_{2q-j-1}, \underset{\sim}{n-1}_j) \subset \tau'_n \subset \tau^*_n$. For

$$\tau(a_{q-j}, \ldots, a_{2q-j-1}, \underset{\sim}{n-1}_j) = \sum_{i=0}^{3} \sum_{e=0}^{1} c_{ie} \omega_e(\underset{\sim}{n-1}_j, a_{q-j+1}) \quad ,$$

and $\omega_e(\underset{\sim}{n-1}_j, a_{q-j+1}) \subset \xi_{2q-1,e}(\underset{\sim}{n-2}_{i+1})$ by the definition of $\xi_{2q-1,e}$. Now we show that if the sum is proper, then τ'_n is Ξ-closed. By Sublemma 1 $\xi_{2q-1,e}(\underset{\sim}{n-2}_{i+1}) \cap \xi_{i'e'}(t) \neq \phi_2$ only in case (a), when $t \oplus i' = \underset{\sim}{n-2}_{i+1} + 2q-1 = \underset{\sim}{n}_i$, and since $i' \leq 2q-1$ we have in this case that $\xi_{i'e'}(t) \subset \xi_{2q-1,e}(\underset{\sim}{n-2}_{i+1})$. Finally, the sum is proper, since by Sublemma 1 again $\xi_{2q-1,e}(\underset{\sim}{n-2}_{i+1}) \cap \xi_{2q-1,e}(\underset{\sim}{n'-2}_{i'+1}) \neq \phi$ only if $\underset{\sim}{n'-2}_{i'+1} = \underset{\sim}{n-2}_{i+1}$, but since $0 \leq i \leq 3 < q$ this requires $n = n'$, $i = i'$. Hence, $\tau'_n = \tau^*_n$. Moreover, this establishes that the τ^*_n are disjoint; and by Sublemma 1 and the explicit formula above for σ^*, the τ^*_n are disjoint from σ^*. Hence Condition (θ_0) is satisfied, since the sum defining θ_0 is proper.

By an argument virtually identical to that for τ^*_n it follows that for $i = 0, \ldots, r-1$, $n \in \mathbb{N}$,

$$\rho^*_{in} = \xi_{2q-1,1}(\underset{\sim}{n-2}_{2i+6}) \qquad \text{for} \quad n \geq 2$$

$$= \xi_{qn+2i+5,\,1}(0) \qquad \text{for} \quad n \leq 1 \quad ,$$

that Envelope Condition (ρ) is satisfied, and that the ρ^*_{in} are disjoint from each other, from the τ^*_n, and from σ^*. In particular, note that there are no n, n', i, i' with $0 \leq i \leq 3$ and $0 \leq i' \leq r-1$ such that $\underset{\sim}{n-2}_{i+1} = \underset{\sim}{n'-2}_{2i'+6}$.

To complete the proof of the Envelope Conditions, consider the π^*_{mn}. Clearly the sum in Envelope Condition (π) is proper. We claim that

(i) (a) $\pi^*_{mn} = \sum_{i=1}^{q} \omega(\underset{\sim}{m}_0, \underset{\sim}{n-1}_i)$, if $n \neq m, m+1$ and $n > 0$

(b) $\pi^*_{m0} = \omega(\underset{\sim}{m}_0, 0)$, if $m \neq 0$

(ii) (a) $\pi^*_{mm} = \sum_{i=1}^{q} (\underset{\sim}{m}_0, \underset{\sim}{m-1}_i) + \xi_{2q-1,0}(\underset{\sim}{m-2}_0)$, if $m \geq 2$

(b) $\pi^*_{11} = \sum_{i=0}^{q} \omega(\underset{\sim}{1}_0, \underset{\sim}{0}_i) + \xi_{q-1,0}(0)$

(c) $\pi^*_{00} = \omega(0,0)$

(iii) (a) $\pi^*_{m,m+1} = \sum_{i=1}^{q} \omega(\underset{\sim}{m}_0, \underset{\sim}{m}_i) + \xi_{2q-1,1}(\underset{\sim}{m-2}_1)$, if $m \geq 2$

(b) $\pi^*_{m,m+1} = \sum_{i=1}^{q} \omega(\underset{\sim}{m}_0, \underset{\sim}{m}_i) + \xi_{qm,1}(0)$, if $m \leq 1$.

We treat just the cases (i), (iia), (iiia); the others are similar. By the definition of ξ_{je}, $\omega(k,\ell) \subset \xi_{je}(t)$, where $t, k, \ell \in \mathbb{N}$, if and only if $k = \ell+1 = t+j+1$ and $e = 0$ or $\ell = k+1 = t+j+1$ and $e = 1$. Thus, $\omega(\underset{\sim}{m}_0, \underset{\sim}{n-1}_i)$ $(m, n \geq 0,\ 1 \leq i \leq q)$ fails to be disjoint from $\xi_{je}(t)$ only if $m = n$, $i = q-1$, $e = 0$, $\underset{\sim}{m}_0 = t+j+1$, or $n = m+1$, $i = 1$, $e = 1$, $\underset{\sim}{m}_1 = t+j+1$. Hence, in case (i) the sum is Ξ-closed, and in cases (iia), (iiia) the Ξ-envelope is obtained by taking $j = 2q-1$ and adding $\xi_{je}(t)$ to the sum.

By Sublemma 1 again the π^*_{mn} as just presented are pairwise disjoint and are disjoint from the ρ^*_{in} presented earlier. This establishes Disjointness Condition (a). The π^*_{mn} are disjoint from σ^*, but not from the τ^*_n, since $\xi_{2q-1}(\underset{\sim}{m-2}_1) \subset \pi^*_{m,m+1}$ and $c_{01}\xi_{2q-1,1}(\underset{\sim}{m-2}_1) \subset \tau^*_m$. But since $c_{01} = -1$ the sum in Condition (θ_1) is proper. Moreover, this settles Disjointness Condition (b).

Finally, Disjointness Condition (c) holds because by the formula in Condition (θ_1) for θ_1 and by the disjointness of σ^* and the τ^*_n from the π^*_{mn},

$$\chi_{\theta_1}(\eta) \cdot \chi_{\pi^*_{mn}}(\eta) = e \neq 0$$

only if $|m-n| \leq 1$, in which case $-\pi^*_{mn} \subset \chi_{\theta_1}(\eta)$ so $e = -1 = \chi_{\alpha_1}(\omega(m,n))$.

This settles everything except for the Extraction Condition and Condition (σ). So let μ be a minimal Φ-completion of θ_1. We consider Condition (σ) first. By Conditions σ^*, (θ_0), and (θ_1), $\sigma_i(a_0, \ldots, a_i) \subset \sigma^* \subset \theta_0 \subset \theta_1 \subset \mu$ for $i = 0, \ldots, 2q-1$, so it remains only to show that if $\sigma_i(a_0, \ldots, a_{i-1}, t) \subset \mu$, then $t = a_i$.

$i = 0$: $\sigma_0(t) = \omega(t,t)$. If $t \in \mathbb{N}$ then $-\omega(t,t) \subset \pi^*_{tt} \subset \theta_1 \subset \mu$, and if $t \in D^*_4 - \mathbb{N} - \{a_0\}$, then $-\omega(t,t) \subset \mu$ by minimality.

$i = 1$: $\sigma_1(a_0, t) = \omega(t, a_0) - \omega(a_0, t)$.

(a) $t \neq a_0$ since $\omega(a_0, a_0) = \sigma_0(a_0) \subset \mu$;

(b) $t \neq a_i$ for $i = 2, \ldots, 2q-1$ since $\omega(a_0, a_i) \subset \sigma_i(a_0, \ldots, a_i) \subset \sigma^* \subset \mu$ (since $d_{i01} = 1$)

(c) $t \neq \underset{\sim}{n}_i$ for $i = 0, 2$, or 3 since $-\omega(\underset{\sim}{n}_i, a_0) \subset c_{i0}\xi_{i0}(\underset{\sim}{n}_0) \subset \tau^*_n$ (since $c_{00} = c_{20} = c_{30} = -1$)

(d) $t \neq \underset{\sim}{n}_1$ since $\omega(a_0, \underset{\sim}{n}_1) \subset c_{11}\xi_{11}(\underset{\sim}{n}_0) \subset \tau^*_n$ (since $c_{11} = 1$)

(e) $t \neq \underset{\sim}{n}_i$ for $i = 4, \ldots, q-1$, and $t \notin D^*_4 - \{a_0, \ldots, a_{2q-1}\} - \mathbb{N}$, since $-\omega(t, a_0) \subset \mu$ by minimality in these cases.

$i = 2$: $\sigma_2(a_0, a_1, t) = -\omega(t, a_0) + \omega(a_0, t) - \omega(t, a_1) + \omega(a_1, t)$

(a) $t \neq a_0, a_1$ because $\sigma_2(a_0, a_1, t)$ is not proper in these cases.

(b) $t \neq a_i$ for $i = 3, \ldots, 2q-1$ since $\omega(a_i, a_0) \subset \sigma_i(a_0, \ldots, a_i)$ (since $d_{i00} = 1$)

(c) $t \neq \underset{\sim}{n}_i$ for $i = 0$ or 2 since $-\omega(a_0, \underset{\sim}{n}_i) \subset c_{i1}\xi_{i1}(\underset{\sim}{n}_0) \subset \tau^*_n$ (since $c_{01} = c_{21} = -1$)

(d) $t \neq \underset{\sim}{n}_1$ since $\omega(\underset{\sim}{n}_1, a_0) \subset c_{10}\xi_{10}(\underset{\sim}{n}_0) \subset \tau^*_n$ ($c_{10} = 1$)

(e) $t \neq \underset{\sim}{n}_i$ for $i = 3, \ldots, q-2$, since either $-\omega(a_0, \underset{\sim}{n}_i) \subset \mu$ (if i is odd) or $-\omega(a_1, \underset{\sim}{n}_i) \subset \mu$ (if i is even) by minimality. That is, for $j = 0, \ldots, r-1$, and for $e = 0, 1$, $e \cdot \xi_{2j+4,1}(\underset{\sim}{n}_0)$ is not a subgraph of σ^*, any τ^*_n, any π^*_{mn}, or (in particular) any ρ^*_{in}.

(f) $t \neq \underset{\sim}{n}_{q-1}$, since $-\omega(a_1, \underset{\sim}{n}_{q-1}) \subset -\xi_{q,1}(\underset{\sim}{n}_0) \subset \tau^*_{n+1}$ (since $c_{01} = -1$)

(g) $t \notin D^*_4 - \{a_0, \ldots, a_{2q-1}\} - \mathbb{N}$ by minimality.

$i = 3$: $\sigma_3(a_0, a_1, a_2, t) = \omega(t, a_0) + \omega(a_0, t) - \omega(t, a_1) - \omega(a_1, t) + \omega(t, a_2) - \omega(a_2, t)$.

(a) $t \neq a_0$, since $\omega(a_1, a_0) \subset \sigma^*$

(b) $t \neq a_1$, since $-\omega(a_0, a_1) \subset \sigma^*$

(c) $t \neq a_2$, since $-\omega(a_2, a_0) \subset \sigma^*$

(d) $t \neq a_i$ for $i = 4, \ldots, 2q-1$, since $-\omega(a_i, a_2) \subset \sigma^*$

(e) $t \neq \underset{\sim}{n}_i$ for $i = 0$ or $i = 2, \ldots, 2q-1$, since $-\omega(\underset{\sim}{n}_i, a_0) \subset \mu$

(f) $t \neq \underset{\sim}{n}_1$, since $-\omega(\underset{\sim}{n}_1, a_2) \subset \tau^*_n$

(g) $t \notin D^*_4 - \{a_0, \ldots, a_{2q-1}\} - \mathbb{N}$ by minimality.

$i = 4, \ldots, 2q-1.$

$$\sigma_i(a_0, \ldots, a_{i-1}, t) = \omega(t, a_0)+\omega(a_0, t) - \sum_{j=1}^{i-2} (\omega(t, a_j)+\omega(a_j, t)) +\omega(t, a_{i-1}) +\omega(a_{i-1}, t)$$

(a) $t \neq a_0, a_1$ since $-\omega(a_0, a_1) \subset \sigma^*$ $(d_{101} = -1)$

(b) $t \neq a_2$ since $-\omega(a_2, a_0) \subset \sigma^*$ $(d_{200} = -1)$

(c) $t \neq a_3$ since $\omega(a_3, a_2) \subset \sigma^*$ $(d_{320} = 1)$

(d) $t \neq a_j$ for $j = 4, \ldots, i-1$, since $\omega(a_j, a_{j-1}) \subset \sigma^*$ but $-\omega(t, a_{j-1}) \subset \sigma(a_0, \ldots, a_{i-1}, t)$

(e) $t \neq a_j$ for $j = i+1, \ldots, 2q-1$, since $-\omega(a_{i-1}, a_j) \subset \sigma^*$ but $\omega(a_{i-1}, t) \subset \sigma(a_0, \ldots, a_{i-1}, t)$

(f) $t \neq \underset{\sim}{n}_i$ for $i = 0$ or $i = 2, \ldots, 2q-1$, since $-\omega(\underset{\sim}{n}_i, a_0) \subset \mu$

(g) $t \neq \underset{\sim}{n}_1$, since $\omega(a_2, \underset{\sim}{n}_1) \subset \tau_n^*$

(h) $t \notin D_4^* - \{a_0, \ldots, a_{2q-1}\} - \mathbb{N}$ by minimality.

This completes the proof of Condition (σ). ∎

Finally we consider the Extraction Condition. By Envelope Condition (τ) and Conditions (θ_0) and (θ_1),

$$\tau(a_{q-i}, \ldots, a_{2q-i-1}, \underset{\sim}{n}_i) \subset \tau_{n+1}^* \subset \theta_0 \subset \theta_1 \subset \mu ,$$

so it remains only to show that if $\tau(a_{q-i}, \ldots, a_{2q-i-1}, t) \subset \mu$ then $t = \underset{\sim}{n}_i$ for some **n**. Now

$$\tau(a_{q-i}, \ldots, a_{2q-i-1}, t) = -\omega(t, a_{q-i}) - \omega(a_{q-i}, t) + \omega(t, a_{q-i-1}) +\omega(a_{q-i+1}, t)$$

$$-\omega(t, a_{q-i+2}) - \omega(a_{q-i+2}, t) - \omega(t, a_{q-i+3}) +\omega(a_{q-i+3}, t) .$$

(a) $t \neq a_j$ for any $j > q-i+2$ since there are no k, ℓ, $k+2 < \ell$ such that $-\omega(a_\ell, a_k) - \omega(a_k, a_\ell) +\omega(a_\ell, a_{k+1}) +\omega(a_{k+1}, a_\ell) - \omega(a_\ell, a_{k+2}) - \omega(a_{k+2}, a_\ell) \subset \mu$

(b) $t \neq a_{q-i+2}$ since $\omega(a_{q-i+1}, a_{q-i+2}) +\omega(a_{q-i+2}, a_{q-i+1}) \subset \mu$ only if $q-i+2 \geq 4$; but then $\omega(a_{q-i+2}, a_{q-i+3}) \subset \mu$

(c) $t \neq a_{q-i+1}$ since $-\omega(a_{q-i+1}, a_{q-i+1}) \subset \mu$ by minimality

(d) $t \neq a_j$ for any $j \leq q-i$ since $\omega(a_j, a_{q-i+1}) +\omega(a_{q-i+1}, a_j) \subset \mu$ only if either $j = 0$, $q-i+1 \geq 3$ or $j = q-i$, $q-i+1 \geq 4$; in either case $-\omega(a_j, a_{q-i+3}) \subset \mu$

(e) $t \neq \underset{\sim}{n}_j$ for any $j \neq i$, $0 \leq j \leq q-1$, since $\omega(\underset{\sim}{n}_j, a_{q-i+1}) +\omega(a_{q-i+1}, \underset{\sim}{n}_j) \subset \mu$ only if $\omega(\underset{\sim}{n}_0, a_{q+j-i+1}) +\omega(a_{q+j-i+1}, \underset{\sim}{n}_0) \subset \mu$ only if $q+j-i+1 \equiv 1$ (mod q), i.e., only if $i = j$

(f) $t \notin D_4^* - \{a_0, \ldots, a_{2q-1}\} - \mathbb{N}$, by minimality.

This completes the proof of the Construction Lemma and the proof of the Unbounded-Existentials Theorem (II). ∎

Historical References. The study of prefix and similarity classes dates from the earliest investigations of the first-order predicate calculus. The solvability of the monadic class, i.e., the (∞) similarity class, goes back to Löwenheim (1915); simpler proofs were given by Skolem (1919) and Behmann (1922). The solvability of the $\exists^*\forall^*$ prefix class was first shown by Bernays and Schönfinkel (1928); that of the $\exists^*\forall\exists^*$ class by Ackermann (1928), Skolem (1928), and Herbrand (1931); and that of the $\exists^*\forall\forall\exists^*$ class by Gödel (1932, 1933), Kalmár (1933), and Schütte (1934). Proofs of the solvability of these classes may also be found in Ackermann (1954), Goldfarb (1974), and Dreben and Goldfarb (1979); proofs of some are also in Church (1956).

On the negative side, the earliest results are not of unsolvability as such, since they antedate any formal notion of recursive unsolvability, but are rather proofs that certain classes are reduction classes, i.e., classes to whose decision problem that for the full predicate calculus can be effectively reduced. Thus Löwenheim (1915) showed that the $(0, \infty)$ similarity class is a reduction class; Herbrand (1931) improved Löwenheim's proof, and sharpened it by showing that the $(0, 3)$ and $(0, 0, 1)$ similarity classes are reduction classes; and finally, Kalmár (1936) (announced in 1932) achieved the reduction to a single dyadic predicate letter.

The earliest reduction result for the form of the quantifier prefix is the reduction to $\forall^*\exists^*$ by Skolem (1920). Gödel (1933) sharpened Skolem's result to the $\forall\forall\forall\exists^*$ case. Cases in which the number of existential quantifiers is bounded were considered by Ackermann (1928), who showed the $\forall\exists \wedge \exists\forall^*$ extended prefix class, and hence the $\forall\exists\exists\forall^*$, $\exists\forall\exists\forall^*$, and $\exists\forall^*\exists$ prefix classes, to be reduction classes; and by Pepis (1938), who showed the $\forall^* \wedge \forall\forall\exists$ extended prefix class, and hence the $\forall^*\exists$ and $\forall\forall\exists\forall^*$ prefix classes, to be reduction classes. Surányi (1950) bounded the total number of quantifiers of either kind, showing the $\forall\forall\exists \wedge \forall\forall\forall$ extended prefix class, and hence the $\forall\forall\forall\exists$ and $\forall\forall\exists\forall$ prefix classes, to be unsolvable. Cases in which no existential quantifier is governed by more than one universal were then the only ones to be settled; at this point, the best result in this direction were Ackermann's (1928) $\forall\exists\exists\forall^*$ and $\exists\forall\exists\forall^*$ classes. Surányi (1951), still using only reduction methods, limited the number of universal quantifiers in Ackermann's classes to three, thus obtaining the $\forall\exists\exists\forall\forall$ and $\exists\forall\exists\forall\forall$ classes. In the early 1960's Büchi (1962) and Wang (1961) for the first time brought the syntactic characterization of satisfiability provided by the expansion theorem directly together with the rapidly developing theory of computability by mathematical machines; this combination yielded much sharper results than the reduction methods. Büchi (1962) showed that the $\exists \wedge \forall\exists\forall$ extended prefix class, and hence the $\exists\forall\exists\forall$, $\forall\exists\exists\forall$, and $\forall\exists\forall\exists$ prefix classes, are unsolvable; he also observed that a similar construction established the unsolvability of the $\forall\exists\forall\forall$ class. The last gap was filled by Kahr, Moore, and Wang (1962), who showed the unsolvability of the $\forall\exists\forall$ class (and hence the $\forall\exists \wedge \forall\forall\forall$ class).

Bounds on the number of predicate letters for schemata of a fixed prefix class were obtained by Kalmár, who showed (1932) the $\exists^*\forall\forall\exists\forall^*(0, 0, 1)$ and (1936) the $\exists^*\forall\forall\exists\forall^*(0, 1)$ classes to be reduction classes. Reduction classes due to Pepis (1938) include $\forall\forall\exists \wedge \forall^*(1, 0, 1)$ (and hence $\forall^*\exists(1, 0, 1)$ and $\forall\forall\exists\forall^*(1, 0, 1)$); $\forall\forall\exists \wedge \forall^*(1, 2)$ (and hence $\forall^*\exists(1, 2)$ and $\forall\forall\exists\forall^*(1, 2)$); $\exists\forall\exists\forall^*(1, 1, 1)$ and $\exists\forall\exists\forall^*(1, 3)$; and $\forall\forall\forall\exists^*(0, 0, 1)$. Prefix classes of schemata with a single dyadic predicate letter were shown to be reduction classes in the following sequence (sharpest results are underlined):

$\exists^*\forall\forall\exists\exists\forall^*(0, 1)$	(Kalmár 1932)
$\exists^*\forall\forall\exists\forall^*(0, 1)$	(Kalmár 1936)
$\exists\forall\exists\forall^*(0, 1)$	(Kalmár 1939)
$\underline{\forall\forall\forall\exists^*(0, 1)}$	(Kalmár and Surányi 1947)
$\forall\forall\exists^*\forall(0, 1)$ and $\forall\exists^*\forall\forall(0, 1)$	(Kalmár 1950, 1951)
$\forall\forall\exists\forall^*(0, 1)$ and $\underline{\forall^*\exists(0, 1)}$	(Kalmár and Surányi 1950)
$\underline{\exists^*\forall\forall\forall\exists(0, 1)}$	(Surányi 1959)
$\underline{\exists^*\forall\exists\forall(0, 1)}$	(Surányi 1959)
$\underline{\forall\exists\forall^*(0, 1)}$	(Denton 1963)
$\underline{\forall\exists^*\forall(0, 1)}$	(Kostryko 1964)
$\underline{\forall\exists\forall\exists^*(0, 1)}$	(Gurevich 1966a, 1966b)

The idea of fixing the number of dyadic predicate letters and the length of the prefix, leaving the number of monadic predicate letters as the only parameter, is presented by Surányi (1951) whose $\exists\forall\exists\forall\forall$ and $\forall\exists\exists\forall\forall$ reduction results apply to the $(\infty, 7)$ similarity class. Surányi (1959) also showed that the $\forall\forall\exists \wedge \forall\forall\forall\ (\infty, 1)$ class is a reduction class; this class becomes $\forall\forall\exists\forall(\infty, 1)$ or $\forall\forall\forall\exists\ (\infty, 1)$ by prenexing. Kahr (1962) refined the Kahr, Moore, and Wang proof for the $\forall\exists\forall$ class to the $\forall\exists\forall(\infty, 1)$ subclass.

The possibility of a complete classification by prefix and similarity type is discussed by Kostryko (1964) and is carried out in Gurevich (1966b), from which the underlying ideas for several of our constructions were derived. General properties of similar classifications are discussed by Gurevich (1969).

The prefix-similarity type classification has been extended in various directions. The first is by consideration of the extended prefix classes. If extended prefix alone is considered, the optimal results are the unsolvability of the $\forall\exists\forall$ and $\forall\exists \wedge \forall\forall\forall$ classes. If similarity type is also considered, there are, in addition to the $(\infty, 1)$ similarity subclasses of these two classes, the following possible reduction classes: $\forall\exists \wedge \forall^*(0,1)$, $\forall\forall\forall\exists \wedge \exists^*(0,1)$, $\forall\exists^* \wedge \forall\forall\forall(0,1)$, $\forall\exists \wedge \exists^*\forall\forall\forall(0,1)$, $\exists^*\forall\exists \wedge \forall\forall\forall(0,1)$, and $\forall\exists\forall \wedge \exists^*(0,1)$. Except for the first of these, which was shown unsolvable by Denton (1963), none of these is known to be a reduction class. But Gurevich (1966b, 1966c, 1970), who proposed this classification, has shown that for some fixed k_0, the $(k_0, 1)$ similarity subclass of each of the above extended prefix classes is unsolvable. His technique is to encode the computations of a universal Turing or two-counter machine on various input words, and the size of k_0 is tied to the number of states of the machine. Reducing k_0 to 0 or some other small number appears to be extremely difficult.

Gurevich (1969, 1976) also extends the classification by introducing function signs into the logical language. Another possibility is to sharpen these results by proving that all the reduction classes are conservative. A reduction class is conservative if there is an effective mapping that associates with each first-order formula F a formula G in the class, such that F is satisfiable if and only if G is satisfiable, and F has a finite model if and only if G has a finite model. All the reduction classes of this chapter are conservative.

In another direction, one may place special restrictions on the kinds of models a formula may have. Thus one may ask of a subclass of the (0, 1) similarity class, whether there is a reduction that associates with every first-order formula F a formula G in the class, such that F is satisfiable if and only if G is satisfiable, and G is satisfiable only if there is a model for G in which the interpretation of the unique dyadic predicate letter is a symmetric, or reflexive, or irreflexive relation. (Such considerations appear, in a technical way, in our proofs in the form of the $\mathscr{L}_\delta(\alpha)$ classes.) These questions have been considered in several places; see, for example, Surányi (1959), Gurevich (1965), Surányi (1971), and Gurevich and Turashvili (1973).

Finally, one may consider the effect of expanding the language to include the identity sign. In this case, the $\forall^*\exists^*$ and $\exists^*\forall\exists^*$ prefix classes are still decidable, as is the (∞) similarity classes. But the solvability of the $\exists^*\forall\forall\exists^*$ prefix class with identity remains an open problem. For a complete discussion, see Dreben and Goldfarb (1979).

Chapter IID

RESTRICTIONS ON THE STRUCTURE OF ATOMIC SUBFORMULAS

Two atomic subformulas of a schema F are said to be coinstantiable if they have identical Herbrand instances, i. e., if the instance of one in some Herbrand instance F_1 of F is identical to the instance of the other in some Herbrand instance F_2 of F. (This is the notion that is called unifiability by Robinson (1965) and in the subsequent computer science literature.) The prefix classification is of intrinsic interest because the variety of possible patterns of coinstantiable atomic subformulas of a schema is related directly to the quantificational structure of the schema: in schemata with few variables, or few alternations of quantifiers, the simplicity of the possible coinstantiations can sometimes be exploited to yield decision procedures. Even when the structure of the prefix permits coinstantiations complex enough to yield unsolvability, decidable subclasses may be identified by restricting the way in which variables may appear together in atomic subformulas. To take a simple example, if no variable may occur with any other (i. e., each atomic subformula has the form Pv. . . v), then the schema may be reduced to a monadic schema whose satisfiability may be decided by an effective procedure. To take a slightly more complex example, if every y-variable appears only with an x-variable that it governs, then again a solvable class is obtained.

In this section we consider the simplest unsolvable prefix classes, $\forall\exists\forall$ and $\forall^*\exists^*$, and attempt to discover the minimal sets of atomic subformulas necessary for unsolvability. Here we group together as atomic subformulas of the same form any two atomic formulas that may be derived from each other simply by changing the predicate letter. Thus we are not concerned with the number of predicate letters of a given degree, only with the sequences of variables such predicate letters may have as arguments. Whether any of these unsolvability results can be strengthened to refer to the $(\infty, 1)$ or $(0, 1)$ similarity class is a question that has not been studied.

This type of investigation has led in two directions. For the $\forall\exists\forall$ case a complete characterization is easy to obtain, thanks to the proof of the unsolvability of the linear sampling problem. Thus the success of the analysis for this case owes more to the discovery of a closely matched type of combinatorial system than to any new insight about logical syntax. For the $\forall^*\exists^*$ case, on the other hand, we are able to identify certain general properties of the patterns in which variables may appear that suffice for, and in some cases are necessary for, unsolvability. However, we also point out below that there is much about this classification we do not understand as yet.

IID.1 <u>The ∀∃∀ Subclasses</u>

If attention is restricted just to monadic and dyadic predicate letters, the atomic subformulas of a prenex schema with prefix $\forall y_1 \exists x \forall y_2$ can assume twelve possible forms:

Monadic	Dyadic
y_1	$y_1 y_1$
y_2	$y_1 y_2$
x	$y_1 x$
	$y_2 y_1$
	$y_2 y_2$
	$y_2 x$
	$x\, y_1$
	$x\, y_2$
	$x\, x$

By considering these in combinations we derive $2^{12}-1$ nonempty subsets, to each of which corresponds the class of schemata whose atomic subformulas are of only the specified forms. The decision problem for these classes of schemata is completely settled by the following:

∀∃∀ <u>Subclass Theorem</u>. For schemata with prefix $\forall y_1 \exists x \forall y_2$, any three forms including the pair $y_1 y_2$, $y_2 x$ or the pair $y_2 y_1$, $x y_2$ yield an unsolvable class; any combination not including such a triple yields a solvable class.

<u>Proof</u>. The solvability part may be found in Dreben, Kahr, and Wang (1962) or Dreben and Goldfarb (1979); we present just the unsolvability proof.

We first note that it suffices to prove the result just for the pair $y_1 y_2$, $y_2 x$, since the result for the other pair will follow by symmetry. Also, the forms $y_1 y_1$, $y_2 y_2$, xx need not be considered separately, since the case in which y_1, y_2, or x is present may be reduced to that with $y_1 y_1$, $y_2 y_2$, or xx by introducing a new dyadic letter P' for each monadic letter P and replacing Pv by $P'vv$ throughout, for whichever is the appropriate variable v. Thus we are left with proving that the pair $y_1 y_2$, $y_2 x$ in conjunction with y_1, y_2, x, $y_1 x$, $y_2 y_1$, $x y_1$, or $x y_2$ yields an unsolvable class.

<u>y_1</u>: By the Bounded Prefix Theorem the forms $y_1 y_2$, $y_2 x$, y_1, y_2, x suffice for unsolvability. Let F be any schema of this type. For each monadic predicate letter P, introduce two dyadic predicate letters P_1, P_2. Conjoin to the matrix the clauses

$$(Py_1 \leftrightarrow P_1 y_1 y_2) \wedge (P_1 y_2 x \leftrightarrow P_2 y_1 y_2) \;,$$

which have the effect of forcing P_1 to depend only on its first argument and P_2 only on its second. Then replace throughout the matrix Py_2 by P_1y_2x, and Px by P_2y_2x. The modified schema G is of the desired type; the proof that F is satisfiable if and only if G is satisfiable is straightforward.

(The "new dyadic letter" in each of the following reductions should be considered as distinct in each step.)

$\underline{y_2}$: Starting from a schema containing only the forms y_1y_2, y_2x, and y_1, introduce a dyadic predicate letter P_1 for each monadic letter P, conjoin $(Py_2 \leftrightarrow P_1y_2x)$ to the matrix, and replace Py_1 by $P_1y_1y_2$ throughout.

$\underline{x}$: Starting from a schema containing only the forms y_1y_2, y_2x, and y_2, introduce a dyadic letter P_1 for each monadic letter P, conjoin $(Px \leftrightarrow P_1y_2x)$ to the matrix, and replace Py_2 by $P_1y_1y_2$ throughout.

$\underline{y_1x}$: Starting from a schema containing only the forms y_1y_2, y_2x, and x, introduce a new dyadic letter P_1 for each monadic letter P, and replace Px by P_1y_1x throughout.

$\underline{xy_1}$: Same as the previous case, except replacing Px by P_1xy_1 throughout.

$\underline{y_2y_1}$: Starting from a schema containing only the forms y_1y_2, y_2x, and y_2, introduce a new dyadic letter P_1 for each monadic letter P, conjoin $(P_1y_2y_1 \leftrightarrow P_1y_2x)$ to the matrix, and replace Py_2 by $P_1y_2y_1$ throughout. Once again, the truth-value of P_1 in any verifying truth-assignment depends only on the first argument.

$\underline{xy_2}$: Starting from a schema containing only the forms y_1y_2, y_2x, and y_2, introduce a new dyadic letter P_1 for each monadic letter P, conjoin $(P_1y_1y_2 \leftrightarrow P_1xy_2)$ to the matrix, and replace Py_2 by $P_1y_1y_2$ throughout.

This exhausts the cases and completes the proof of the $\forall\exists\forall$ Subclass Theorem. ∎

IID.2 Skolem Subclasses: Minimal Bases

In this section we undertake an analysis, similar to that carried out above for the $\forall\exists\forall$ case, for the Skolem prefix $\forall^*\exists^*$. This study is prompted by the ease with which cases of the problem may be described (there being only a single quantifier alternation in the prefix, the variables are of one of two kinds, and the criteria for coinstantiability are simple), and by the broad classes of schemata in Skolem form that have been shown to be solvable. Although these positive results, combined with the negative results presented in this section, provide a first step towards a complete analysis, we fall well short of our goal of exhaustive classification. First, the classes considered here do not lend themselves to fine distinctions according to the atomic forms that may and may not occur; that is, the cases that arise in that way do not seem to reflect a natural combinatorial substructure. Instead we consider classes determined by the sets of variables that can occur together (rather than the sequences of variables). Taking this broader perspective, we are able to delineate some general conditions sufficient for unsolvability, which combine with the solvability results to settle all cases involving only a small number of variables. Still, there remain many unsettled cases, the simplest of which are discussed at the end of this section.

We begin with some definitions. A basis set is a finite set of finite sets of variables among the particular variables $y_1, y_2, \ldots, x_1, x_2, \ldots$. If F is any schema, then the basis of F is the set of all sets $\{v_1, \ldots, v_k\}$ such that $Pv_1 \ldots v_k$ is an atomic subformula of F for some predicate letter P. Thus the basis of a schema is a basis set, provided that the variables are among $y_1, y_2, \ldots,$ $x_1, x_2, \ldots$. A member of a basis set is monadic, dyadic, or triadic if it contains exactly one, two, or three variables, respectively, and polyadic if it contains two or more variables. A basis set is dyadic if it contains only dyadic sets.

Our analysis proceeds by considering, for various basis sets S, the set B(S) of all schemata F with prefix $\forall y_1 \ldots \forall y_m \exists x_1 \ldots \exists x_n$ whose basis is S, where $\{y_1, \ldots, y_m, x_1, \ldots, x_n\}$ is the set of variables of S. Two crucial notions in the analyses of such basis sets are those of tying and cleaving. Call a set of variables a y-set if it is a subset of $\{y_1, y_2, \ldots\}$, i.e., a set of y-variables of a schema with the given prefix. Then two y-sets V, W in S are S-tied if and only if either (a) V = W or recursively (b) there are sets U_1, $U_2 \in S$ with the same sets of x-variables such that V is the set of y-variables in U_1, and the set of y-variables in U_2 is S-tied to W. For example, $\{y_1\}$ and $\{y_2, y_3\}$ are S-tied in $S = \{\{y_1\}, \{y_1, x_1, x_2\}, \{y_4, y_5, x_1, x_2\}, \{y_4, y_5, x_3\}, \{y_2, y_3, x_3\}, \{y_2, y_3\}\}$, because of the "alternating chain" obtained by taking the sets in the illustrated order: the first and second sets have the same y-variables, the second and third the same x-variables, the third and fourth the same y-variables, etc. Intuitively, if V and W are S-tied then there is a schema with basis S that implies that all substituents for members of $V \cup W$ are identical in any Herbrand instance of F; we shall see examples of such schemata later (p. 151).

A variable v cleaves a basis set S into sets S_1 and S_2 if and only if (S_1, S_2) is a partition of S with v the only variable in common between the parts; that is, if and only if $S_1 \cup S_2 = S$, $S_1 \cap S_2 = \emptyset$, and $(\cup S_1) \cap (\cup S_2) = \{v\}$. A schema whose basis is cleaved by a y-variable has a "structural weakness" which makes it possible to reduce it to a schema whose matrix is a conjunction of two parts that share no variable at all. In certain cases this reduction simplifies the problem of determining satisfiability.

Dreben and Goldfarb (1979) exploit such structural weakness in various forms in order to yield solvability. For example, a monadic set can always be added to a basis set their methods apply to without impairing solvability. And if S contains only a single polyadic y-set, then B(S) is a solvable class. Thus two or more polyadic y-sets must be in the basis in order to obtain unsolvability. The following general result, proved by Dreben and Goldfarb (1979), indicates how far it is possible to go in accommodating such polyadic y-sets in the basis set and still preserve solvability. Say that two y-variables are S-tied if there is a pair of S-tied y-sets, each containing one of the variables. Then B(S) is solvable for basis sets S with the following property: For all distinct polyadic y-sets Y_1 and Y_2 in S, either

(1) no variable in Y_1 is S-tied with any variable in Y_2, or

(2) some y-variable cleaves S into sets S_1 and S_2, such that $Y_1 \in S_1$ and $Y_2 \in S_2$.

In contradistinction to the above, we next present certain minimal basis sets that yield unsolvable classes of Skolem schemata. We then give some rules by which these basis sets can be expanded to yield other unsolvable classes. Instead of simply listing the basis sets, we use a suggestive graphical representation. A small circular dot represents a variable, with solid black dots for x-variables and unfilled dots for y-variables; a line between two dots represents a dyadic member of the basis set; and a closed curve surrounding several dots represents a polyadic member of the basis set. (Just as monadic sets do not impact the solvability results, they are not needed for the unsolvability results.) This graphical representation has several advantages: basis sets that differ only in the names of their variables are pictured in the same way, and any diagram formed from one representing an unsolvable class by adding more dots or lines also represents an unsolvable class.

Minimal Skolem Basis Theorem. Each of the diagrams illustrated in Figures 22a, 22b, 23a, 23b, 23c, and 24 represents a basis set S whose associated class B(S) of schemata in Skolem form is unsolvable.

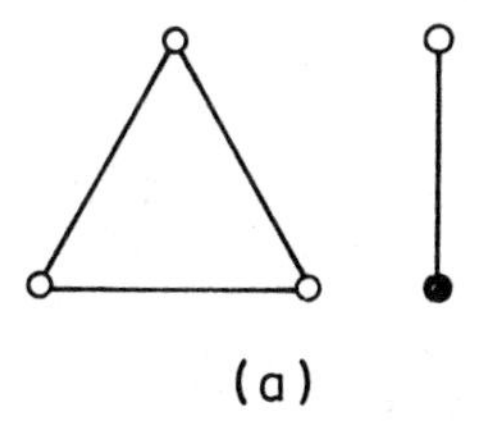

(a)

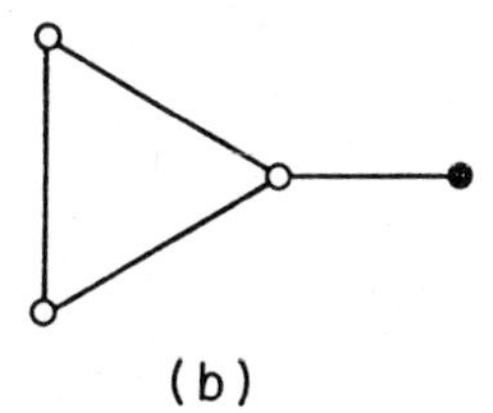

(b)

Figure 22

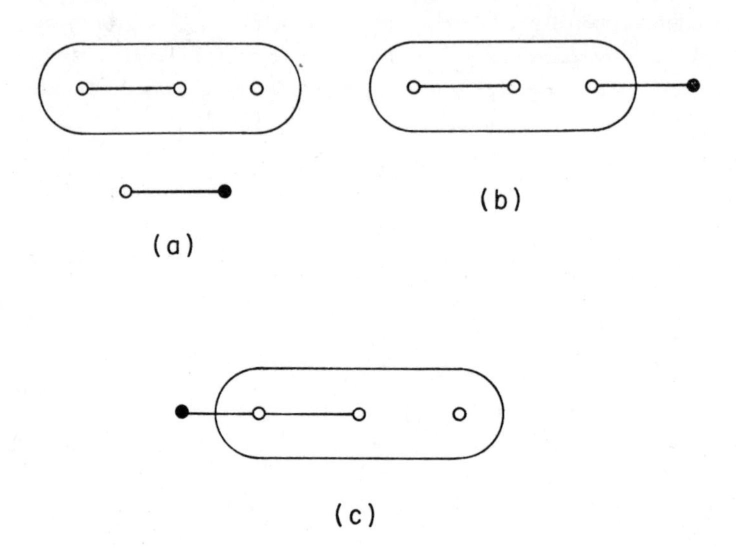

Figure 23

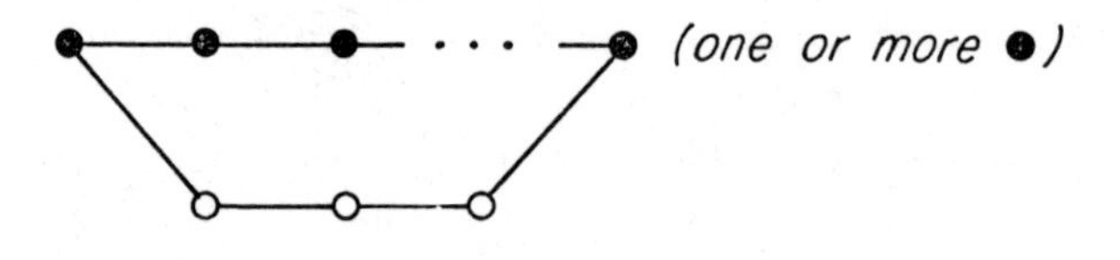

Figure 24

(For example, the class corresponding to Figure 22a includes all schemata $\forall y_1 \forall y_2 \forall y_3 \forall y_4 \exists x_1\ F^M$ with basis $\{\{y_1, y_2\}, \{y_2, y_3\}, \{y_1, y_3\}, \{y_4, x\}\}$; that of Figure 23b, schemata with prefix $\forall y_1 \forall y_2 \forall y_3 \exists x_1$ and basis $\{\{y_1, y_2\}, \{y_1, y_2, y_3\}, \{y_3, x_1\}\}$. Note that, by the result cited on p. 143, any basis set obtained by removing a set of variables from one of the illustrated basis sets corresponds to a solvable class of schemata.)

Proof.

Figure 22: The unsolvability of this class follows from that of the $\forall\forall\forall \wedge \forall\exists$ extended prefix class (Bounded Prefix Theorem, p. 97). The only problem is to eliminate monadic atomic formulas in favor of dyadic atomic formulas with the specified sets of arguments. But this is easily done; given a schema

$F = \forall y_1 \forall y_2 \forall y_3 M_1 \wedge \forall y_4 \exists x_1 M_2$, where M_1 and M_2 are quantifier-free, associate with each monadic letter P of F a dyadic letter P'; replace in the matrix $M_1 \wedge M_2$ each atomic formula Pv by an atomic formula $P'vw$, where w is some variable such that $\{v, w\}$ is in the basis set, thus obtaining a new matrix $M_1' \wedge M_2'$; and let G be

$$\forall y_1 \forall y_2 \forall y_3 \forall y_4 \exists x_1 (M_1' \wedge \bigwedge (P' y_1 y_2 \leftrightarrow P' y_1 y_3) \wedge M_2'),$$

the conjunction over all the newly introduced predicate letters P'. The middle conjunct guarantees that in any truth-assignment verifying $E(G)$ the truth-value of atomic formulas with predicate letter P' depends only on the first argument. The basis of G is as illustrated in Figure 22a; to obtain a schema whose basis is as shown in Figure 22b, let H be $\forall y_1 \forall y_2 \forall y_3 \exists x_1 (G^M[y_4/y_3])$. Then each of G, H is satisfiable if and only if F is satisfiable, so the unsolvability follows.

Figure 23: The unsolvability of this class follows from the previous construction. This time we need to show how the three triangularly linked dyadic y-sets can be replaced by a dyadic y-set and a triadic y-set of which the dyadic y-set is a subset. Again, the trick is to introduce predicate letters with arguments that "don't matter." Let G be the schema obtained just above from a schema F in $\forall\forall\forall \wedge \forall\exists$; thus the matrix of G may be written $M \wedge M'$, where M contains no variable but y_1, y_2, y_3 and M' no variable but y_4, x_1. The basis of G is as illustrated in Figure 22a, with y_1, y_2, y_3 the three y-variables forming the triangle. Introduce a new triadic letter P' for each dyadic letter P, and conjoin to the matrix of G the conjunction $\bigwedge (Py_1y_2 \leftrightarrow P'y_1y_2y_3)$ which forces the truth-values of atomic formulas with the new predicate letters P' to depend only on their first two arguments. Then each atomic subformula Py_iy_j in the matrix of G may be replaced by $P'y_iy_jy_k$, where y_k is whichever of y_1, y_2, y_3 is neither y_i nor y_j. The basis of the resulting schema is as shown in Figure 23a, and the other forms (Figures 23b and 23c) may be obtained by substitutions in the matrix.

Figure 24: It is convenient to break this case into two subcases, depending on the number n of x-variables. The construction for the case $n > 1$ is essentially a more complex version of that for $n = 1$, but the $n = 1$ case presents minor complications of its own not present in the $n > 1$ case, due to the coinstantiability of atomic formulas of the forms y_1x_1 and y_3x_1 (if $n > 1$ then y_1x_1 and y_3x_n are *not* coinstantiable).

We make free use in our construction of monadic predicate letters, even though according to our convention monadic argument sets are disallowed. Any monadic letter P can, however, be supplanted by a dyadic letter P' by replacing

Pv by P'vw throughout, where $\{v, w\}$ is in the basis set, and conjoining the clause $(P'y_2y_1 \leftrightarrow P'y_2y_3)$ to the matrix.

Our unsolvability proof is by reduction of the origin-constrained tiling problem (p. 7). Let $\mathscr{D} = (D, D_0, \overline{H}, \overline{V})$ be a Δ_0-system; we construct a schema F with the illustrated basis such that F is satisfiable if and only if $\mathscr{D}$ accepts some tiling. The universe of the intended model $\mathfrak{A}$ for F (in case $\mathscr{D}$ accepts some tiling) is $\mathbb{N} \times \mathbb{N}$. F has a monadic predicate letter Q_d for each tile $d \in D$, and also dyadic predicate letters H and V, which are to represent the relations of a point on the plane to the one to its right or above it. Thus if $\mathscr{D}$ accepts a tiling τ then the following interpretations $H^{\mathfrak{A}}$, $V^{\mathfrak{A}}$, $Q_d^{\mathfrak{A}}$ are intended for the letters H, V, and Q_d:

$$H^{\mathfrak{A}}(p, q) \quad \text{if and only if} \quad q = p + \langle 1, 0 \rangle$$

$$V^{\mathfrak{A}}(p, q) \quad \text{if and only if} \quad q = p + \langle 0, 1 \rangle$$

$$Q_d^{\mathfrak{A}}(p) \quad \text{if and only if} \quad \tau(p) = d \quad (d \in D) \ .$$

<u>n = 1 Case</u>. F also has an additional monadic letter E, whose intended interpretation is

$$E^{\mathfrak{A}}(\langle p_1, p_2 \rangle) \qquad \text{if and only if} \qquad p_1 \ \text{is even.}$$

F is the schema $\forall y_1 \forall y_2 \forall y_3 \exists x_1 F^M$, where the matrix F^M is the conjunction of the following subformulas:

(1) $$\big((Ey_1 \leftrightarrow Ey_2) \wedge (Ey_2 \leftrightarrow Ey_3)\big) \rightarrow \big((Ex_1 \leftrightarrow \neg Ey_1) \wedge Hy_1x_1\big)$$

(2) $$(\neg Ey_1 \wedge Ey_2 \wedge \neg Ey_3) \rightarrow \Big(Ex \wedge \Big(\bigvee_{d \in D_0} Q_d x_1\Big)\Big)$$

(3) $$(Ey_1 \wedge \neg Ey_2 \wedge Ey_3) \rightarrow (Ex_1 \wedge Vy_1x_1)$$

(4) $$(Vy_2y_1 \wedge Hy_2y_3) \rightarrow (Hy_1x_1 \wedge Vy_3x_1 \wedge (Ex \leftrightarrow Ey_3))$$

(5) $$Hy_1y_2 \rightarrow \bigvee_{\langle d, d' \rangle \in \overline{H}} (Q_d y_1 \wedge Q_{d'} y_2)$$

(6) $$Vy_1y_2 \rightarrow \bigvee_{\langle d, d' \rangle \in \overline{V}} (Q_d y_1 \wedge Q_{d'} y_2)$$

(7) $\mathop{\triangledown}_{d \in D} Q_d y_1$, where $\triangledown$ denotes "exclusive or".

Clearly the basis of F is as illustrated in Figure 24, except for monadic sets, which may be eliminated as described above.

First suppose that τ is a tiling of $\mathbb{N} \times \mathbb{N}$ is accepted by $\mathscr{D}$; then we claim that F is true in the model $\mathfrak{A}$ over $\mathbb{N} \times \mathbb{N}$ specified above. Clearly (5), (6) and

(7), which involve only universal quantification, hold in $\mathfrak{A}$. Now for any values of y_1, y_2, y_3, the antecedent of at most one of (1)-(4) holds in $\mathfrak{A}$; for letting + denote a pair $\langle i,j\rangle$ such that i is even and - a pair such that i is odd, the antecedent of (1) requires that $\langle y_1, y_2, y_3\rangle$ be $\langle +, +, +\rangle$ or $\langle -, -, -\rangle$; (2) requires $\langle -, +, -\rangle$; (3) requires $\langle +, -, +\rangle$; and (4) requires $\langle +, +, -\rangle$ or $\langle -, -, +\rangle$. Thus we need only show that if an antecedent is verified, there is some choice for x_1 that makes the consequent of that conjunct true. Here is how to choose: if the antecedent of (1) holds, let $x_1 = \langle i+1, j\rangle$, where $y_1 = \langle i,j\rangle$; if that of (2), let $x_1 = \langle 0,0\rangle$; if that of (3), let $x_1 = \langle i, j+1\rangle$, where $y_1 = \langle i,j\rangle$; and if that of (4), let $x_1 = \langle i+1, j+1\rangle$, where $y_2 = \langle i,j\rangle$ (and hence $y_1 = \langle i, j+1\rangle$ and $y_3 = \langle i+1, j\rangle$).

Next suppose that F is satisfiable, let $\mathscr{A}$ be a truth-assignment such that $\mathscr{A} \models E(F)$, and let f be the triadic indicial correlate of x_1. By (1) there are s_0, $s_1 \in D(F)$ such that $\mathscr{A} \models Es_0 \wedge \neg Es_1$, namely either $s_0 = \text{ɨ}$ and $s_1 = f(\text{ɨ ɨ ɨ})$ or vice versa. Define a mapping $\gamma : \mathbb{N} \times \mathbb{N} \to D(F)$ inductively as follows:

$$\gamma(0,0) = f(s_1\ s_0\ s_1)$$

$$\gamma(0,j+1) = f(\gamma(0,j)\ s_1\ s_0)$$

$$\gamma(i+1,0) = f(\gamma(i,0)\ \gamma(i,0)\ \gamma(i,0))$$

$$\gamma(i+1,j+1) = f(\gamma(i,j+1)\ \gamma(i,j)\ \gamma(i+1,j))$$

The mapping γ provides a term in D(F) corresponding to each lattice point $\langle i,j\rangle$. From (1-4) it follows that $\mathscr{A} \models E\gamma(i,j)$ if and only if i is even, $\mathscr{A} \models H\gamma(i,j)\gamma(i+1,j)$, and $\mathscr{A} \models V\gamma(i,j)\gamma(i,j+1)$ for all $i, j \in \mathbb{N}$.

Now define a tiling $\tau : \mathbb{N} \times \mathbb{N} \to D$ by

$$\tau(i,j) = \text{that } d \in D \text{ such that } \mathscr{A} \models Q_d\gamma(i,j)\,.$$

Then clauses (2), (5), (6) and (7) guarantee that τ is well-defined and is a solution of $\mathscr{D}$.

<u>$n > 1$ Case</u>. In this case the monadic letter E is not needed, but a dyadic letter I is used instead, whose intended interpretation is

$$I^{\mathfrak{A}}(p,q) \quad \text{if and only if} \quad p = q.$$

The schema F is $\forall y_1 \forall y_2 \forall y_3 \exists x_1 \ldots \exists x_n F^M$, where F^M is the conjunction of the following clauses. Let C be the formula

$$Ix_1x_2 \wedge \ldots \wedge Ix_{n-1}x_n\ .$$

(1) $\neg Hy_1y_1 \wedge Iy_1y_1 \wedge \left(Iy_1y_2 \rightarrow \bigwedge_{d\in D} (Q_dy_1 \leftrightarrow Q_dy_2)\right)$

(2) $(\neg Hy_2y_1 \wedge Hy_2y_3) \rightarrow \left(\bigvee_{d\in D_0} Q_dx_1\right)$

(3) $(\neg Hy_2y_1 \wedge \neg Hy_2y_3) \rightarrow (Hy_1x_1 \wedge Vy_3x_n)$

(4) $(Hy_2y_1 \wedge Vy_2y_3) \rightarrow (Vy_1x_1 \wedge C \wedge Hy_3x_n)$

(5) $(Hy_2y_1 \wedge Iy_2y_3) \rightarrow (Iy_1x_1 \wedge C \wedge Hy_3x_n)$

(6) $Hy_1y_2 \rightarrow \bigvee_{\langle d,d'\rangle\in\overline{H}} (Q_dy_1 \wedge Q_{d'}y_2)$

(7) $Vy_1y_2 \rightarrow \bigvee_{\langle d,d'\rangle\in\overline{V}} (Q_dy_1 \wedge Q_{d'}y_2)$

(8) $\mathop{\triangledown}_{d\in D} Q_dy_1$, where $\triangledown$ denotes "exclusive or".

If $\mathscr{D}$ accepts a tiling τ, then the intended interpretation $\mathfrak{A}$ is a model for F. Clearly (1), (6), (7), and (8) hold in $\mathfrak{A}$. And for any values for y_1, y_2, y_3, the antecedent of at most one of (2)-(5) can hold in $\mathfrak{A}$, so it suffices to show how the x_i can be chosen to satisfy the corresponding consequent. If the antecedent of (2) holds, let $x_1 = \langle 0,0\rangle$; if that of (3), let $x_1 = \langle i,j+1\rangle$, where $y_1 = \langle i,j\rangle$, and let $x_n = \langle k+1,\ell\rangle$, where $y_3 = \langle k,\ell\rangle$; if that of (4), let $x_1 = \ldots = x_n = \langle i+1,j+1\rangle$, where $y_2 = \langle i,j\rangle$ (and $y_1 = \langle i+1,j\rangle$, $y_3 = \langle i,j+1\rangle$); if that of (5), let $x_1 = \ldots = x_n = \langle i,j\rangle$, the value of y_1. In each case it is easily checked that the appropriate consequent holds in $\mathfrak{A}$.

Conversely, suppose that F is satisfiable and let $\mathscr{A} \models E(F)$. To prove that $\mathscr{D}$ accepts some tiling, it suffices to show the existence of a mapping $\gamma: \mathbb{N}\times\mathbb{N} \rightarrow D(F)$ with the following properties:

(a) $\mathscr{A} \models Q_d\gamma(0,0)$, for some $d \in D_0$;

(b) For all i, j in $\mathbb{N}$, $\mathscr{A} \models Q_d\gamma(i,j)$ for exactly one $d \in D$;

(c) For all i, j in $\mathbb{N}$, $\mathscr{A} \models H\gamma(i,j)\ \gamma(i+1,j)$;

(d) For each i, j in $\mathbb{N}$, there are a $q > 0$ and terms $t_1,\ldots,t_q$ in $D(F)$ such that $\mathscr{A} \models V\gamma(i,j)t_1 \wedge It_1t_2 \wedge \ldots \wedge It_{q-1}t_q \wedge It_q\gamma(i,j+1)$.

For given a mapping γ fulfilling (a)-(d), an accepted tiling $\tau: \mathbb{N}\times\mathbb{N} \rightarrow D$ may be obtained by defining $\tau(i,j)$ to be that tile $d \in D$ such that $\mathscr{A} \models Q_d\gamma(i,j)$. By conjuncts (1), (6), (7), (8) τ will be a well-defined solution of $\mathscr{D}$.

To define γ inductively, let f_i be the triadic indicial correlate of x_i for $i = 1, \ldots, n$. First note that by (1) and (3) there are $s_1, s_2, s_3 \in D(F)$ such that $\mathscr{A} \models \neg Hs_2 s_1 \wedge Hs_2 s_3$, for example $s_1 = s_2 = \text{ł}$ and $s_3 = f_1(\text{ł ł ł})$. Then let

$$\gamma(0,0) = f_1(s_1\, s_2\, s_3)$$

$$\gamma(0,j+1) = f_n(\gamma(0,j)\gamma(0,j)\gamma(0,j)) \qquad \text{for each} \quad j \geq 0$$

$$\gamma(i+1,0) = f_1(\gamma(i,0)\gamma(i,0)\gamma(i,0)) \qquad \text{for each} \quad i \geq 0 \quad .$$

Before proceeding, we observe that (a) holds by conjunct (2), that (b) holds by conjunct (8), that (c) holds for $j = 0$ by conjunct (3), and also that (d) holds for $i = 0$ by conjuncts (3) and (1) (since $\mathscr{A} \models V\gamma(0,j)\gamma(0,j+1) \wedge I\gamma(0,j+1)\gamma(0,j+1)$, so that for $i = 0$ we may take $q = 1$ and $t_1 = \gamma(0,j+1)$ in (d)).

Now suppose that for some i and j, γ has been defined on $\langle i,j\rangle$, $\langle i+1,j\rangle$, and $\langle i,j+1\rangle$ so as to fulfill (c) and (d) for $\langle i,j\rangle$. We define $\gamma(i+1,j+1)$. By (c) and (d) $\mathscr{A}$ verifies both $H\gamma(i,j)\gamma(i+1,j)$ and, for some $t_1, \ldots, t_q \in D(F)$,

$$V\gamma(i,j)t_1 \wedge It_1t_2 \wedge \ldots \wedge It_{q-1}t_q \wedge It_q\gamma(i,j+1) \quad .$$

For notational convenience let $t_{q+1} = \gamma(i,j+1)$. Then by conjunct (4) there are terms $t_1^1, \ldots, t_1^n$ such that

$$\mathscr{A} \models V\gamma(i+1,j)t_1^1 \wedge It_1^1t_1^2 \wedge \ldots \wedge It_1^{n-1}t_1^n \wedge Ht_1t_1^n \quad ,$$

namely, for each j, let $t_1^j = f_j(\gamma(i+1,j)\gamma(i,j)t_1)$. Then by conjunct (5) there are $t_2^1, \ldots, t_2^n$ such that

$$\mathscr{A} \models It_1^nt_2^1 \wedge It_2^1t_2^2 \wedge \ldots \wedge It_2^{n-1}t_2^n \wedge Ht_2t_2^n \, ;$$

namely $t_2^j = f_j(t_1^nt_1t_2)$. By q more Herbrand instances of (5) there are $t_i^1, \ldots, t_i^n$ for each $i \leq q+1$ such that $\mathscr{A} \models It_i^jt_i^{j+1}$ for each $j < n$ and $\mathscr{A} \models It_i^nt_{i+1}^1$ for each $i < q+1$, and also $\mathscr{A} \models Ht_it_i^n$. Let $\gamma(i+1,j+1) = t_{q+1}^n$. Then condition (d) is fulfilled. So is condition (c), since $\mathscr{A} \models Ht_{q+1}\gamma(i+1,j+1)$, that is $H\gamma(i,j+1)\gamma(i+1,j+1)$.

This completes the proof of the Minimal Skolem Basis Theorem. ∎

IID.3 Skolem Subclasses: Basis Expansion Theorems

We next consider certain systematic ways in which the minimal basis sets illustrated in Figures 22, 23, and 24 may be expanded so as to produce larger basis sets whose corresponding classes of schemata are unsolvable.

Skolem Basis Expansion Theorem (I). Suppose S is a basis set containing a set as illustrated in Figure 25(a), or two sets as illustrated in Figure 25(b).

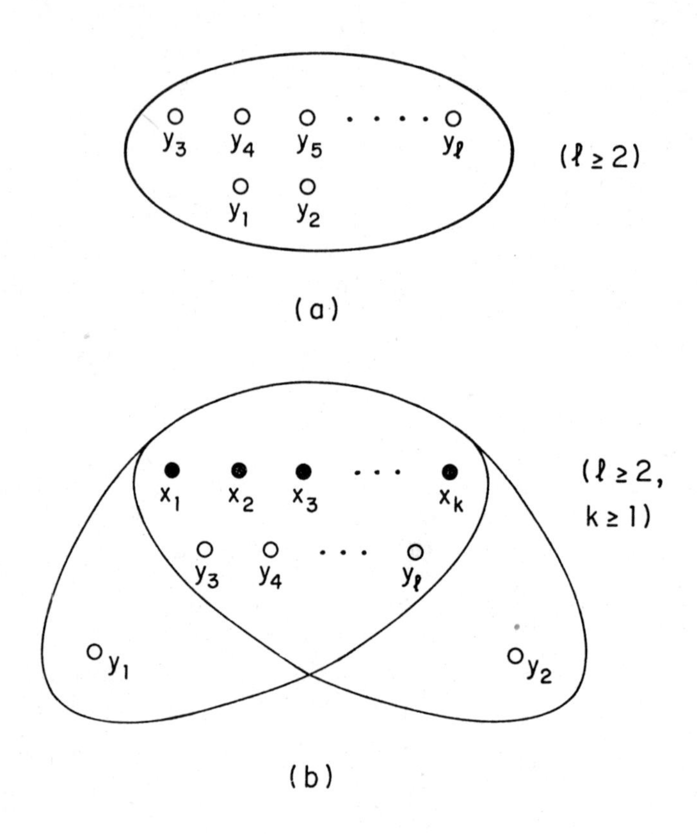

Figure 25

Let S' result from S by collapsing the two y-variables y_1, y_2, i.e., by replacing y_1 by y_2 everywhere y_1 occurs in S. Then B(S) is unsolvable provided that B(S') is unsolvable.

Proof. The method in each case is to start from a schema F in B(S') and to derive a schema in B(S) that has a subformula that guarantees the equality of the values of the two y-variables. Let $F = \forall y_2 \ldots \forall y_m \exists x_1 \ldots \exists x_n F^M$ be a schema with basis S'. For each j-place predicate letter R in F let R' be a new 2j-place predicate letter, and let G_1 be the quantifier-free formula formed from F^M by replacing subformulas as follows:

(1) If $Rv_1 \ldots v_j$ is an atomic subformula of F^M such that $y_2 \in \{v_1, \ldots, v_j\}$ and $\{v_1, \ldots, v_j, y_1\} \in S$, then replace $Rv_1 \ldots v_j$ by $R'v_1 \ldots v_j w_1 \ldots w_j$, where $\langle w_1, \ldots, w_j \rangle$ is the result of replacing y_2 by y_1 in $\langle v_1, \ldots, v_j \rangle$;

(2) if $Rv_1 \ldots v_j$ is an atomic subformula of F^M such that $y_2 \in \{v_1, \ldots, v_j\}$, $\{v_1, \ldots, v_j, y_1\} \notin S$, but $\{v_1, \ldots, v_j\} - \{y_2\} \cup \{y_1\} \in S$, then replace $Rv_1 \ldots v_j$ by $R'w_1 \ldots w_j\, w_1 \ldots w_j$, where $\langle w_1, \ldots, w_j \rangle$ is as in (1);

(3) otherwise, if $Rv_1 \ldots v_j$ is an atomic formula falling under neither (1) nor (2), then replace $Rv_1 \ldots v_j$ by $R'v_1 \ldots v_j\, v_1 \ldots v_j$.

For example, if $\{y_1, y_2, y_3\} \in S$, $\{y_1, y_2, y_4\} \notin S$, $\{y_1, y_4\} \in S$, and $\{y_2, y_4\} \in S$, then by (1) Ry_2y_3 would be replaced by $R'y_2y_3y_1y_3$, by (2) Ry_2y_4 would be replaced by $R'y_1y_4y_1y_4$, and by (3) Ry_3y_4 would be replaced by $R'y_3y_4y_3y_4$.

The next steps depend on whether the basis set of Figure 25a or 25b is desired. Let P be a new $(\ell+1)$-place predicate letter in case (a), or $(k+\ell-1)$-place in case (b). Then G is the schema $\forall y_1 \ldots \forall y_m \exists x_1 \ldots \exists x_n (A_1 \wedge (A_2 \rightarrow G_1))$, where

in case (a), A_1 is $Py_1y_1y_2y_3 \ldots y_\ell$

A_2 is $Py_1y_2y_2y_3 \ldots y_\ell$; and

in case (b), A_1 is $Py_2y_3y_4 \ldots y_\ell x_1 \ldots x_k$ and

A_2 is $Py_1y_3y_4 \ldots y_\ell x_1 \ldots x_k$.

Clearly G is in B(S). It remains to show that G is satisfiable if and only if F is satisfiable.

Suppose F is satisfiable. By vacuous quantification F implies $F_0 = \forall y_1 \ldots \forall y_m \exists x_1 \ldots \exists x_n F^M$, so F_0 is satisfiable. Let $\mathscr{A}$ be a truth-assignment verifying $E(F_0)$. Note that $D(F_0) = D(G)$. Let $\mathscr{B}$ be the truth-assignment defined as follows:

(i) For any $t_1, \ldots, t_{2j}$ in D(G)

$\mathscr{B} \models R't_1 \ldots t_{2j}$ if and only if $\mathscr{A} \models Rt_1 \ldots t_j$.

(ii) In case (a), for any $t_1, t_2, \ldots, t_{\ell+1}$,

$\mathscr{B} \models Pt_1 \ldots t_{\ell+1}$ if and only if $t_1 = t_2$;

In case (b), for any $t_1, \ldots, t_{k+\ell-1}$,

$\mathscr{B} \models Pt_1 \ldots t_{k+\ell-1}$ if and only if for some $s_1, \ldots, s_m$, $t_i = s_{i+1}$ for $1 \leq i \leq \ell-1$ and $t_i = f_{i-\ell+1}(s_1 \ldots s_m)$ for $\ell \leq i \leq k+\ell-1$.

We claim that $\mathscr{B}$ verifies $E(G)$. For if $A_1' \wedge (A_2' \rightarrow G_1') = G^*[y_1/t_1, \ldots, y_m/t_m]$ is any Herbrand instance in $E(G)$, then $\mathscr{B} \models A_1'$ by definition of $\mathscr{A}$ and A_1. And if $\mathscr{B} \models A_2'$ then, in either case (a) or case (b), $t_1 = t_2$. But then, by the construction of G_1 from F^M, $\mathscr{B} \models G_1'$ just in case $\mathscr{A} \models F_0^*[y_1/t_2, y_2/t_2, \ldots, y_m/t_m]$. This is a member of $E(F_0)$ and hence verified by $\mathscr{A}$.

Conversely, G implies $\forall y_1 \ldots \forall y_m \exists x_1 \ldots \exists x_n (A_1[y_1/y_2] \wedge (A_2[y_1/y_2] \rightarrow G_1[y_1/y_2])$, and hence $G_0 = \forall y_1 \ldots \forall y_m \exists x_1 \ldots \exists x_n G_1[y_1/y_2]$. Thus it suffices to show that if G_0 is satisfiable then F is satisfiable. But G_0 is just like F, except in having $R'v_1 \ldots v_j v_1 \ldots v_j$ wherever F has $Rv_1 \ldots v_j$; the result follows easily.

This completes the proof of the Skolem Basis Expansion Theorem (I). ■

The technique used to prove this theorem can be extended to prove the following more general result:

Skolem Basis Expansion Theorem (II). Suppose S is a basis set containing a set as illustrated in Figure 26a, or two sets as illustrated in Figure 26b.

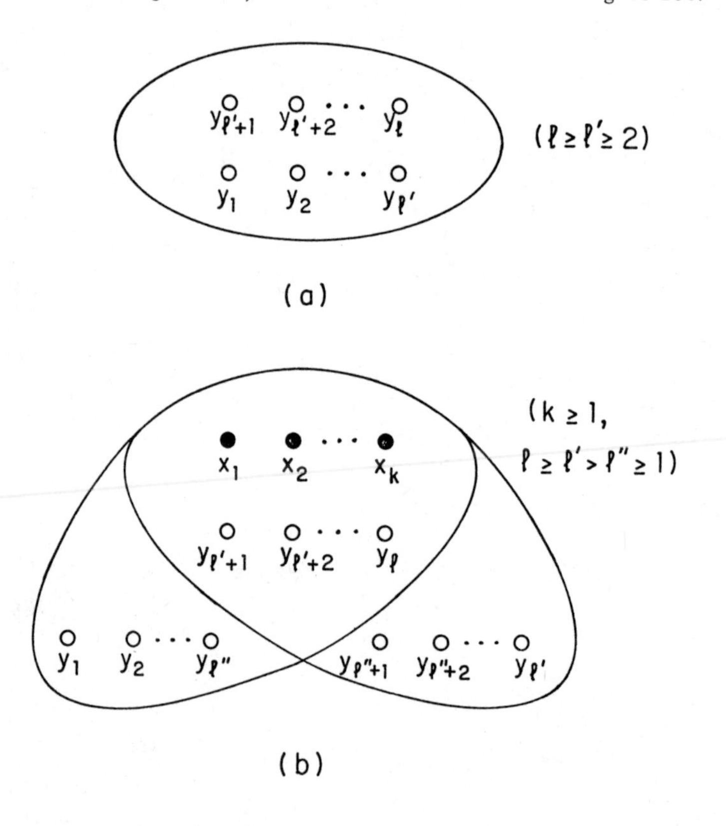

Figure 26

Let S' result from S by collapsing all the variables $y_1, \ldots, y_{\ell'}$, i.e., replacing each of $y_2, \ldots, y_{\ell'}$ by y_1 everywhere it occurs in S; then $B(S)$ is unsolvable provided that $B(S')$ is unsolvable.

(Note that the previous theorem is the $\ell = \ell' = 2$ case of this one.)

We omit the proof of this fact, which presents no special problems. The principal variations on the previous construction are that a j-place predicate letter R is replaced by a $j\ell'$-place predicate letter R', and that the atomic formulas A_1 and A_2 must be adjusted slightly.

With the aid of the two Skolem Basis Expansion Theorems it immediately follows that several broad classes are unsolvable. First, the "cycle" in Figure 22a or 22b may be increased in size, so as to give, for example, the basis set of Figure 27a. By the same token we may derive from Figure 24 such patterns as those illustrated in Figures 27b and 27c. In fact $B(S)$ is unsolvable, if the diagram of the basis set S contains any cycle consisting of a run of ●—● edges, followed by any combination of segments of the forms ○—●—○ and ○—○ that contains at least two of the latter type, and terminating at the beginning of the ●—● edges.

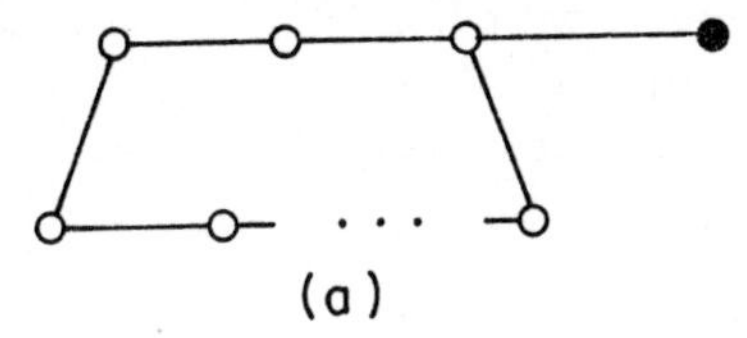

(a)

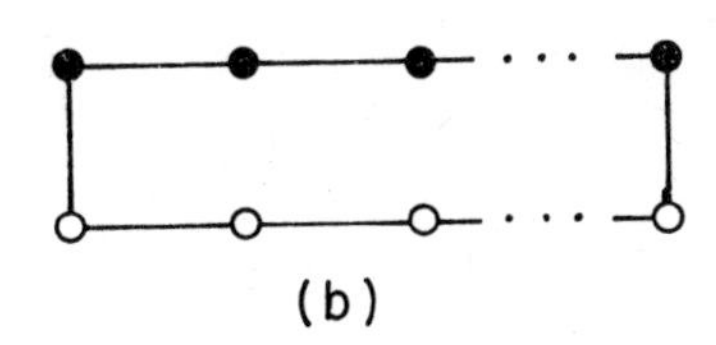

(b)

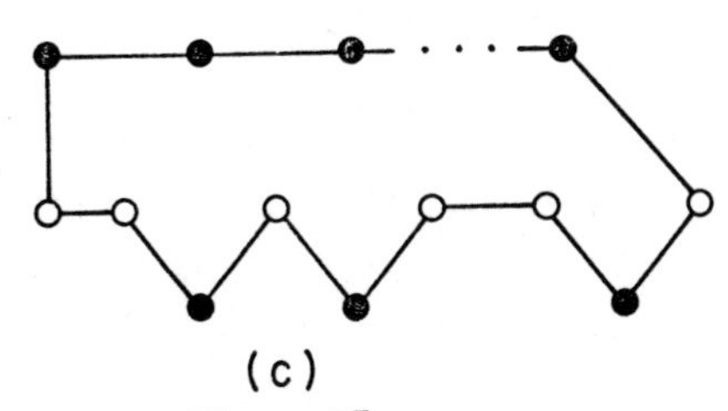

(c)

Figure 27

Generalizing in another direction, from these theorems we obtain the result that if only the y-sets are taken into account (every set containing an x-variable may occur) then the presence of any two distinct polyadic y-sets leads to unsolvability. This result meshes nicely with the solvability result cited on p. 144; if only one such y-set is present a solvable class results.

Unfortunately these results do not cover all the cases, and even for certain dyadic basis sets S the decision problem for B(S) remains open. One such basis set is illustrated in Figure 28.

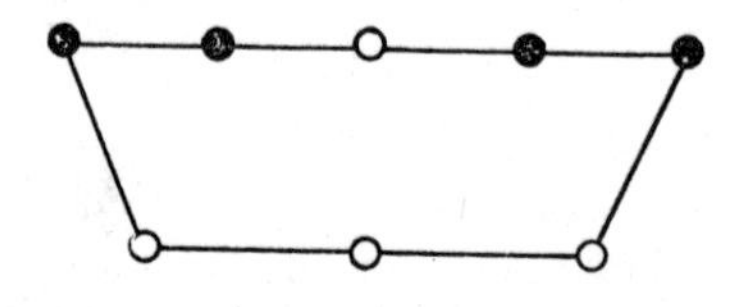

Figure 28

Here there is a cycle but the run of x-variables is interrupted by a y-variable; moreover the x-variables cannot be used to collapse the interpolated y-variable since there are two x-variables in a row.

Another problematical case arises when "collapsing" results in a solvable class, for example as shows in Figure 29a. No subset of this basis set corresponds to an unsolvable class. When the first Skolem Basis Expansion Theorem is applied to variable x and the two neighboring y-variables, the basis set of Figure 29b results, which is solvable since the central y-variable cleaves the basis into two parts, each containing one of the y-sets.

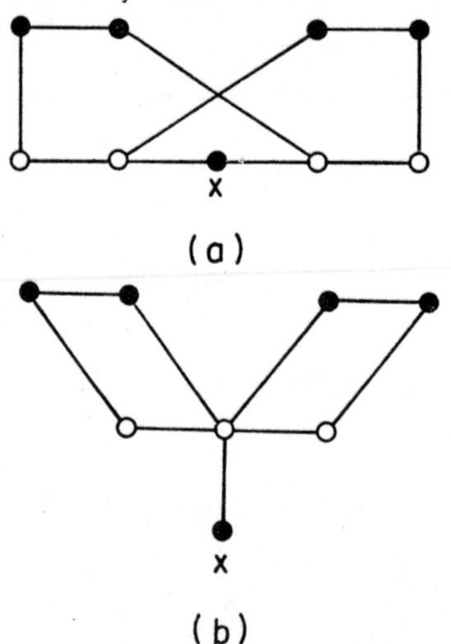

Figure 29

We conjecture that both these examples yield solvable classes, and that the methods of Dreben and Goldfarb (1979) can be extended to furnish decision procedures.

Historical References. The classification of $\forall\exists\forall$ schemata by atomic forms was suggested by Büchi (1962), Wang (1962), and Dreben, Kahr, and Wang (1962). Kostryko (1966) proved the unsolvability of the $\forall\exists\forall(\infty, 1)$ subclass with only monadic forms and any three of the four dyadic forms y_1y_2, y_2y_1, xy_2, y_2x. Aanderaa's thesis (1966), which led to the formulation of the linear sampling problem, was motivated by the study of the y_1y_2, y_2x, y_1y_1 class, which Dreben, Kahr, and Wang had conjectured to be solvable, but Aanderaa showed to be unsolvable.

Schemata in Skolem form and restricted as to the forms of their atomic subformulas have been considered by Skolem (1928), Church (1951), Dreben (1961), Friedman (1963), Maslov (1966, 1967, 1968), Joyner (1976), and Dreben and Goldfarb (1979) from the point of view of solvability; the unsolvability of such classes is discussed by Goldfarb and Lewis (1975) and by Goldfarb (1974).

Chapter IIE

RESTRICTIONS ON TRUTH-FUNCTIONAL STRUCTURE

The decision problem for first-order logic is unsolvable because of quantification, in the sense that the underlying truth-functional logic without quantification has a solvable decision problem. The restrictions considered in Chapters IIC and IID are primarily limitations on the extent to which quantification can be used to create complex patterns of coinstantiation in the Herbrand expansion. In this chapter and the next such quantificational restrictions are combined with restrictions on the form of the matrix: in this chapter, limitation of the truth-functional form to various restricted conjunctive and disjunctive normal forms, and in the next chapter, bounds on the number of atomic subformulas a schema may have.

The starting point for investigation in these directions is the observation, due to Herbrand (1931), that the decision problem is solvable for schemata whose matrices are conjunctions of signed atomic formulas: such a schema is unsatisfiable if and only if some negated and some unnegated conjunct have coinstantiable atomic formulas. (Herbrand's result is actually the dual, for provability when the matrix is a disjunction of signed atomic formulas.) Attempts to generalize Herbrand's class may view it as the $n = 1$ case in either of two series: the classes $\mathscr{D}_n$ of schemata in disjunctive normal form with at most n disjuncts, or the classes $\mathscr{C}_n$ of schemata in conjunctive normal form with at most n disjuncts per conjunct. (By truth-functional moves alone, each formula in $\mathscr{D}_n$ is equivalent to one in $\mathscr{C}_n$, so $\mathscr{D}_n$ may be viewed as a subset of $\mathscr{C}_n$.) We show in this chapter that both $\mathscr{D}_2$ and $\mathscr{C}_2$ (which we call <u>Krom</u>, after M. R. Krom, who studied it) are unsolvable. In fact we show that certain similarity and prefix subclasses of these two classes are unsolvable. For Krom we have almost complete analyses by prefix and by similarity type (though not by these two features considered jointly); for $\mathscr{D}_2$ the state of our knowledge is much poorer. In neither case is the result the same as for schemata without the truth-functional restriction, since the $\forall\exists\forall$ subclasses of Krom and $\mathscr{D}_2$ are both solvable. It is worth noting that $\mathscr{C}_3$, the schemata with matrices in conjunctive normal form and at most three disjuncts per conjunct, has the same solvable and unsolvable subclasses of schemata as full first-order logic, if only prefix is taken into account: any conjunct $(A_1 \vee \ldots \vee A_n)$ with $n \geq 4$ signed atomic disjuncts may be "split" by introducing a new k-place predicate letter P, where k is the total number of y-variables in the schema; and replacing the conjunct by the two conjuncts $(A_1 \vee A_2 \vee Py_1 \ldots y_k) \wedge (\neg Py_1 \ldots y_k \vee A_3 \vee \ldots \vee A_n)$, where $y_1, \ldots, y_k$ are the y-variables. This transformation preserves satisfiability (though obviously not similarity type) and, when repeated, ultimately results in a schema in $\mathscr{C}_3$.

We mention one other class: <u>Horn</u> (named after Alfred Horn), the schemata in conjunctive normal form with at most one negated disjunct per conjunct. Though

Krom and Horn are incomparable in the sense of set inclusion, they have closely related decision problems: in each case in which a similarity or prefix subclass of Krom is known to be solvable or unsolvable, the corresponding Horn subclass has the same property. Actually our understanding of Horn is somewhat more extensive than that of Krom, since the $\forall\exists\forall\exists$ prefix subclass of Horn is known to be unsolvable, but the corresponding question for Krom remains open.

IIE.1 Prefix Subclasses of Krom and Horn

Following the conventions presented in Section IIC. 1, we use $\forall\exists\forall$ (for example) to denote the class of all prenex schemata with prefixes of the form $\forall v_1 \exists v_2 \forall v_3$ for some variables v_1, v_2, v_3. Then $\forall\exists\forall \cap$ Krom (for example) denotes the class of all schemata in Krom with such prefixes.

This section is devoted to the proof of:

Krom and Horn Prefix Theorem. The following are unsolvable:

$\exists\forall\exists\forall \cap$ Krom $\cap$ Horn
$\forall\exists\exists\forall \cap$ Krom $\cap$ Horn
$\forall\exists\forall\forall \cap$ Krom $\cap$ Horn
$\forall\forall\exists\forall \cap$ Krom $\cap$ Horn
$\forall\exists\forall\exists \cap$ Horn
$\forall\forall\forall\exists \cap$ Horn .

In fact, the narrower subclasses of these classes are unsolvable that contain only schemata whose matrices have no conjuncts of the form $(\neg A_1 \vee \neg A_2 \vee \ldots \vee \neg A_n)$ for $n > 1$, where the A_i are atomic.

(The slightly stronger form of the theorem is used in Chapter IF; it restricts the conjuncts of Krom $\cap$ Horn schemata to the three forms A, $\neg A$, and $A \rightarrow B$.)

This result should be seen in contrast to the solvability results cited on p. 92 for prenex schemata without truth-functional restrictions, and the solvability of the following special cases:

$\forall\exists\forall \cap$ Krom (Aanderaa and Lewis, 1973)
$\forall\exists\forall \cap$ Horn (Goldfarb, 1974; Dreben and Goldfarb, 1979)
$\exists^*\forall^*\exists^* \cap$ Krom (Maslov, 1964; Dreben and Goldfarb, 1979).

In combination these give an exhaustive prefix classification of Krom and Horn, except for the classes $\forall\exists\forall\exists^n \cap$ Krom for $n \geq 1$. Whether any of these classes, or the union class $\forall\exists\forall\exists^* \cap$ Krom, is unsolvable remains open. Our strong conjecture is that all these classes are solvable.

Proof of Krom and Horn Prefix Theorem. The proof is in four parts, the tool being in each case the special two-counter machines of Section IF2; the Special Two-Counter Theorem (Inputless Version) (p. 62) implies the unsolvability of each class.

(i) $\exists\forall\exists\forall \cap$ Krom $\cap$ Horn and $\forall\exists\exists\forall \cap$ Krom $\cap$ Horn.

These cases are treated at once by proving that $\exists\forall \wedge \forall\exists\forall \cap$ Krom $\cap$ Horn is unsolvable; by prenexing either of the prefix types $\exists\forall\exists\forall$ or $\forall\exists\exists\forall$ may be obtained. Let $\mathcal{M} = \langle I_1, \ldots, I_n \rangle$ be a special two-counter machine, and let $P_1, \ldots, P_n$ be dyadic predicate letters. We construct a formula $F_{\mathcal{M}} = \exists z \forall y\, M_1 \wedge \forall y_1 \exists x \forall y_2\, M_2$ in Krom $\cap$ Horn that is unsatisfiable if and only if $(1,0,0) \overset{*}{\vdash}_{\mathcal{M}} (n,0,0)$. Let c be the 0-place indicial correlate of z and f the monadic indicial correlate of x; thus $D(F) = \{f^p(c) \mid p \geq 0\}$ has the same structure as the natural numbers with successor. If $F_{\mathcal{M}}$ is satisfiable then the following truth-assignment $\mathcal{A}$ will verify $E(F_{\mathcal{M}})$:

$$\mathcal{A} \models P_i f^p(c) f^q(c) \quad \text{if and only if} \quad (1,0,0) \overset{*}{\vdash}_{\mathcal{M}} (i,p,q) \quad .$$

The definitions of M_1 and M_2 are as follows:

$$M_1 = P_1 zz \wedge \neg P_n zz \wedge \bigwedge (P_i zy \rightarrow P_j yz)$$

(the conjunction over all i such that I_i is $S(k,j)$ for some k)

$$M_2 = \bigwedge (P_i y_2 y_1 \rightarrow P_j y_2 x)$$

(the conjunction over all i such that I_i is $A(j)$)

$$\wedge \bigwedge (P_i x y_2 \rightarrow P_j y_1 y_2)$$

(the conjunction over all i such that I_i is $S(j,k)$ for some k).

If $\mathcal{M}$ does not $\vdash$-halt, i.e., it is not the case that $(1,0,0) \overset{*}{\vdash}_{\mathcal{M}} (n,0,0)$, then the truth-assignment $\mathcal{A}$ defined above verifies $E(F_{\mathcal{M}})$. For in any Herbrand instance of F the substituent for z is $c = f^0(c)$, and that for x is $f^{p+1}(c)$, if that for y_1 is $f^p(c)$. Then the Herbrand instances of $P_1 zz$ and $\neg P_n zz$ are verified by $\mathcal{A}$, since the only such instances are $P_1 cc$ and $\neg P_n cc$, and $(1,0,0) \overset{*}{\vdash}_{\mathcal{M}} (1,0,0)$ by the reflexiveness of $\overset{*}{\vdash}_{\mathcal{M}}$, but not $(1,0,0) \overset{*}{\vdash}_{\mathcal{M}} (n,0,0)$ by hypothesis. $\mathcal{A}$ verifies each Herbrand instance of a conjunct $P_i zy \rightarrow P_j yz$, i.e.,

$$P_i c f^q(c) \rightarrow P_j f^q(c) c$$

where $q \in \mathbb{N}$ and I_i is $S(k,j)$ for some j, since if $\mathcal{A} \models P_i c f^q(c)$ then $(1,0,0) \overset{*}{\vdash}_{\mathcal{M}} (i,0,q)$ by definition of $\mathcal{A}$, $(i,0,q) \vdash_{\mathcal{M}} (j,q,0)$ by definition of $\vdash_{\mathcal{M}}$ (p. 61), $(1,0,0) \overset{*}{\vdash}_{\mathcal{M}} (j,q,0)$ by definition of $\overset{*}{\vdash}_{\mathcal{M}}$, and hence $\mathcal{A} \models P_j f^q(c)c$ by definition of $\mathscr{A}$ again. Similarly, $\mathscr{A}$ verifies each Herbrand instance of the other two types of conjunct, i.e.,

$$P_i f^q(c) f^p(c) \rightarrow P_j f^q(c) f^{p+1}(c)$$

where $p, q \in \mathbb{N}$ and I_i is $A(j)$, or

$$P_i f^{p+1}(c) f^q(c) \rightarrow P_j f^p(c) f^q(c)$$

where $p, q \in \mathbb{N}$ and I_i is $S(j, k)$ for some k.

Conversely, if $F_{\mathscr{M}}$ is satisfiable then let $\mathscr{A}$ be any truth-assignment verifying $E(F_{\mathscr{M}})$. Then a straightforward induction establishes that if $(1, 0, 0) \vdash^{*}_{\mathscr{M}} (i, p, q)$ then $\mathscr{A} \models P_i f^p(c) f^q(c)$. Since also $\mathscr{A} \models \neg P_n cc$, it follows that not $(1, 0, 0) \vdash^{*}_{\mathscr{M}} (n, 0, 0)$.

(ii) $\forall\exists\forall\forall \cap$ Krom $\cap$ Horn and $\forall\forall\exists\forall \cap$ Krom $\cap$ Horn.

For these classes we again encode special two-counter machines, this time using the $\Vdash_{\mathscr{M}}$ relation between extended instantaneous descriptions rather than the $\vdash_{\mathscr{M}}$ relation between instantaneous descriptions. There is no term in $D(F)$, for F in one of these classes, that might naturally play the role of 0 in the way that the constant c does in the construction above. So instead we use the third y-variable to represent a base relative to which the contents of the two counters are encoded. This representation requires use of a triadic predicate letter for each machine state, rather than a dyadic letter as above.

Specifically, let $\mathscr{M} = \langle I_1, \ldots, I_n \rangle$ be a special two-counter machine, and let $M_{\mathscr{M}}$ be the matrix

$$P_1 y_3 y_3 y_3 \wedge \neg P_n y_3 y_3 y_3 \wedge \bigwedge_{I_i = A(j)} (P_i y_2 y_3 y_1 \rightarrow P_j y_2 y_3 x)$$

$$\wedge \bigwedge_{I_i = S(j,k)} ((P_i y_2 x y_3 \rightarrow P_j y_2 y_1 y_3) \wedge (P_i y_2 y_2 y_3 \rightarrow P_k y_2 y_3 y_2))$$

and let $F_{\mathscr{M}}$ be $\forall y_1 \exists x \forall y_2 \forall y_3 M_{\mathscr{M}}$. Then $F_{\mathscr{M}}$ is unsatisfiable if and only if $(1, 0, 0) \Vdash^{*}_{\mathscr{M}} (n, 0, 0)$. The proof is almost identical to that for the previous case; a truth-assignment verifying $E(F_{\mathscr{M}})$, in case $\mathscr{M}$ does not $\Vdash$-halt from $(1, 0, 0)$, is

$$\mathscr{A} \models P_i f^p(\text{ł}) f^q(\text{ł}) f^r(\text{ł}) \quad \text{if and only if} \quad (1, 0, 0) \Vdash^{*}_{\mathscr{M}} (i, q-p, r-p) \ .$$

Thus the first argument plays the role of the base, the second and third arguments correspond to the counters. Note that there is no way to restrict the applicability of a conjunct

$$P_i y_2 x y_3 \rightarrow P_j y_2 y_1 y_3$$

to cases in which the number corresponding to x is greater than that corresponding to y_2. It is for this reason that extended instantaneous descriptions, with their possibly negative counter values, must be used, and that special two-counter machines must be encoded instead of standard machines.

Also, let $G_{\mathcal{M}}$ be $\forall y_1 \forall y_2 \exists x \forall y_3 M_{\mathcal{M}}$. If it is not the case that $(1,0,0) \Vdash^*_{\mathcal{M}} (n,0,0)$ then $F_{\mathcal{M}}$ is satisfiable, and since $F_{\mathcal{M}}$ implies $G_{\mathcal{M}}$, $G_{\mathcal{M}}$ is satisfiable. To prove the converse, suppose that $G_{\mathcal{M}}$ is satisfiable and let $\mathcal{A}$ be any truth-assignment verifying $E(G_{\mathcal{M}})$. Let g be the dyadic indicial correlate of x, and define a mapping

$$\gamma : \mathbb{N} \to D(G_{\mathcal{M}}) \quad \text{by} \quad \gamma(0) = \ell$$
$$\gamma(i+1) = g(\gamma(i)\ell), \quad \text{for each} \quad i \geq 0 .$$

Suppose that $(1,0,0) \Vdash^*_{\mathcal{M}} (n,0,0)$; then there is a sequence of extended instantaneous descriptions $(i_1, p_1, q_1), \ldots, (i_k, p_k, q_k)$ for some $k \geq 1$ such that $(i_1, p_1, q_1) = (1,0,0)$, $(i_k, p_k, q_k) = (n,0,0)$, $(i_j, p_j, q_j) \Vdash_{\mathcal{M}} (i_{j+1}, p_{j+1}, q_{j+1})$ for each $j < k$, and such that $p_j \geq 0$ and $q_j \geq 0$ for each j. Then a straightforward induction on j proves that $\mathcal{A} \models P_{i_j} \ell\gamma(p_j)\gamma(q_j)$ for each j. Taking $j = k$ yields a contradiction, since also $\mathcal{A} \models \neg P_n \ell\gamma(0)\gamma(0)$ since $\neg P_n y_3 y_3 y_3$ is a conjunct of $M_{\mathcal{M}}$. ∎

Note that it is crucial in this last argument that $\forall y_3$ follow $\exists x$ in the prefix, so that the indicial term supplanting x in $G^*_{\mathcal{M}}$ does not have y_3 as an argument, even though it does have y_2 (the "base") as an argument. Thus this proof does not extend to show that $(1,0,0) \Vdash^*_{\mathcal{M}} (n,0,0)$ if and only if $\forall y_1 \forall y_2 \forall y_3 \exists x M_{\mathcal{M}}$ is satisfiable. Indeed, this assertion is false, and as we have mentioned $\exists^*\forall^*\exists^* \cap$ Krom is solvable.

One oddity of the proof for $\forall\exists\forall\forall \cap$ Krom (but not that for $\forall\forall\exists\forall \cap$ Krom) may be noted before continuing. Unlike nearly every other unsolvability proof in Part II, this one cannot, it would appear, be extended easily to establish the <u>conservative</u> unsolvability of the class (p. 138), and indeed the conservative unsolvability of this class remains an open problem.

(iii) <u>$\forall\exists\forall\exists \cap$ Horn</u>

To show this class unsolvable, we start from an arbitrary schema $F = \exists z \forall y M_1 \wedge \forall y_1 \exists x \forall y_2 M_2$ in the class $\exists\forall \wedge \forall\exists\forall \cap$ Horn, which was proved unsolvable in (i) above. We transform such a schema into a schema in the class $\exists \wedge \forall\forall \wedge \forall\exists\forall \cap$ Horn that is satisfiable if and only if F is satisfiable. By judicious prenexing this schema may be transformed into a prenex schema with prefix type $\forall\exists\forall\exists$.

Let Z be a new monadic predicate letter and let y' be a new variable, and let G be

$$\exists z\, Zz \wedge \forall y \forall y'(Zy' \to M_1[z/y']) \wedge \forall y_1 \exists x \forall y_2 M_2 \ .$$

By distributing the antecedent Zy' over the conjuncts of $M_1[z/y']$ the schema G may be transformed into a Horn schema as required. Note that $D(F) = D(G)$. Any truth-assignment $\mathscr{A}$ verifying $E(F)$ can be extended to one verifying $E(G)$ by defining

$$\mathscr{A} \models Zt \quad \text{if and only if} \quad t = c, \text{ the 0-place indicial correlate of } z.$$

Conversely, if $\mathscr{A}$ is any truth-assignment verifying $E(G)$ then $\mathscr{A}$ also verifies $E(F)$. For let F' be any Herbrand instance $F^*[y/t, y_1/t_1, y_2/t_2]$ in $E(F)$. Then F' is a conjunction $M_1' \wedge M_2'$, where the Herbrand instance $G' = G^*[y/t, y'/c, y_1/t_1, y_2/t_2]$ in $E(G)$ is $Zc \wedge (Zc \rightarrow M_1') \wedge M_2'$. Since $\mathscr{A} \models G'$, also $\mathscr{A} \models M_1' \wedge M_2'$, i.e., $\mathscr{A} \models F'$.

(iv) <u>$\forall\forall\forall\exists \cap$ Horn</u>

It follows from the proof of (iii) that $\exists \wedge \forall\exists\forall \cap$ Horn is unsolvable, since the class $\exists \wedge \forall\forall \wedge \forall\exists\forall$ may be reduced to $\exists \wedge \forall\exists\forall$ by quantificational maneuvers. (This makes all the more remarkable the solvability of $\forall\exists\forall \cap$ Horn.) To prove $\forall\forall\forall\exists \cap$ Horn unsolvable, we start from an arbitrary schema $F = \exists z M_1 \wedge \forall y_1 \exists x \forall y_2 M_2$ in $\exists \wedge \forall\exists\forall \cap$ Horn, and transform F into a schema in the class $\forall\forall\forall\exists \wedge \forall \wedge \forall\forall\forall \cap$ Horn that is satisfiable if and only if F is satisfiable; this schema has a prenex equivalent with prefix $\forall\forall\forall\exists$.

Let E and S be new dyadic predicate letters and let Z be a new monadic predicate letter. Let G be the schema

$$\begin{aligned} &\forall y_1' \forall y_2' \forall y_3' \exists x' (E y_1' y_1' \wedge (E y_2' y_3' \rightarrow S y_1' x') \wedge (S y_2' y_3' \rightarrow Z x')) \\ &\qquad \wedge \forall y (Zy \rightarrow M_1[z/y]) \\ &\qquad \wedge \forall y_1 \forall y_2 \forall y_3 (S y_1 y_3 \rightarrow M_2[x/y_3]) \quad . \end{aligned}$$

Let c be the 0-place indicial correlate of z, let f be the monadic indicial correlate of x, and let g be the triadic indicial correlate of x'.

Suppose F is satisfiable, and let $\mathscr{A} \models E(F)$. Define a mapping $\gamma: D(G) \rightarrow D(F)$ as follows:

$$\gamma(\text{ł}) = c$$

$$\gamma(g(t_1 t_2 t_3)) = \begin{cases} c, & \text{if} \quad t_2 \neq t_3 \\ f(\gamma(t_1)), & \text{if} \quad t_2 = t_3 \end{cases} \quad .$$

Then define a truth-assignment $\mathscr{B}$ thus: For a k-place predicate letter P appearing in F,

$$\mathscr{B} \models P t_1 \ldots t_k \qquad \text{if and only if} \qquad \mathscr{A} \models P\gamma(t_1) \ldots \gamma(t_k);$$

and

$$\mathscr{B} \models E t_1 t_2 \qquad \text{if and only if} \qquad t_1 = t_2,$$

$$\mathscr{B} \models St_1t_2 \quad \text{if and only if} \quad \gamma(t_2) = f(\gamma(t_1)) \, ,$$

$$\mathscr{B} \models Zt \quad \text{if and only if} \quad \gamma(t) = c \, .$$

Then $\mathscr{B} \models E(G)$. For let $G' \in E(G)$. Then for some t, t_1, t_2, t_3, and some quantifier-free formula C, G' is

$$C \wedge (Zt \rightarrow M_1[z/t]) \wedge (St_1t_3 \rightarrow M_2[y_1/t_1, y_2/t_2, y_3/t_3]) \, .$$

Clearly $\mathscr{B} \models C$. If $\mathscr{B} \models Zt$ then $\gamma(t) = c$ and

$$\mathscr{B}(M_1[z/t]) = \mathscr{A}(M_1[z/\gamma(t)]) = \mathscr{A}(M_1[z/c]) = 1 \, ,$$

since $M_1[z/c]$ is an Herbrand instance of F. Similarly, if $\mathscr{B} \models St_1t_3$ then $\gamma(t_3) = f(\gamma(t_1))$ and $\mathscr{B} \models M_2[y_1/t_1, y_2/t_2, x/t_3]$ since $\mathscr{A} \models M_1[y_1/\gamma(t_1), y_2/\gamma(t_2), x/f(\gamma(t_1))]$.

Conversely, suppose G is satisfiable and let $\mathscr{B} \models E(G)$. Define a mapping $\delta : D(F) \rightarrow D(G)$ by

$$\delta(c) = f(\ast \ast f(\ast \ast \ast))$$

$$\delta(f(t)) = f(\delta(t) \ast \ast) \qquad \text{for each} \quad t \in D(F).$$

Then $\mathscr{B} \models Z\delta(c)$ and $\mathscr{B} \models S\delta(t)\,\delta(f(t))$ for each $t \in D(F)$, whence it follows readily that the truth-assignment $\mathscr{A}$ such that

$$\mathscr{A} \models Pt_1 \ldots t_k \qquad \text{if and only if} \qquad \mathscr{B} \models P\delta(t_1) \ldots \delta(t_k) \, ,$$

where P is a k-place predicate letter of F and $t_1, \ldots, t_k \in D(F)$, verifies $E(F)$.

This completes the proof of the Krom and Horn Prefix Theorem. ∎

IIE.2 The (0, 1) Similarity Class

In this section we prove:

Krom ∩ Horn (0, 1) Theorem. Schemata whose matrices are conjunctions of formulas Pv_1v_2, $\neg Pv_1v_2$, and $Pv_1v_2 \rightarrow Pv_3v_4$, where $v_1, \ldots, v_4$ range over variables but P is a fixed dyadic predicate letter, comprise an unsolvable class.

The proof is long and extremely technical. There are three main steps. The first step (Unsolvability Lemma) is to reduce the Post Correspondence Problem to a decision problem for a class of schemata with function signs. These formulas are not the functional forms of formulas without function signs. The second step (Sufficient Conditions Lemma) is to use the language of bigraphs to state conditions on certain "ρ-closed" sets of bigraphs under which it is possible to reduce the class of schemata of the Unsolvability Lemma to class of the schemata of the theorem. Here "ρ-closed" is a new notion, related to closure of a set of formulas under logical implication. Finally we show (Construction Lemma) that the sufficient conditions can be met. What makes the proof so technical throughout is its dependence on very special features of the construction; thus the Sufficient Conditions cannot be relaxed or generalized so as to make them simpler and more comprehensible, since such a generalization would demand more powerful constructions than we are able to provide.

We introduce two definitions before beginning the proof. Recall that ω is the "identity bigraph", so that $\omega = \omega(0,1) = \langle \{\langle 0,1\rangle\}, \phi, 2\rangle$. Then let $\omega_0(0,1) = \omega(0,1)$, and $\omega_1(0,1) = \omega(1,0)$. Also, for $m = 0, 1$, let $\bar{m} = 1 - m$. Thus, $\omega_m(a_0, a_1) = \omega(a_m, a_{\bar{m}})$, $\omega_m(a_m, a_{\bar{m}}) = \omega(a_0, a_1)$, etc.

We regularly use ι to denote the identity mapping in any domain.

Unsolvability Lemma. Let S be a finite set of monadic function signs, and let S_0 be a subset of S. Let D be the domain generated by the function signs in S and the constant c, and let D_0 be the domain generated by c and the function signs in S_0. For any

$$T \subset \{1\} \times S \times S_0 \cup \{-1\} \times S \times S,$$

let

$$\eta_T = \sum_{\langle e,f,g\rangle \in T} \; \sum_{t \in D_0} e \cdot \omega(f(t), g(t)) ;$$

and for any $Q \subset \{0,1\} \times S^3$, let

$$F_Q = \bigwedge_{\langle m,f,g,h\rangle \in Q} (\hat{\omega}_m(f(x), g(y)) \rightarrow \hat{\omega}_m(x, h(y))) .$$

Then the problem of determining, given S, S_0, T, and Q, whether or not $\mu(\iota) \models E(F_Q, D_0)$ for some completion μ of η_T, is unsolvable.

Before proceeding to the proof it is worth making a few general comments. T and Q are finite objects, but η_T is (in general) infinite. That T is not presented simply as a subset of $\{-1, 1\} \times S^2$ is due to the sort of annoying technicality mentioned above; we need the stronger, if less natural, statement presented here in order to complete the construction. The function signs in $S - S_0$ enter into the definitions of η_T and F_Q only as the outermost function signs of terms; they encode "states" in a way that would more naturally be done with predicate letters, were more than one predicate letter available. Finally, observe the manner in which it could happen that there would be no completion μ of η_T such that $\mu(\iota) \models E(F_Q, D_0)$. This would happen if there were, for $i = 1, \ldots, k-1$ for some k, conjuncts $\hat{\omega}(s_i, t_i) \rightarrow \hat{\omega}(s_{i+1}, t_{i+1})$ of members of $E(F_Q, D_0)$, such that $\omega(s_1, t_1) - \omega(s_k, t_k) \subset \eta_T$. This would require in turn that $s_1 = f_1(s_1')$, $t_1 = g_1(t_1')$,

$s_k = f_k(s_k'),\ t_k = f_k(t_k')$ for some $f_1, f_k, g_k \in S$, $g_1 \in S_0$, and $s_1', t_1', s_k', t_k' \in D_0$, such that $\langle 1, f_1, g_1\rangle$ and $\langle -1, f_k, g_k\rangle$ are members of T.

Proof of the Unsolvability Lemma. We associate with each correspondence system $\mathscr{P}$ (p. 55) sets S, S_0, T and Q such that the stated expansion of F_Q has a verifying truth-assignment as described if and only if $\mathscr{P}$ has a solution. So let Σ be an alphabet and let $\mathscr{P} \subset \Sigma^+ \times \Sigma^+$. Let ℓ be the length of the longest word appearing as either component of a pair in $\mathscr{P}$. For any k, let $\Sigma^{(k)}$ be the set of words in Σ^* of length at most k. The set S consists of a function sign $f_{\langle v, w\rangle}$ for each pair $\langle v, w\rangle \in \Sigma^{(\ell)} \times \Sigma^{(\ell)}$, and a function sign f_a for each symbol $a \in \Sigma$. The set S_0 is the subset of S containing only the function signs f_a for $a \in \Sigma$. Then with D and D_0 defined as in the statement of the lemma, there is a natural mapping $\delta : D_0 \to \Sigma^*$, namely $\delta(f_{a_1}(f_{a_2}(\ldots(f_{a_n}(c))\ldots))) = a_1a_2\ldots a_n$, for any $a_1, \ldots, a_n \in \Sigma (n \geq 0)$. Now let Γ be a set of bigraphs,

$$\Gamma = \{\omega(f_{\langle a, w\rangle}(s), t) \mid a \in \Sigma,\ w \in \Sigma^{(\ell)},\ s,\ t \in D_0\}$$

$$\cup \{\omega(s, f_{\langle v, w\rangle}(t)) \mid v \in \Sigma^{(\ell)},\ w \in \Sigma^{(\ell)} - \{\varepsilon\},\ s,\ t \in D_0\} .$$

Each member of Γ corresponds to a pair in $\Sigma^* \times \Sigma^*$ as follows:

$$\gamma(\omega(f_{\langle a, w\rangle}(s), t)) = \langle a\delta(s),\ w\delta(t)\rangle$$

$$\gamma(\omega(s, f_{\langle v, w\rangle}(t))) = \langle v\delta(s),\ w\delta(t)\rangle .$$

Conversely, let each member of $\Sigma^+ \times \Sigma^+$ correspond to a member of Γ as follows: for every $a, b \in \Sigma$, $v, w \in \Sigma^*$, let

$$\gamma_0(av, bw) = \omega(f_{\langle a, \varepsilon\rangle}(\delta^{-1}(v)),\ \delta^{-1}(bw)) .$$

Then for each pair $\langle v, w\rangle \in \Sigma^+ \times \Sigma^+$, let $\theta_{v,w}$ be the sum of all bigraphs in $\gamma^{-1}(v, w)$; note that $\gamma_0(v, w) \subset \theta_{v,w}$. Also, let $\Theta = \{\theta_{v,w} \mid \langle v, w\rangle \in \Sigma^+ \times \Sigma^+\}$.

We next show that there is a set $Q_0 \subset \{0, 1\} \times S^3$ such that the formula F_{Q_0} constructed as described in the statement of the Lemma has the following properties. If μ is a complete bigraph of dimension 2 with $V(\mu) = D$, then

(a) If $\mu(\iota) \models E(F_{Q_0}, D_0)$, then for each pair $\langle v, w\rangle \in \Sigma^+ \times \Sigma^+$, if $\gamma_0(v, w) \subset \mu$ then $\theta_{v,w} \subset \mu$; and

(b) If μ is Θ-closed, then $\mu(\iota) \models E(F_{Q_0}, D_0)$.

Thus F_{Q_0} has the properties that if $\mu(\iota) \models E(F_{Q_0}, D_0)$ and the one member $\gamma_0(v, w)$ of $\gamma^{-1}(v, w)$ is a subgraph of μ, then all the other members of $\gamma^{-1}(v, w)$ are also subgraphs of μ; and if, for each $\langle v, w\rangle \in \Sigma^+ \times \Sigma^+$, either $\theta_{v,w} \subset \mu$ or $-\theta_{v,w} \subset \mu$, then $\mu(\iota) \models E(F_{Q_0}, D_0)$.

Instead of constructing Q_0 we construct F_{Q_0} directly. Let F_{Q_0} be the conjunction of all subformulas

$$\hat{\omega}(f_{\langle a, w\rangle}(x), f_b(y)) \to \hat{\omega}(f_{\langle a, wb\rangle}(x), y) ,$$

$$\hat{\omega}(f_{\langle a, w\rangle}(x), f_b(y)) \to \hat{\omega}(x, f_{\langle a, wb\rangle}(y)) ,$$

and

$$\hat{\omega}(f_a(x), f_{\langle v, wb\rangle}(y)) \to \hat{\omega}(x, f_{\langle va, wb\rangle}(y)) ,$$

for all $a, b \in \Sigma$ and $v, w \in \Sigma^{(\ell-1)}$. (Clearly this formula is F_{Q_0} for some set $Q_0 \subset \{0,1\} \times S^3$.) The proof of (a) is an easy induction. Intuitively, conjuncts of the first kind remove function signs from the right hand term and add the corresponding symbols to the right hand component of the pair $\langle v, w\rangle$; conjuncts of the third kind remove from and add to the left hand sides; and conjuncts of the second kind make the transition between operations on the right side and operations on the left. Note that only instances of F_{Q_0} over D_0 (not D) are needed. The proof of (b) is immediate from the fact that if $\hat{\omega}(s,t) \to \hat{\omega}(s', t')$ is an instance over D_0 of one of the conjuncts of F_{Q_0}, then $\omega(s,t)$ and $\omega(s', t')$ are members of Γ and $\gamma(\omega(s,t)) = \gamma(\omega(s',t'))$, so that $\omega(s,t)$ and $\omega(s',t')$ are subgraphs of the same member of Θ.

Now to complete the construction of the formula F_Q, conjoin to F_{Q_0} conjuncts

$$\hat{\omega}(f_a(x), f_{\langle v, w\rangle}(y)) \to \hat{\omega}(f_{\langle a, \epsilon\rangle}(x), y)$$

for each pair $\langle v, w\rangle \in \mathscr{P}$ and each $a \in \Sigma$. Since for $\langle v, w\rangle \in \mathscr{P}$ and any $v', w' \in \Sigma^*$, the set $\gamma^{-1}(vav', ww')$ contains $\omega(f_a(\delta^{-1}(v')), f_{\langle v, w\rangle}(\delta^{-1}(w')))$, it follows from property (a) of F_{Q_0} that if $\mu(\iota) \models E(F, D_0)$, then for any $\langle v, w\rangle \in \mathscr{P}$, $v' \in \Sigma^+$, and $w' \in \Sigma$, if $\gamma_0(vv', ww') \subset \mu$ then $\gamma_0(v', w') \subset \mu$. That is, if $\gamma_0(p)$ is a subgraph of μ for some pair of words p, then so is $\gamma_0(p')$ for each pair p' which can be obtained by removing words forming a pair in $\mathscr{P}$ from the beginnings of the words forming the pair p.

Finally, let T be such that

$$\eta_T = \sum_{a\in\Sigma} \sum_{t\in D_0} \omega(f_{\langle a, \epsilon\rangle}(t), f_a(t)) - \sum_{\substack{a\in\Sigma \\ \langle va, w\rangle \in \mathscr{P}}} \sum_{t\in D_0} \omega(f_a(t), f_{\langle v, w\rangle}(t));$$

note that $T \subset \{1\} \times S \times S_0 \cup \{-1\} \times S \times S$.

Then if $\mu(\iota) \models E(F, D_0)$ and $\eta_T \subset \mu$, then $\gamma_0(v, w) \subset \mu$ whenever $v, w \in \Sigma^+$ and for some presolution $\langle v', w'\rangle$ of $\mathscr{P}$, $v'v = w'w$. For if $v' = w' = \varepsilon$ and $v'v = w'w$, then $v = w$ and $\gamma_0(v, w) \subset \eta_T$; and the induction step follows from the fact that if $\langle v'', w''\rangle \in \mathscr{P}$ and $\gamma_0(v''v, w''w) \subset \mu$ then $\gamma_0(v, w) \subset \mu$. But then $\mathscr{P}$ has no solution. For if $\langle v'v, w'w\rangle$ were a solution and $\langle v, w\rangle \in \mathscr{P}$ (i.e., $\langle v, w\rangle$ is the last pair in used in forming the solution), then $\gamma_0(v, w) \subset \mu$, and hence by property (a) of F_{Q_0} also $\omega(f_a(c), f_{\langle v'', w\rangle}(c)) \subset \mu$, where $v = v''a$ and $a \in \Sigma$. But this is impossible since $-\omega(f_a(c), f_{\langle v'', w\rangle}(c)) \subset \eta_T$ by the definition of η_T.

Conversely, if $\mathscr{P}$ has no solution then let μ be the minimal Θ-completion of $\Sigma\, \gamma_0(v, w)$, the sum over all $v, w \in \Sigma^+$ such that for some presolution $\langle v', w'\rangle$ of $\mathscr{P}$, $v'v = w'w$. Then $\mu(\iota) \models E(F_{Q_0}, D_0)$ by property (b) of F_{Q_0}; $\mu(\iota)$ verifies each instance over D_0 of the additional conjuncts added to obtain F_Q from F_{Q_0}, because of the way presolutions give rise to other presolutions; and $\eta_T \subset \mu$ since $\langle \epsilon, \epsilon\rangle$ is a presolution but $\mathscr{P}$ has no solution.

This completes the proof of the Unsolvability Lemma. ■

Define a graph to be a sum of bigraphs $\omega(a, b)$ for various a, b. If ρ is a reflexive, transitive relation on graphs, then say that a bigraph ξ is ρ-closed if and only if $\beta \subset \xi$ whenever $\alpha \subset \xi$ and $\rho(\alpha, \beta)$, and call a set of bigraphs ρ-closed if each member is ρ-closed. This notion of closure is essentially a "one-way" version of the notion of Ξ-closure (for Ξ a set of bigraphs) used extensively in Chapter IIC; its introduction here arises from the use of simple conditionals, rather than biconditionals, in constructing formulas.

Sufficient Conditions Lemma. Let S, S_0, c, D, and D_0 be as stated in the Unsolvability Lemma; let d be a new constant and let

$$\lambda = \sum_{t \in D_0} \omega(t, d) .$$

For $m, e \in \{0, 1\}$ and function signs $f, g, h \in S$, let H_{mf}, J_{mefgh}, and J^*_{mefgh} be new monadic function signs. Let S_1 be a set of monadic function signs containing all the members of S, and also all the H_{mf}, J_{mefgh}, and J^*_{mefgh}. Let D_1 be the domain generated by the constants c, d and the function signs in S_1. Let ρ be a reflexive, transitive relation on graphs and let Ξ be a ρ-closed set of graphs with vertices in D_1. Let $\bar{\lambda}$ and θ be ρ-closed graphs, with $\lambda \subset \bar{\lambda}$, and for $m = 0, 1$, $h \in S$, $s, t \in D - \{c\}$, and $u \in D$, let π_{msu}, τ_{m0sth}, and τ_{m1uth} be members of Ξ. Suppose that the following conditions are satisfied.

Disjointness. All the graphs $\bar{\lambda}$, θ, π_{msu}, τ_{m0sth}, and τ_{m1sth} (as m, e range over $\{0, 1\}$, s, t over $D - \{c\}$, u over D, and h over S) are pairwise disjoint.

Subgraph. For any $s, t \in D$, $m, e \in \{0, 1\}$, and $f, g, h \in S$,

(a) $\omega_m(H_{mf}(s), t) \subset \pi_{mf(s)t}$

(b) $\omega_m(J_{mefgh}(s), J^*_{mefgh}(t)) \subset \pi_{mf(s)g(t)}$

(c) $\omega_m(J^*_{m0fgh}(t), s) \subset \tau_{m0f(s)g(t)h}$

(d) $\omega_m(J^*_{m1fgh}(t), s) \subset \tau_{m1sg(t)h}$

(e) $\rho(\omega_m(H_{mf}(s), g(t)), \pi_{mf(s)g(t)})$

Extraction. Let μ be any minimal Ξ-completion of $\overline{\lambda} + \theta$. Then for any $m, e \in \{0, 1\}$, $f, g, h \in S$, and $s, t \in D_1$,

(a) if $\omega_m(J_{mefgh}(s), t) \subset \mu$ then $s \in D$ and $t = J^*_{mefgh}(u)$ for some $u \in D$;

(b) if $\omega_m(J^*_{mefgh}(t), s) \subset \mu$ then either $s, t \in D$, or else $\omega_m(s, H_{\overline{m}k}(t)) \subset \theta$ for each $k \in S$;

(c) if $\omega(t, d) \subset \mu$ then $t \in D_0$.

Schema. There is a prenex schema G_0 in $\exists\exists\forall\exists^*\forall$ such that

(a) the indicial function signs of G_0^* are c, d, and the members of S_1;

(b) the functional form of G_0 is a conjunction of subformulas $\hat{\omega}(t_1, t_2) \rightarrow \hat{\omega}(t_3, t_4)$, where $t_1, \ldots, t_4$ are among c, d, x, y, and f(x), for $f \in S_1$ and x, y two fixed variables; and

(c) for any complete bigraph μ of dimension 2 with $V(\mu) = D_1$, $\mu(\iota) \models E(G_0^*, D_1)$ if and only if μ is ρ-closed.

Then for any sets T and Q as specified in the Unsolvability Lemma, there is a schema G of the form specified by the theorem such that the indicial function signs of G are c, d, and the members of S_1, and G is satisfiable if and only if $\nu(\iota) \models E(F_Q, D_0)$ for some completion ν of η_T.

Two comments are in order before beginning the proof. First, the Extraction Conditions are, except for (c), not definable by means of the formal notion of "extraction" of Chapter IIB. However, we use the term here because it is suggestive of the same phenomenon; certain conditions on μ single out certain vertices. The other comment is about the various domains relevant to this lemma. There are really four: D_0, D, D_1, and D(G), which are generated, respectively, by c and the function signs in S_0; c and those in S; c, d and those in S_1; and ł, c, d and those in S_1. Thus $D_0 \subset D \subset D_1 \subset D(G)$. Since (p. 69) G is satisfiable if and only if $E(G, D_1)$ is satisfiable, we henceforth use D_1 instead of D(G). Also, the relevance of D_0 in the construction of G is only through the definition of λ and Extraction Condition (c); the other parts of the lemma refer only to D or D_1. Finally, the only terms in D_1 relevant to the construction will be those of the form f(t), for $f \in S_1 - S$ and $t \in D$. That is, the outermost function signs are drawn from a larger set than the others. This situation is much like that in the proof of the Unsolvability Lemma; there the function signs in $S - S_0$ enter the proof only as the outermost function signs of terms.

Proof of the Sufficient Conditions Lemma. First let

$$M_0 = G_0^* \wedge \hat{\omega}(c, d) \wedge \bigwedge_{f \in S_0} (\hat{\omega}(x, d) \rightarrow \hat{\omega}(f(x), d)) .$$

If $\mu(\iota) \models E(M_0, D_1)$ then μ is ρ-closed by the Schema Condition and $\lambda \subset \mu$ by the definition of λ. Also, if μ is any minimal Ξ-completion of $\overline{\lambda} + \theta$ then $\mu(\iota)$ verifies each instance over D_1 of each conjunct of M_0. For (i) μ is ρ-closed since $\bar{\lambda}, \theta$, and each member of Ξ are ρ-closed, so $\mu(\iota) \models E(G_0^*, D_1)$ by the Schema Condition; (ii) $\lambda \subset \mu$ so that $\omega(c, d) \subset \mu$; and (iii) if $t \in D_1$ and $\omega(t, d) \subset \mu$ then $t \in D_0$ by Extraction Condition (c) and so $\omega(f(t), d) \subset \lambda \subset \mu$ for each $f \in S_0$. Hence

(M_0-A) if μ is any minimal Ξ-completion of $\overline{\lambda} + \theta$ then $\mu(\iota) \models E(M_0, D_1)$; and

(M_0-B) if $\mu(\iota) \models E(M_0, D_1)$ then $\lambda \subset \mu$ and μ is ρ-closed.

Now given any T and Q as stated in the Unsolvability Lemma, let $\eta = \eta_T$ and $F = F_Q$ be as defined there, and let G be that schema with the same prefix as G_0 whose functional form is

$$\begin{aligned}
&M_0 \\
&\wedge \bigwedge_{\langle 1, f, g\rangle \in T} (\hat{\omega}(g(x), d) \rightarrow \hat{\omega}(H_{0f}(x), g(x))) && (1) \\
&\wedge \bigwedge_{\substack{m=0,1 \\ f,g,h \in S}} (\hat{\omega}_m(J_{m0fgh}(x), y) \rightarrow \hat{\omega}_m(y, x)) && (2) \\
&\wedge \bigwedge_{\langle m, f, g, h\rangle \in Q} (\hat{\omega}_m(J^*_{m0fgh}(x), y) \rightarrow \hat{\omega}_m(y, H_{\overline{m}h}(x))) && (3) \\
&\wedge \bigwedge_{\substack{m=0,1 \\ f,g,h \in S}} (\hat{\omega}_m(J_{m1fgh}(x), y) \rightarrow \hat{\omega}_m(y, f(x))) && (4) \\
&\wedge \bigwedge_{\substack{m=0,1 \\ f,g,h \in S}} (\hat{\omega}_m(J^*_{m1fgh}(x), y) \rightarrow \hat{\omega}_m(y, H_{\overline{m}g}(x))) && (5) \\
&\wedge \bigwedge_{\langle -1, f, g\rangle \in T} \neg \hat{\omega}(H_{0f}(x), g(x)) && (6)
\end{aligned}$$

We must show that G is satisfiable if and only if $\nu(\iota) \models E(F, D_0)$ for some completion ν of η.

(F $\triangleright$ G). First suppose that $\nu(\iota) \models E(F, D_0)$, where ν is a completion of η. Let μ be the minimal completion of

$$\overline{\lambda} + \theta + \sum_{m,s,t} \chi_\nu(\omega_m(s, t)) \cdot \pi_{mst} + \sum_{m,e,s,t,h} \chi_\nu(\omega_m(s, t)) \cdot \tau_{mesth} ,$$

the summations being over all $m, e \in \{0, 1\}$, $s, t \in D$, and $h \in S$, for which π_{mst} and τ_{mesth} are defined. The sum is proper by the Disjointness Condition; and since each π_{mst} and τ_{mesth} is a member of Ξ, $\mu(\iota) \models E(M_0, D_1)$ by $(M_0\text{-}A)$. We next show that $\mu(\iota)$ verifies each instance over D_1 of each conjunct (1)-(6).

(1) If $\langle 1, f, g\rangle \in T$ and $\omega(g(t), d) \subset \mu$, then $g(t) \in D_0$ by Extraction Condition (c) and $\omega(f(t), g(t)) \subset \eta \subset \nu$. Hence $\pi_{0f(t)g(t)} \subset \mu$; but $\omega(H_{0f}(t), g(t)) \subset \pi_{0f(t)g(t)}$ by Subgraph Condition (a).

(2) If $\omega_m(J_{m0fgh}(s), t) \subset \mu$ then $s \in D$ and $t = J^*_{m0fgh}(u)$ for some $u \in D$ by Extraction Condition (a). Hence by Subgraph Condition (b), $\omega_m(J_{m0fgh}(s), t) \subset \pi_{mf(s)g(u)} \subset \mu$. Hence $\tau_{m0f(s)g(u)h} \subset \mu$; but by Subgraph Condition (c), $\omega_m(t, s) = \omega_m(J^*_{m0fgh}(u), s) \subset \tau_{m0f(s)g(u)h} \subset \mu$.

(3) By Extraction Condition (b), if $\omega_m(J^*_{m0fgh}(t), s) \subset \mu$ then either $\omega_m(s, H_{\overline{m}h}(t)) \subset \theta \subset \mu$, or else $s, t \in D$ and by Subgraph Condition (c), $\tau_{m0f(s)g(t)h} \subset \mu$. In the latter case, by the definition of μ, $\omega_m(f(s), g(t)) \subset \nu$; and since $\hat{\omega}_m(f(x), g(y)) \to \hat{\omega}_m(x, h(y))$ is a conjunct of F, also $\omega_m(s, h(t)) \subset \nu$, that is, $\omega_{\overline{m}}(h(t), s) \subset \nu$. Then $\pi_{\overline{m}h(t)s} \subset \mu$ by the definition of μ, and by Subgraph Condition (a), $\omega_{\overline{m}}(H_{\overline{m}h}(t), s) \subset \mu$, i.e., $\omega_m(s, H_{\overline{m}h}(t)) \subset \mu$.

(4) By Extraction Condition (a) again, if $\omega_m(J_{m1fgh}(s), t) \subset \mu$ then $s \in D$ and $t = J^*_{m1fgh}(u)$ for some $u \in D$, whence $\pi_{mf(s)g(u)} \subset \mu$. Then by Subgraph Condition (d), $\omega_m(t, f(s)) = \omega_m(J^*_{m1fgh}(u), f(s)) \subset \tau_{m1f(s)g(u)h} \subset \mu$.

(5) By Extraction Condition (b) again, if $\omega_m(J^*_{m1fgh}(t), s) \subset \mu$ then either $\omega_m(s, H_{\overline{m}g}(t)) \subset \theta \subset \mu$, or else $s, t \in D$ so that by Subgraph Condition (d), $\tau_{m1sg(t)h} \subset \mu$ and hence $\omega_m(s, g(t)) = \omega_{\overline{m}}(g(t), s) \subset \nu$. Then $\pi_{\overline{m}g(t)s} \subset \mu$ so by Subgraph Condition (a), $\omega_{\overline{m}}(H_{\overline{m}g}(t), s) = \omega_m(s, H_{\overline{m}g}(t)) \subset \mu$.

(6) Let $\langle -1, f, g\rangle \in T$ and $t \in D_1$. If $t \in D_1 - D$ then $-\omega(H_{0f}(t), g(t)) \subset \mu$ by the minimality of μ. Otherwise, by Subgraph Condition (e) and the fact that μ is ρ-closed, $\omega(H_{0f}(t), g(t)) \subset \mu$ only if $\pi_{0f(t)g(t)} \subset \mu$, i.e., (by the definition of μ) only if $t \in D_0$ and $\omega_0(f(t), g(t)) \subset \nu$. But $-\omega_0(f(t), g(t)) \subset \nu$ for all $t \in D_0$ since $\neg\,\hat{\omega}(f(x), g(x))$ is a conjuct of F, so $\mu(\iota) \models \neg\,\hat{\omega}(H_{0f}(t), g(t))$.

(<u>G ▷ F</u>) Now suppose that $\mu(\iota) \models E(G)$; then by $(M_0\text{-}B)$, $\lambda \subset \mu$ and μ is ρ-closed. Let ν be the minimal completion of $\Sigma\,\omega(s, t)$, the sum being over all s, t in D such that $\pi_{0st} \subset \mu$ or $\pi_{1ts} \subset \mu$. We first show that in fact $\chi_\mu(\pi_{0f(s)g(t)}) = \chi_\mu(\pi_{1g(t)f(s)})$ for any $f(s), g(t) \in D$. For if $\pi_{mf(s)g(t)} \subset \mu$ then $\omega_m(J_{m1fgh}(s), J^*_{m1fgh}(t)) \subset \mu$ for each $h \in S$ by Subgraph Condition (b); since $\mu(\iota) \models E(G)$, $\omega_m(f(s), H_{\overline{m}g}(t)) \subset \mu$ from conjuncts (4) and (5); and $\pi_{\overline{m}g(t)f(s)} \subset \mu$ by Subgraph Condition (e) and the fact that μ is ρ-closed.

Next we show that ν is a completion of η. Since $\lambda \subset \mu$, $\omega(t, d) \subset \mu$ for each $t \in D_0$. Thus if $\omega(f(t), g(t)) \subset \eta$ then $\omega(g(t), d) \subset \mu$ (here we use the fact that if $\langle 1, f, g\rangle \in T$ then $g \in S_0$) and by conjunct (1), $\omega(H_{0f}(t), g(t)) \subset \mu$. Then by Subgraph Condition (e) $\pi_{0f(t)g(t)} \subset \mu$ and hence $\omega(f(t), g(t)) \subset \nu$. Similarly, using conjunct (6), if $-\omega(f(t), g(t)) \subset \eta$ then $-\omega(H_{0f}(t), g(t)) \subset \mu$ so $-\omega(f(t), g(t)) \subset \nu$.

Finally, to show that $\nu(z) \models E(F, D_0)$, let $\hat{\omega}_m(f(s), g(t)) \rightarrow \hat{\omega}_m(s, h(t))$ be an instance over D_0 of a conjunct of F, and suppose that $\omega_m(f(s), g(t)) \subset \nu$. Then $\pi_{mf(s)g(t)} \subset \mu$ so by Subgraph Condition (b) $\omega_m(J_{m0fgh}(s), J^*_{m0fgh}(t)) \subset \mu$ and by conjuncts (2) and (3), $\omega_m(s, H_{\overline{m}h}(t)) = \omega_{\overline{m}}(H_{\overline{m}h}(t), s) \subset \mu$ so that $\pi_{\overline{m}h(t)s} \subset \mu$ by Subgraph Condition (e). Hence $\omega_{\overline{m}}(h(t), s) = \omega_m(s, h(t)) \subset \nu$ as required.

This completes the proof of the Sufficient Conditions Lemma. ∎

To complete the proof of the theorem we must define ρ, Ξ, $\overline{\lambda}$, θ, π_{mst}, and τ_{mesth} so as to meet the sufficient conditions. Actually, we do this in a backwards fashion; we first define the schema G_0 described in the Sufficient Conditions Lemma, then define the other objects in terms of G_0. Then the Schema Condition and the Subgraph Conditions will follow easily; the major difficulties will lie with the Disjointness and Extraction Conditions.

Let S, S_0, D, D_0 be as in the Unsolvability and Sufficient Conditions Lemmata. Let S contain p function signs $f_1, f_2, \ldots, f_p$, and let $R = \{3^i \cdot 5^j \cdot 7^k \cdot 11^m \cdot 13^e \mid 1 \leq i, j, k \leq p,\ m, e = 0, 1\}$. Here $3^i \cdot 5^j \cdot 7^k \cdot 11^m \cdot 13^e$ is meant to furnish two encodings of the formula $\hat{\omega}_m(f_i(x), f_j(y)) \rightarrow \hat{\omega}_m(x, f_k(y))$ as $e = 0, 1$. Let $r = \max R$, and for $1 \leq i \leq p$, $0 \leq j < r$, and $m \in \{0, 1\}$ let g_{ijm} and g^*_{ijm} be new monadic function signs which are the monadic indicial correlates of certain variables. Then let $S_1 = S \cup \{g_{ijm}, g^*_{ijm} \mid 1 \leq i \leq p,\ 0 \leq j < r,\ m \in \{0, 1\}\}$. In particular, let

$$H_{mf_i} = g_{i0m}$$

$$J_{mef_if_jf_k} = g_{i\ell m}, \qquad \text{where } \ell = 3^i \cdot 5^j \cdot 7^k \cdot 11^m \cdot 13^e - 1$$

$$J^*_{mef_if_jf_k} = g^*_{j\ell\overline{m}}, \qquad \text{where } \ell = 3^i \cdot 5^j \cdot 7^k \cdot 11^m \cdot 13^e - 1 .$$

<u>Construction Lemma.</u> Let G_0 be a schema whose functional form is

$$\bigwedge_{i=1}^{p} \bigwedge_{m=0}^{1} \bigwedge_{j=0}^{r-2} (\hat{\omega}_m(g_{ijm}(x), y) \rightarrow \hat{\omega}_m(y, g_{i,j+1,m}(x))) \qquad (1)$$

$$\wedge \bigwedge_{j=1}^{r-2} (\hat{\omega}_m(g^*_{ij\overline{m}}(x), y) \rightarrow \hat{\omega}_m(y, g^*_{i,j+1,\overline{m}}(x))) \qquad (2)$$

$$\wedge\ (\hat{\omega}_m(f_i(x), y) \rightarrow \hat{\omega}_m(y, g^*_{i1\overline{m}}(x))) \quad . \qquad (3)$$

Define $\rho(\alpha,\beta)$, where α and β are graphs, as follows: $\rho(\alpha,\beta)$ if and only if, for each $\omega(a,b) \subset \beta$, there is an $\omega(a',b') \subset \alpha$, such that $\hat{\omega}(a',b') \to \hat{\omega}(a,b)$ is a logical consequence of $E(G_0^*, D)$. Let $\overline{\rho}(\alpha) = \Sigma_{\rho(\alpha,\beta)} \beta$, and define $\Xi = \{\overline{\rho}(\omega(s,t)) \mid s,t \in D_1\}$. Let $\overline{\lambda} = \overline{\rho}(\lambda)$;

$$\pi_{mf(s)t} = \overline{\rho}(\omega_m(H_{mf}(s),t)) \quad ;$$

$$\tau_{m0f(s)g(t)h} = \overline{\rho}(\omega_m(J^*_{m0fgh}(t), s)) \quad ;$$

$$\tau_{m1sg(t)h} = \sum_{i=1}^{p} \overline{\rho}(\omega_m(J^*_{m1f_igh}(t), s)) \quad ;$$

$$\theta = \sum_{s,t \in D_0} \sum_{m=0}^{1} \sum_{i,i'=1}^{p} \sum_{\substack{j,j'=0 \\ j+1 \notin R \\ j'+1 \notin R}}^{r-1} \overline{\rho}(\omega_m(g_{ijm}(s), g_{i'j'\overline{m}}(t)) + \omega_m(g_{i0m}(s), g^*_{i'j'm}(t))) .$$

Then $\overline{\lambda}$, θ, the π_{mst}, and the τ_{mesth} are as stated in the Sufficient Conditions Lemma.

Proof. Clearly ρ is reflexive and transitive, and Ξ is ρ-closed. Subgraph Conditions (a), (c), (d), and (e) follow immediately from the definitions. For Subgraph Condition (b), note that the following is a sequence $A_1, A_2, \ldots, A_q$ of atomic formulas such that $A_i \to A_{i+1}$ is a conjunct of some member of $E(G_0)$ for $1 \le i < q$ (the parentheses indicate from which conjunct each pair is derived):

$\hat{\omega}_m\left(H_{mf_i}(s), f_j(t)\right)$, i.e., $\hat{\omega}_m(g_{i0m}, f_j(t))$,

(from (1))

$\hat{\omega}_m(f_j(t), g_{i1m}(s))$,

(from (2))

$\hat{\omega}_m(g_{i1m}(s), g^*_{j1\overline{m}}(t))$,

(from (1))

$\hat{\omega}_m(g^*_{j1\overline{m}}(t), g_{i2m}(s))$,

(from (3))

$\hat{\omega}_m(g_{i2m}(s), g^*_{j2m}(t))$,

...

(from (1) and (3))

$\hat{\omega}_m(g_{i\ell m}(s), g^*_{j\ell m}(t))$ (where $\ell = 3^i \cdot 5^j \cdot 7^k \cdot 11^m \cdot 13^e - 1$)

i.e., $\hat{\omega}_m\left(J_{mef_if_jf_k}(s), J^*_{mef_if_jf_k}(t)\right)$.

Also, the Schema Condition is satisfied by virtue of the way ρ is defined from G_0.

Before proceeding to prove that the Disjointness and Extraction Conditions are satisfied, we present explicit values for $\overline{\lambda}$, the π_{mst} and τ_{mesth}, and θ. Checking that these sums are correct is a tedious but straightforward task, which requires proving, for each sum σ, that if $\omega(s,t) \subset \sigma$ and $\hat{\omega}(s,t) \to \hat{\omega}(s',t')$ is a conjunct of some member of $E(G_0)$ then $\omega(s',t') \subset \sigma$. This in turn requires matching $\omega(s,t)$ against the antecedent of each conditional in the matrix of G_0^*. We omit the details and simply state the results.

$$\overline{\lambda} = \lambda + \sum_{\substack{s \in D_0 \\ 1 \le i \le p}} \omega(d, g_{i11}^*(s))$$

$$\pi_{mf_i(s)c} = \omega_m(g_{i0m}(s), c) + \omega_m(c, g_{i1m}(s))$$

$$\pi_{mf_i(s)f_j(t)} = \omega_m(g_{i0m}(s), f_j(t)) + \omega_m(f_j(t), g_{i1m}(s)) + \sum_{k=1}^{r-1} \omega_m(g_{ikm}(s), g_{jk\overline{m}}^*(t))$$

$$+ \sum_{k=1}^{r-2} \omega_m(g_{jk\overline{m}}^*(t), g_{j,k+1,m}(s))$$

The graphs τ_{mesth} are defined in terms of the graphs $\overline{\rho}(\omega_m(J_{mefgh}^*(t), s))$ for various m, e, f, g, h, so we simply give those values. Let $\ell = 3^i \cdot 5^j \cdot 7^k \cdot 11^m \cdot 13^e - 1$; then

$$\overline{\rho}\left(\omega_m\left(J_{mef_if_jf_k}^*(t), c\right)\right) = \omega_m(g_{i\ell\overline{m}}^*(t), c) + \omega_m(c, g_{i,\ell+1,\overline{m}}^*(t)) \quad ;$$

and

$$\overline{\rho}\left(\omega_m\left(J_{mef_if_jf_k}^*(t), f_{i'}(u)\right)\right) = \omega_m(g_{j\ell\overline{m}}^*(t), f_{i'}(u)) + \omega_m(f_{i'}(u), g_{j,\ell+1,\overline{m}}^*(t))$$

$$+ \sum_{k=1}^{r-\ell-2} (\omega_m(g_{j,\ell+k,\overline{m}}^*(t), g_{i'k\overline{m}}^*(u))$$

$$+ \omega_m(g_{i'k\overline{m}}^*(u), g_{j,\ell+k+1,\overline{m}}^*(t)))$$

$$+ \omega_m(g_{i'1m}^*(u), g_{j\ell\overline{m}}^*(t)) \quad .$$

Finally,

$$\theta = \sum_{s,t \in D_0} \sum_{m=0}^{1} \sum_{i,i'=1}^{p} \sum_{\substack{j,j'=0 \\ j+1 \notin R \\ j'+1 \notin R}}^{r-1} (\omega_m(g_{ijm}(s), g_{i'j'\overline{m}}(t)) + \omega_m(g_{i'j'\overline{m}}(t), g_{i,j+1,m}(s))$$

$$+ \omega_m(g_{i0m}(s), g_{i'j'm}^*(t)) + \omega_m(g_{i'j'm}^*(t), g_{i1m}(s))) \; .$$

Proof of Disjointness Conditions: We must show that $\overline{\lambda}$, θ, the π_{mst}, and the τ_{mesth} are pairwise disjoint. Clearly, $\overline{\lambda}$ is disjoint from the others since if $\omega(s,t) \subseteq \overline{\lambda}$ then either $s = d$ or $t = d$. Also, θ is disjoint from the τ_{mesth}, since θ is a sum of graphs of the forms $\omega(g(-), g(-))$, $\omega(g(-), g^*(-))$, and $\omega(g^*(-), g(-))$, while each τ_{mesth} is a sum of graphs of the forms $\omega(g^*(-), c)$ $\omega(c, g^*(-))$, $\omega(g^*(-), f(-))$, $\omega(f(-), g^*(-))$, and $\omega(g^*(-), g^*(-))$. Also θ is disjoint from the π_{mst}; the only conflicts would be on subgraphs of the form $\omega_m(g_{ijn}(-), g^*_{i'j'n'}(-))$, but here θ would have $n = n'$ and π_{mst} would have $n = \overline{n}'$. The π_{mst} are disjoint from the $\tau_{m'es't'h}$ by a similar argument, as are the π_{mst} from each other and the τ_{mesth} from each other.

Proof of Extraction Conditions: (a) For any ℓ such that $\ell+1 \in R$, $\omega_m(g_{i\ell m}(-), -)$ is not a subgraph of any $\tau_{m'esth}$, or of $\overline{\lambda}$; nor is it a subgraph of θ, since $\ell+1 \in R$. The result then follows from inspection of the $\pi_{m'st}$.

(b) Similarly, $\omega_m(g^*_{j\ell\overline{m}}(-), -)$, where $\ell+1 \in R$, is not a subgraph of $\overline{\lambda}$ or of θ. If $\omega_m(g^*_{j\ell\overline{m}}(t), s) \subseteq \pi_{m's't'}$ for some m', s', t' then $s = g_{j,\ell+1,m}(s'')$ for some $s'' \in D$, and for any i, $1 \le i \le p$, $\omega_m(s, H_{\overline{m}f_i}(t)) = \omega_m(g_{j,\ell+1,m}(s''),$ $g_{i0\overline{m}}(t)) = \omega_{\overline{m}}(g_{i0\overline{m}}(t), g_{j,\ell+1,m}(s''))$, which is a subgraph of θ since $\ell+2 \notin R$ if $\ell+1 \in R$. And if $\omega_m(g^*_{i\ell\overline{m}}(t), s) \subseteq \tau_{m'es't'h}$ then either $s = f_{i'}(u)$ for some i', $1 \le i' \le p$, and some $u \in D$, or else $s = g^*_{jk\overline{m}}(u)$ for some $u \in D$, where either $k = \ell' + \ell + 1$ for some ℓ' such that $\ell'+1 \in R$, or $k = \ell - \ell'$ for some ℓ' such that $\ell'+1 \in R$. In either of the last two cases $\omega_m(s, H_{\overline{m}f_i}(t)) = \omega_{\overline{m}}(g_{i0\overline{m}}(t),$ $g_{jk\overline{m}}(u)) \subseteq \theta$ for any i, $1 \le i \le p$, since if $\ell+1 \in R$ and $\ell'+1 \in R$ then $\ell'+\ell+2 \notin R$ and $\ell-\ell'+1 \notin R$. For $\ell, \ell' \equiv 0 \pmod 2$, so $\ell+\ell'+2 \equiv 0 \pmod 2$; and $\ell, \ell' \equiv 2 \pmod 3$, so $\ell - \ell'+1 \equiv 1 \pmod 3$. But $k \in R$ only if $k \equiv 1 \pmod 2$ and $k \equiv 0 \pmod 3$.

(c) follows immediately by inspection of the sums.

This completes the proof of the Construction Lemma and the proof of the Krom $\cap$ Horn (0, 1) Theorem. ■

IIE.3 Joint Classification by Prefix and Similarity Type

The results of Section IIE.1 and Section IIE.2 are, except for the open question of $\forall\exists\forall\exists^* \cap$ Krom, the strongest possible for prefix and similarity subclasses of Krom or Horn, when prefix and similarity type are considered separately. When these parameters are considered jointly only much weaker results are known, but it is worth noting what kinds of methods have been applied along these lines.

(a) Bounded Prefixes with Unbounded Similarity Types

Scrutiny of the proofs of Section IIE. 1 reveals that they establish the unsolvability of these subclasses of Krom and Horn:

$\exists\forall\exists\forall\ (0,\infty) \cap \text{Krom} \cap \text{Horn}$

$\forall\exists\exists\forall\ (0,\infty) \cap \text{Krom} \cap \text{Horn}$

$\forall\exists\forall\forall\ (0,0,\infty) \cap \text{Krom} \cap \text{Horn}$

$\forall\forall\exists\forall\ (0,0,\infty) \cap \text{Krom} \cap \text{Horn}$

$\forall\exists\forall\exists\ (0,\infty) \cap \text{Horn}$

$\forall\forall\forall\exists\ (0,\infty) \cap \text{Horn}$

In each case the method was to reduce the halting problem for "inputless" special two-counter machines by associating a dyadic or triadic predicate letter with each state. The unlimited number of states of the machines thus gives rise to unlimited numbers of such predicate letters in the schemata. Now if the two-counter machines are regarded not as inputless, but as having an input supplied as a number initially on the counters, then there are single machines with unsolvable halting problems, in the sense that their behavior with various inputs cannot be effectively predicted as a function of the input (Special Two-Counter Theorem, Universal Version, p. 62).

This observation can be used to sharpen the above results on Krom and Horn schemata to the extent of replacing $(0,\infty)$ by (∞,k) and $(0,0,\infty)$ by $(0,\infty,k)$ in the list, where k is some fixed (but large) number--the number of states in the smallest special two-counter machine with an unsolvable halting problem. (Sharper results, with $k = 1$ or 2, for example, would require other methods; but no solvability results in this direction are known, so it is possible that k can in each case be reduced to 1.) For example, the $\exists\forall \wedge \forall\exists\forall\ (\infty,k) \cap \text{Krom} \cap \text{Horn}$ class can be shown unsolvable by choosing a fixed "universal" machine $\mathcal{M}$ and encoding it as in part (i) of the proof in Section IIE. 1 (p. 158), except that the conjunct P_1zz in the $\exists z\forall y$ part of the matrix, which "starts" the machine, is omitted and replaced by an atomic formula A_0z. Then to get an infinite class of schemata with an unsolvable decision problem, introduce, for various n, monadic letters $A_1,\ldots,A_n$, and add to the $\forall y_1\exists x\forall y_2$ part of the matrix the conjunction

$$\bigwedge_{i=0}^{n-1} (A_iy_1 \rightarrow A_{i+1}x) \wedge (A_ny_1 \rightarrow P_1y_1y_1)$$

which in effect counts to n and then starts the machine with n on both counters. A similar method works for the $\forall\exists\forall\forall$ and $\forall\forall\exists\forall$ prefixes, using dyadic letters to represent the input relative to a base. The two Horn classes $\forall\exists\forall\exists\ (\infty,k)$ and $\forall\forall\forall\exists\ (\infty,k)$ may be obtained, as before, by reducing certain Krom classes (though k may have to be increased slightly).

(b) Unbounded-Existentials Classes with Bounded Similarity Type

Another way of "preparing input" is to introduce initial existential variables $z_0, \ldots, z_n$ and to include a subformula

$$\exists z_0 \ldots \exists z_n \forall y_1 \exists x \forall y_2 \left(P z_0 z_0 \wedge \left(\bigwedge_{i=0}^{n-1} (P y_1 z_i \rightarrow P x z_{i+1}) \right) \wedge \ldots \right)$$

which counts in the first argument of a dyadic letter. This method, applied to the $\exists\forall\exists\forall$ prefix type and using a universal machine, yields the unsolvability of $\exists^*\forall\exists\forall\ (0,k) \cap \text{Krom} \cap \text{Horn}$ for some k. A similar method would work for the $\forall\exists\exists\forall$ prefix type, but here the construction of Section IIE. 2 yields a stronger result. The schemata constructed there can, by careful prenexing, be given the prefix type $\exists\forall\exists^*\forall$ instead of $\exists\exists\forall\exists^*\forall$ as noted on p. 167; thence directly, or by straightforward reduction, follow the unsolvability of

$\exists\forall\exists^*\forall\ (0,1) \cap \text{Krom} \cap \text{Horn}$

$\forall\exists^*\forall\ (1,1) \cap \text{Krom} \cap \text{Horn}$

$\forall\exists^*\forall\ (0,2) \cap \text{Krom} \cap \text{Horn}$.

The class $\forall\forall\forall\exists \wedge \exists^*\ (0,k) \cap \text{Horn}$ can be shown unsolvable by starting from the construction in Section IIE. 1 that yields a schema in $\forall\forall\forall\exists \cap \text{Horn}$ for each special two-counter machine (p. 161). This schema has a predicate letter S, which represents the successor relation between numbers. Apply this construction to a universal machine; introduce four new dyadic predicate letters P, Q, Q', R and conjoin $\exists z_0 \ldots \exists z_n\ (\bigwedge_{i=0}^{n-1} P z_i z_{i+1}) \wedge Q z_0 z_0 \wedge Q' z_n z_n$ to the schema; and then conjoin to the matrix of the $\forall y_1 \forall y_2 \forall y_3 \exists x$ part of the schema the conjuncts

$$(Q y_1 y_2 \wedge S y_1 y_3 \rightarrow R y_3 y_2) \wedge (R y_1 y_2 \wedge P y_2 y_3 \rightarrow Q y_1 y_3) \quad ,$$

which have the effect of "counting" in the first argument of Q. Finally, replace the subformula $P_1 vv$, which "starts the machine", by

$$(Q y_1 y_2 \wedge Q' y_2 y_2 \rightarrow P_1 y_1 y_1) \quad .$$

The class $\forall\exists\forall\exists^*\ (0,k) \cap \text{Horn}$ appears to be more difficult; perhaps the methods of Section IIC.5 can be adapted.

Although the problem of reducing k to 1 in any of these cases appears to be difficult, it is an easy matter to use x-variables to allow replacement of several predicate letters by a single predicate letter of higher degree. Thus $\mathscr{P}(0,\infty) \cap \text{Krom}$ is reducible to $\exists^*\mathscr{P}(0,0,1) \cap \text{Krom}$ for any prefix class $\mathscr{P}$, which implies the unsolvability of $\exists^*\forall\exists\forall\ (0,0,1)$; other results of this kind are also possible.

(c) Unbounded-Universals Classes with Bounded Similarity Type

There remain the cases with bounded similarity type but an unlimited number of $\forall$'s. For the class $\forall^*\exists(0,k) \cap$ Horn a construction is easily obtained, using a clause such as

$$(Sy_1y_2 \wedge Sy_2y_3 \wedge \cdots \wedge Sy_{n-1}y_n \rightarrow P_1y_ny_n)$$

to prepare a large input. But this method does not work for Krom, or for any $\mathscr{C}_n$, for that matter. In fact, we conjecture that no results along these lines are possible.

Conjecture. Let σ be a mapping from $\mathbb{N} - \{0\}$ to $\mathbb{N}$ (not $\mathbb{N} \cup \{\infty\}$) and let $\mathscr{P} \subset \{\forall, \exists\}^*$ be such that, for some k, no member of $\mathscr{P}$ contains more than k occurrences of $\exists$. Then for each n, $\mathscr{P}(\sigma) \cap \mathscr{C}_n$ is solvable.

That is, if the size of the clauses is bounded, then in order to capitalize on the availability of an unlimited number of universal quantifiers, either an unlimited number of predicate letters, or predicate letters of unlimited degree, must also be available.

IIE.4 The Two-Disjunct Class

In this section we prove the unsolvability of $\mathscr{D}_2$, the class of schemata in disjunctive normal form whose matrices have only two disjuncts. This result is an easy consequence of the Krom $\cap$ Horn (0,1) Theorem.

Two-Disjunct Theorem. $\mathscr{D}_2$ is unsolvable.

Proof. Let $F = QF^M$ be any schema of the form described by the Krom $\cap$ Horn (0,1) Theorem. That is, F^M is a conjunction of the form

$$\bigwedge_{\langle v_1, v_2\rangle \in S^+} \hat{\omega}(v_1, v_2)$$

$$\wedge \bigwedge_{\langle v_1, v_2\rangle \in S^-} \neg\hat{\omega}(v_1, v_2)$$

$$\wedge \bigwedge_{\langle v_1, \ldots, v_4\rangle \in T} (\hat{\omega}(v_1, v_2) \rightarrow \hat{\omega}(v_3, v_4))$$

where S^+, S^- are sets of pairs of variables and T is a set of quadruples of variables. Then let $z_1, \ldots, z_4$ be variables not occurring in F, and let $G = Q\forall z_1 \forall z_2 \forall z_3 \forall z_4 G^M$, where

$$G^M = \bigwedge_{\langle v_1,\ldots,v_4\rangle \in T} \Bigg(\bigg(\bigwedge_{\langle u_1,u_2\rangle \in S^+} \hat{\omega}(u_1,u_2,v_1,v_2,v_3,v_4)\bigg) \quad (1)$$

$$\wedge \bigg(\bigwedge_{\langle u_1,u_2\rangle \in S^-} \neg\, \hat{\omega}(u_1,u_2,v_1,v_2,v_3,v_4)\bigg) \quad (2)$$

$$\wedge\ (\hat{\omega}(z_1,z_2,z_1,z_2,z_3,z_4) \rightarrow \hat{\omega}(z_3,z_4,v_1,v_2,v_3,v_4))\Bigg). \quad (3)$$

Note first that G may be drawn into disjunctive normal form so as to become a member of $\mathscr{D}_2$; this is evident from the fact that all the conjuncts $A \rightarrow B$ have the same antecedent.

Suppose first that F is satisfiable, and let $\mathscr{A} \models E(F)$. Define a truth-assignment $\mathscr{B}$ on $E(G)$ as follows: $\mathscr{B} \models \hat{\omega}(s_1,s_2,t_1,t_2,t_3,t_4)$, where $s_1,s_2,t_1,\ldots,t_4 \in D(G)$, if and only if $\mathscr{A} \models \hat{\omega}(s_1,s_2)$ and $\hat{\omega}(t_1,t_2) \rightarrow \hat{\omega}(t_3,t_4)$ is a conjunct of some member of $E(F)$. Then clearly $\mathscr{B}$ verifies each Herbrand instance of a conjunct from (1) or (2). And if

$$\hat{\omega}(s_1,s_2,t_1,t_2,t_3,t_4) \rightarrow \hat{\omega}(s_1',s_2',t_1',t_2',t_3',t_4')$$

is an Herbrand instance of some conjunct from (3), then $s_1 = t_1$, $s_2 = t_2$, $t_3 = s_1'$, $t_4 = s_2'$. Now if $\mathscr{B}$ verifies the antecedent, then $\mathscr{A} \models \hat{\omega}(s_1,s_2)$ and $\hat{\omega}(t_1,t_2) \rightarrow \hat{\omega}(t_3,t_4)$ is an Herbrand instance of some conjunct of F^M. But then $\mathscr{A} \models \hat{\omega}(t_3,t_4)$, i.e., $\mathscr{A} \models \hat{\omega}(s_1',s_2')$. Since also $\hat{\omega}(t_1',t_2') \rightarrow \hat{\omega}(t_3',t_4')$ is an Herbrand instance of some conjunct of F^M, it follows that $\mathscr{B}$ verifies the consequent as well.

Conversely, suppose that G is satisfiable and let $\mathscr{B} \models E(G)$. Then let $\mathscr{A} \models \hat{\omega}(s_1,s_2)$ if and only if $\mathscr{B} \models \hat{\omega}(s_1,s_2,t_1,t_2,t_3,t_4)$ for <u>every</u> $t_1,\ldots,t_4$ such that $\hat{\omega}(t_1,t_2) \rightarrow \hat{\omega}(t_3,t_4)$ is an Herbrand instance of some conjunct of F^M. Again, it is easy to see that $\mathscr{A}$ verifies each signed atomic formula that is a conjunct of an Herbrand instance of F. And if $\hat{\omega}(t_1,t_2) \rightarrow \hat{\omega}(t_3,t_4)$ is an Herbrand instance of a conjunct of F^M then

$$\hat{\omega}(t_1,t_2,t_1,t_2,t_3,t_4) \rightarrow \hat{\omega}(t_3,t_4,t_1',t_2',t_3',t_4')$$

is a conjunct of an Herbrand instance of G for <u>every</u> t_1', t_2', t_3', t_4' such that $\hat{\omega}(t_1',t_2') \rightarrow \hat{\omega}(t_3',t_4')$ is a conjunct of an Herbrand instance of F. Thus if $\mathscr{A} \models \hat{\omega}(t_1,t_2)$, then $\mathscr{B} \models \hat{\omega}(t_1,t_2,t_1,t_2,t_3,t_4)$, hence $\mathscr{B} \models \hat{\omega}(t_3,t_4,t_1',t_2',t_3',t_4')$ for all such t_1', t_2', t_3', t_4', and hence $\mathscr{A} \models \hat{\omega}(t_3,t_4)$.

This completes the proof of the Two-Disjunct Theorem. ■

<u>Historical References</u>. The study of Krom schemata was motivated by Herbrand's result on $\mathscr{C}_1$ (1931) discussed in Section IIE. 1, and by the reduction by Chang and Keisler (1962) of any formula to an equivalent one in $\mathscr{C}_3$ (this is a much stronger result than showing that $\mathscr{C}_3$ is an unsolvable or reduction class, since it requires the formula in $\mathscr{C}_3$ to have exactly the same models as the original formula). Krom first found (1966) a formula not equivalent to any formula in $\mathscr{C}_2$; thus the strong reduction theorem of Chang and Keisler cannot be extended. He later showed (1970) that nevertheless, $\mathscr{C}_2$ is unsolvable. Two-counter machines are used by Aanderaa (1971) to provide a simpler proof of the unsolvability of $\mathscr{C}_2$, including the $\exists\forall\exists\forall$ and $\forall\exists\exists\forall$ prefix subclasses; the $\forall\exists\forall\forall$ and $\forall\forall\exists\forall$ subclasses were treated by Aanderaa and Lewis (1973). Börger (1971, 1974) gives many results on subclasses of $\mathscr{C}_2$, including cases in which prefix and similarity type are considered jointly. Orevkov (1971) also has a number of results of this sort. The solvable class $\exists^*\forall^*\exists^* \cap \text{Krom}$ is due to Maslov (1964); the $\forall\exists\forall \cap \text{Krom}$ class to Aanderaa (1971) (see also Aanderaa and Lewis (1973)). Finally, we note that Aanderaa (personal communication) showed the solvability of $\exists \wedge \forall\exists\forall \cap \text{Krom}$.

The unsolvability of $\text{Krom} \cap \text{Horn}(0, 1)$ was shown by Lewis (1974, 1976). That of $\mathscr{D}_2$ is due to Orevkov (1968); the proof presented here is based on that of Lewis (1976).

Chapter IIF

SCHEMATA WITH FEW ATOMIC SUBFORMULAS

In this chapter we consider classes of schemata restricted as to the number of atomic subformulas a schema may have. To be specific, let $\mathscr{N}_n$ be the class of schemata with exactly n distinct atomic subformulas (but possibly containing several occurrences of some of these subformulas). The construction presented here starts with schemata that belong to a fixed prefix class $\mathscr{P}$ and whose matrices are conjunctions of subformulas of the three forms A, $A \rightarrow B$, and $\neg A$, where A and B are atomic. Three lemmata reduce this class of schemata to subclasses of $\mathscr{P}\forall^* \cap \mathscr{N}_6$, $\mathscr{P}\forall^* \cap \mathscr{N}_5$, and $\mathscr{P}\forall^* \cap \mathscr{N}_4$, respectively.

Schemata with prefix type $\forall\exists\forall\forall$ whose matrices are conjuctions of such subformulas are shown to form an unsolvable class by the Krom and Horn Prefix Theorem, so the first part of the following theorem is immediate; the second part is a corollary.

<u>Few Atomic Subformula Theorem.</u> $\forall\exists\forall^* \cap \mathscr{N}_4$ and $\forall^*\exists \cap \mathscr{N}_5$ are unsolvable.

Whether these results can be strengthened to yield the unsolvability of $\mathscr{N}_3$ we do not know; however, $\mathscr{N}_2$ is solvable (Goldfarb 1974, Dreben and Goldfarb 1979).

As for the classification by both prefix and number of atomic subformulas, we know that in one sense these results are the strongest possible: there is no unsolvable class $\mathscr{P} \cap \mathscr{N}_k$ for any k, if $\mathscr{P}$ is a set of prefix types not containing words with an unlimited number of $\forall$'s (Dreben and Goldfarb 1979). Of course this does not rule out the possibility that $\mathscr{P} \cap \mathscr{N}_3$ is unsolvable for some $\mathscr{P}$, but not for $\mathscr{P} = \forall\exists\forall^*$; and there also remains the question of such classes as $\forall^*\exists\forall \cap \mathscr{N}_4$.

Moreover, it is clear that to show $\mathscr{N}_k$ to be unsolvable for some fixed k we must consider schemata with predicate letters of arbitrarily great degree; for to restrict both the number of atomic formulas and the number of arguments is to consider only a finite number of logically inequivalent formulas. Thus questions of similarity type are, in this context, trivial. To deal with predicate letters of unlimited degree we require special notation.

As described on p. 86 , a letter with a wavy underline, such as $\underset{\sim}{x}$, denotes a sequence $x_0, \ldots, x_{n-1}$. A <u>vector</u> is a bigraph $\omega(\underset{\sim}{t})$ for some $\underset{\sim}{t}$, i.e., $\omega(t_0, \ldots, t_{n-1})$ for some not necessarily distinct $t_0, \ldots, t_{n-1}$. To make the dimension explicit, we sometimes write $\omega_n(\underset{\sim}{t})$ instead of $\omega(\underset{\sim}{t})$. If $\eta = \omega_n(\underset{\sim}{t})$ and $\theta = \omega_m(\underset{\sim}{s})$ are vectors, then $\eta^\frown\theta = \omega_{n+m}(t_0, \ldots, t_{n-1}, s_0, \ldots, s_{m-1})$.

Next, let $v_0, \ldots, v_{n-1}$ be distinct variables, let $t_0, \ldots, t_{n-1}$ be terms, and let θ be a bigraph such that $V(\theta) = \{v_0, \ldots, v_{n-1}\}$; then $\theta[v_0/t_0, \ldots, v_{n-1}/t_{n-1}]$, or $\theta[\underset{\sim}{v}/\underset{\sim}{t}]$, is the result of replacing each v_i by t_i in θ. Note that if $\eta = \theta[\underset{\sim}{v}/\underset{\sim}{t}]$, then $\hat{\eta} = \hat{\theta}[\underset{\sim}{v}/\underset{\sim}{t}]$; in this case η is said to be an <u>instance</u> of θ. In particular, η is an <u>Herbrand instance</u> of θ (with respect to a schema F) if $\hat{\theta}$ is an atomic subformula of F and $\hat{\eta}$ is an Herbrand instance of $\hat{\theta}$. Also, if $\theta = \alpha^\frown\beta$, and $\eta = \alpha'^\frown\beta'$ is an Herbrand instance of θ, where α' is an instance of α and β' of β, then α', β' are said to be Herbrand instances of α and β, respectively (always with respect to a particular schema F, which is not, however, usually mentioned explicitly).

One final convention: an $\forall$ or $\exists$ preceding a sequence of variables stands for the same sequence with the $\forall$ or $\exists$ distributed through it; thus $\forall xy\underset{\sim}{z}$, where $\underset{\sim}{z}$ is the sequence $z_0, \ldots, z_{n-1}$, stands for $\forall x \forall y \forall z_0 \ldots \forall z_{n-1}$.

IIF. 1 <u>Six Atomic Formulas</u>

<u>Six Atomic Formula Lemma</u>. Let $\mathscr{P}$ be a prefix type such that an $\exists$ follows an $\forall$ in $\mathscr{P}$, and let F be a schema with prefix type $\mathscr{P}$ whose matrix is a conjuction of subformulas of the three forms A, $\neg$A, and A $\to$ B, where A and B are atomic. Then we can construct a schema $G \in \mathscr{P}\forall^*$ such that G is satisfiable if and only if F is satisfiable and

$$G^M = \hat{\beta}_0 \wedge \neg\hat{\beta}_1 \wedge (\hat{\beta}_2 \to \hat{\beta}_3) \wedge (\hat{\beta}_4 \to \hat{\beta}_5)$$

where $\beta_0, \ldots, \beta_5$ are vectors of the same dimension such that

(1) All the variables of β_2, β_3, β_4 are y-variables of G governing no x-variable of G;

(2) $\beta_2 = \omega(\underset{\sim}{u})$, $\beta_3 = \omega(\underset{\sim}{v})$ for some $\underset{\sim}{u}, \underset{\sim}{v}$ such that $\underset{\sim}{v}$ is a permutation of $\underset{\sim}{u}$;

(3) Neither β_0 nor β_1 has any variable in common with β_2 or β_3;

(4) Each of β_0, β_1, β_2, β_3, is an instance of each of β_2, β_3;

(5) β_4 has no variable in common with β_2;

(6) β_4 is an instance of β_2.

(These special properties of $\beta_0, \ldots, \beta_5$ are crucial for establishing the next two lemmata.)

<u>Proof.</u> First note that by simple reductions we may assume that F^M has the form

$$\hat{\kappa}_0 \wedge \neg\hat{\kappa}_1 \wedge \bigwedge_{i=1}^{n} (\hat{\kappa}_{2i} \to \hat{\kappa}_{2i+1})$$

where the κ_i are tagged bigraphs of the same dimension d, say, $\kappa_i = \langle h_i, \alpha_i \rangle$, where $h_i \in \mathbb{N}$, α_i are bigraphs of dimension d, and α_0 and α_1 contain only y-variables not appearing in $\alpha_2, \ldots, \alpha_{2n+1}$. (These reductions may entail adding universal quantifiers to $\mathscr{P}$, but as the goal is to obtain the prefix class $\mathscr{P}\forall^*$ this is of no consequence. Alternatively, note that for the particular prefix type $\mathscr{P} = \forall\exists\forall\forall$, which is the only one for which this lemma will actually be applied, the proof of the Krom and Horn Prefix Theorem supplies schemata already nearly in this form.)

Now the whole trick is to find a way of encoding the large conjunction in F^M as the arguments of a single atomic formula, in such a way that these arguments may be viewed as a table and the only logic that needs to be captured in G is that for "looking up" the individual conjuncts of F in the table. To construct this table we need a few tools.

Let $h = \max\{h_0, \ldots, h_{2n+1}\}$, and for some k let $\sigma_0, \ldots, \sigma_h$ be vectors which are 2-templates of dimension k such that $\sigma_i(p,q) \neq \sigma_j(p',q')$ whenever $i \neq j$ and $p \neq q$. (For example, we could let $k = h+2$, and $\sigma_i(p,q) = \omega_{i+1}(p,\ldots,p)^\frown\omega(q)^\frown\omega_{h-i}(p,\ldots,p)$.) Let x', y' be variables not appearing in F. Then $\sigma_{h_i}(x',y')^\frown\alpha_i$ may be considered an encoding of κ_i, in the sense that any instance of $\sigma_{h_i}(x',y')^\frown\alpha_i$ with distinct substituents for x' and y' uniquely determines i and an instance of α_i. Let $m = k+d$ be the dimension of such a vector. Next, for $i = 1, \ldots, n$ let

$$\lambda_i = \sigma_{h_{2i}}(x',y')^\frown\alpha_{2i}{}^\frown\sigma_{h_{2i+1}}(x',y')^\frown\alpha_{2i+1} ,$$

which may be considered an encoding of the conjuct $(\hat{\kappa}_{2i} \rightarrow \hat{\kappa}_{2i+1})$ of F^M. Let μ_i $(i = 1, \ldots, n)$ be the vector of dimension $\ell = 2mn$, $\lambda_i{}^\frown\lambda_{i+1}{}^\frown\cdots{}^\frown\lambda_n{}^\frown\lambda_1{}^\frown\ldots{}^\frown\lambda_{i-1}$. Thus each μ_i represents a permutation of the large conjunction in F^M and is a version of the table we are seeking.

Now, since an $\exists$ follows an $\forall$ in $\mathscr{P}$, there are variables y, x of F such that y is a y-variable, x is an x-variable, and y governs x (so that in any Herbrand instance of F the substituents for y and x are distinct). Let $\beta_0, \ldots, \beta_5$ be the following vectors of dimension $m+\ell$, where the variables u_i, v_i, v_i', w_i are all new:

$$\beta_0 = \sigma_{h_0}(x',y')^\frown\alpha_0{}^\frown\mu_1$$

$$\beta_1 = \sigma_{h_1}(x,y)^\frown\alpha_1{}^\frown\omega_\ell(\underset{\sim}{u})$$

$$\beta_2 = \omega_m(\underset{\sim}{v})^\frown\omega_\ell(\underset{\sim}{v'})$$

$$\beta_3 = \omega_m(\underset{\sim}{v})^\frown\omega_\ell(v'_{2m}, \ldots, v'_{\ell-1}, v'_0, \ldots, v'_{2m-1})$$

$$\beta_4 = \omega_m(\underset{\sim}{w})^\frown\omega_\ell(\underset{\sim}{w})$$

$$\beta_5 = \omega(w_m, \dots, w_{2m-1})^\frown \mu_1 .$$

Finally, let $G = Q\forall x'y'\underset{\sim}{u}\underset{\sim}{v}\underset{\sim}{v}'\underset{\sim}{w}(\hat{\beta}_0 \wedge \neg \hat{\beta}_1 \wedge (\hat{\beta}_2 \rightarrow \hat{\beta}_3) \wedge (\hat{\beta}_4 \rightarrow \hat{\beta}_5))$, where Q is the prefix of F. Clearly, G satisfies conditions (1)-(6) above.

($\underline{F \rhd G}$) If F is satisfiable, then let $\theta(\iota) \models E(F)$, where ι is the identity map on the Herbrand domain of F, and let η be the minimal completion of $\Sigma\, \omega_k(\underset{\sim}{r})^\frown \omega_d(\underset{\sim}{s})^\frown \omega_\ell(\underset{\sim}{t})$, the sum being over all $\underset{\sim}{r}$, $\underset{\sim}{s}$, $\underset{\sim}{t}$ such that

(i) $\omega_\ell(\underset{\sim}{t})$ is an Herbrand instance of some μ_i, $i = 1, \dots, n$; and

(ii) if $\omega_k(\underset{\sim}{r}) = \sigma_j(r, r')$ for some j and some <u>distinct</u> terms r, r', then $\langle j, \omega_d(\underset{\sim}{s})\rangle \subset \theta$.

Then $\eta(\iota) \models E(G)$. For let $\hat{\beta}_0' \wedge \neg \hat{\beta}_1' \wedge (\hat{\beta}_2' \rightarrow \hat{\beta}_3') \wedge (\hat{\beta}_4' \rightarrow \hat{\beta}_5')$ be an Herbrand instance of G. Then clearly, $\beta_0' \subset \eta$ and $-\beta_1' \subset \eta$. If β_2' satisfies (ii) and (i) with $i = i_0$, then β_3' satisfies (ii) and (i) with $i \equiv i_0 + 1 \pmod n$. Now let $\beta_4' = \omega_k(\underset{\sim}{r})^\frown \omega_d(\underset{\sim}{s})^\frown \omega_\ell(\underset{\sim}{t})$, $\beta_5' = \omega_k(\underset{\sim}{r}')^\frown \omega_d(\underset{\sim}{s}')^\frown \omega_\ell(\underset{\sim}{t}')$. Then $\omega_k(\underset{\sim}{r})^\frown \omega_d(\underset{\sim}{s}) = \omega_m(t_0, \dots, t_{m-1})$, and $\omega_k(\underset{\sim}{r}')^\frown \omega_d(\underset{\sim}{s}') = \omega_m(t_m, \dots, t_{2m-1})$. Now if $\beta_4' \subset \eta$, then $\omega_\ell(\underset{\sim}{t})$ satisfies (i), so $\omega_{2m}(\underset{\sim}{t}) = \omega_k(\underset{\sim}{r})^\frown \omega_d(\underset{\sim}{s})^\frown \omega_k(\underset{\sim}{r}')^\frown \omega_d(\underset{\sim}{s}')$ is for some i an Herbrand instance of

$$\lambda_i = \sigma_{h_{2i}}(x', y')^\frown \alpha_{2i}{}^\frown \sigma_{h_{2i+1}}(x', y')^\frown \alpha_{2i+1} .$$

Since also $\omega_k(\underset{\sim}{r})$ and $\omega_d(\underset{\sim}{s})$ satisfy (ii), either the substituents for x', y' are identical, in which case $\omega_k(\underset{\sim}{r}')$, $\omega_d(\underset{\sim}{s}')$ satisfy (ii) vacuously, or else $\langle h_{2i}, \omega(\underset{\sim}{s})\rangle \subset \theta$ so that $\langle h_{2i+1}, \omega(\underset{\sim}{s}')\rangle \subset \theta$ (since $\theta(\iota) \models E(F)$) and again, $\omega_k(\underset{\sim}{r}')$, $\omega_k(\underset{\sim}{s}')$ satisfy (ii). Clearly, $\omega_\ell(\underset{\sim}{t}')$ satisfies (i) in any event. Hence, if $\beta_4' \subset \eta$ then $\beta_5' \subset \eta$ and $\eta(\iota) \models \hat{\beta}_4' \rightarrow \hat{\beta}_5'$.

($\underline{G \rhd F}$) Now suppose that G is satisfiable and let $\theta(\iota) \models E(G)$. Define η to be the minimal completion of $\Sigma\langle j, \alpha\rangle$, the sum being over all j, α such that $\sigma_j(p, q)^\frown \alpha^\frown \mu \subset \theta$ whenever p, q are terms in $D(G)$ and μ is an Herbrand instance of some μ_i. Because $\theta(\iota)$ verifies each Herbrand instance of $\hat{\beta}_2 \rightarrow \hat{\beta}_3$, this is equivalent to requiring that $\sigma_j(p, q)^\frown \alpha^\frown \mu \subset \theta$ whenever $p, q \in D(G)$ and μ is an Herbrand instance of μ_1. Now let

$$\hat{\varkappa}_0' \wedge \neg \hat{\varkappa}_1' \wedge \bigwedge_{i=1}^{n} (\hat{\varkappa}_{2i}' \rightarrow \hat{\varkappa}_{2i+1}')$$

be an Herbrand instance of F, where $\varkappa_i' = \langle h_i, \alpha_i'\rangle$ for each i. Then $\varkappa_0' \subset \eta$ since $\theta(\iota)$ verifies each instance of $\hat{\beta}_0$ and neither x', y', nor any of the (y-) variables of α_0 appear in μ_1; and $-\varkappa_1' \subset \eta$. And if $\varkappa_{2i}' \subset \eta$ $(1 \le i \le n)$ then for any $p, q \in D(G)$ and any Herbrand instance μ_1' of μ_1, there is an Herbrand instance $\hat{\beta}_4' \rightarrow \hat{\beta}_5'$ of $\hat{\beta}_4 \rightarrow \hat{\beta}_5$ such that

$$\beta_4' = \sigma_{h_{2i}}(p,q)^\frown\alpha_{2i}'^\frown\sigma_{h_{2i}}(p,q)^\frown\alpha_{2i}'^\frown\sigma_{h_{2i+1}}(p,q)^\frown\alpha_{2i+1}'^\frown\beta'$$

for some β' of dimension ℓ-2m, and

$$\beta_5' = \sigma_{h_{2i+1}}(p,q)^\frown\alpha_{2i+1}'^\frown\mu_1' \quad .$$

Then $\beta_4' \subset \theta$ since $\varkappa_{2i}' \subset \eta$, and $\beta_5' \subset \eta$ since $\theta(z) \models \hat{\beta}_4' \rightarrow \hat{\beta}_5'$. Since this is the case for any p, q, μ_1', it follows that $\varkappa_{2i+1}' \subset \eta$.

This completes the proof of the Six Atomic Formula Lemma. ∎

IIF. 2 Five Atomic Formulas

Five Atomic Formula Lemma. Let $G \in \mathscr{P}\forall^*$ be a schema as described in the Six Atomic Formula Lemma. Then we can construct a schema $H \in \mathscr{P}\forall^*$ such that H is satisfiable if and only if G is satisfiable, and

$$H^M = \hat{\gamma}_0 \wedge \neg\, \hat{\gamma}_1 \wedge (\hat{\gamma}_2 \rightarrow (\hat{\gamma}_3 \wedge \hat{\gamma}_4)) \quad ,$$

where $\gamma_0, \ldots, \gamma_4$ are vectors of the same dimension such that

(1') All the variables of γ_2 and γ_3 are y-variables of H governing no x-variable of H;

(2') $\gamma_2 = \omega(\underset{\sim}{u})$, $\gamma_3 = \omega(\underset{\sim}{v})$ for some $\underset{\sim}{u}, \underset{\sim}{v}$ such that $\underset{\sim}{v}$ is a permutation of $\underset{\sim}{u}$;

(3') Neither γ_0 nor γ_1 has any variable in common with γ_2 or γ_3;

(4') Each of $\gamma_0, \gamma_1, \gamma_2, \gamma_3$ is an instance of each of γ_2, γ_3.

Proof. Let G and $\beta_0, \ldots, \beta_5$ be as in the statement of the Six Atomic Formula Lemma (p. 180), where $\beta_0, \ldots, \beta_5$ are of dimension m. By (1) and (6), β_2 and β_4 contain only y-variables governing no x-variable, and β_4 is an instance of β_2. Let $y_1, \ldots, y_n$ be the (distinct) variables of β_2 and let $\beta_4 = \beta_2[\underset{\sim}{y}/\underset{\sim}{z}]$, where the z_i are not necessarily distinct from each other but are (by (5)) distinct from the y_i. Also, let $\underset{\sim}{u}, \underset{\sim}{v}, \underset{\sim}{w}$ be n-tuples of new variables. Then define $\gamma_0, \ldots, \gamma_4$ as follows:

$$\gamma_0 = \beta_0{}^\frown\omega(\underset{\sim}{u})^\frown\omega(\underset{\sim}{u})$$

$$\gamma_1 = \beta_1{}^\frown\omega(\underset{\sim}{v})^\frown\omega(\underset{\sim}{v})$$

$$\gamma_2 = \beta_2{}^\frown\omega(\underset{\sim}{w})^\frown\omega(\underset{\sim}{w})$$

$$\gamma_3 = \beta_3{}^\frown\omega(\underset{\sim}{w})^\frown\omega(\underset{\sim}{w})$$

(a) $\omega_n(\underset{\sim}{p})$ is not an Herbrand instance of γ_2 or

(b) $-\omega_n(\underset{\sim}{p})^\frown\omega_n(\underset{\sim}{q}) \subset \theta$ for every $\underset{\sim}{q}$.

Let $\underset{\sim}{v}$ be a list of all the variables of the functional form J^* of J, and let $H^* = \hat{\gamma}_0^* \wedge \neg\hat{\gamma}_1^* \wedge (\hat{\gamma}_2^* \rightarrow (\hat{\gamma}_3^* \wedge \hat{\gamma}_4^*))$. Now consider any Herbrand instance of H, say, $H^*[\underset{\sim}{v}/\underset{\sim}{s}] = \hat{\gamma}_0' \wedge \neg\hat{\gamma}_1' \wedge (\hat{\gamma}_2' \rightarrow (\hat{\gamma}_3' \wedge \hat{\gamma}_4'))$.

(i) $\gamma_0' \subset \eta$. For by (1') and (3') there is a $\underset{\sim}{t}$ such that $\gamma_0^*[\underset{\sim}{v}/\underset{\sim}{t}] = \gamma_0^*[\underset{\sim}{v}/\underset{\sim}{s}] = \gamma_0'$ and $\gamma_1^*[\underset{\sim}{v}/\underset{\sim}{t}] = \gamma_1^*[\underset{\sim}{v}/\underset{\sim}{s}] = \gamma_3^*[\underset{\sim}{v}/\underset{\sim}{t}]$; such a $\underset{\sim}{t}$ may be formed from $\underset{\sim}{s}$ by keeping the same substituents for the variables actually occurring in γ_0^* or γ_1^* and choosing substituents for the variables in γ_3^* so as to obtain $\gamma_1^*[\underset{\sim}{v}/\underset{\sim}{s}]$ as an instance of γ_3^* (which is possible by (4')). Moreover, $\underset{\sim}{t}$ may be chosen so that $\omega_n(\underset{\sim}{z})[\underset{\sim}{v}/\underset{\sim}{t}] = \gamma_4[\underset{\sim}{v}/\underset{\sim}{t}]$, since the variables in $\underset{\sim}{z}$ do not appear in γ_0-γ_4. Then $J^*[\underset{\sim}{v}/\underset{\sim}{t}] = (\hat{\zeta}_0' \wedge \neg\hat{\zeta}_1') \vee (\hat{\zeta}_2' \wedge \neg\hat{\zeta}_3')$, where $\zeta_2' = \zeta_3'$ and $\zeta_0' = \gamma_2'^\frown\gamma_0'$ for some γ_2'. Hence $\gamma_2'^\frown\gamma_0' \subset \theta$ and by the sublemma and the definition of η, $\gamma_0' \subset \eta$.

(ii) $-\gamma_1' \subset \eta$. The argument is similar to (i). We can find a $\underset{\sim}{t}$ such that $\gamma_1^*[\underset{\sim}{v}/\underset{\sim}{t}] = \gamma_1'$, and $\gamma_0^*[\underset{\sim}{v}/\underset{\sim}{t}] = \gamma_2^*[\underset{\sim}{v}/\underset{\sim}{t}] = \omega(\underset{\sim}{y})[\underset{\sim}{v}/\underset{\sim}{t}]$. Hence, in $J^*[\underset{\sim}{v}/\underset{\sim}{t}]$ the first disjunct is falsified by $\theta(z)$, so $\gamma_1'^\frown\gamma_4' \subset \theta$ for some γ_4'; hence $-\gamma_1' \subset \eta$.

(iii) If $\gamma_2' \subset \eta$ then $-\gamma_2'^\frown\omega_n(\underset{\sim}{q}) \subset \theta$ for all $\underset{\sim}{q}$. Then $\gamma_1'^\frown\gamma_4'$ - $\gamma_3'^\frown\omega(\underset{\sim}{q}) \subset \theta$ for all $\underset{\sim}{q}$. Hence, $\gamma_3' \subset \eta$. If γ_4' is not an Herbrand instance of γ_2, then $\gamma_4' \subset \eta$, otherwise $\gamma_4' \subset \eta$ by the sublemma.

This completes the proof of the Four Atomic Formula Lemma. ∎

IIF. 4 The Theorem

We can now prove the main result of this chapter.

Few Atomic Subformula Theorem. $\forall\exists\forall^* \cap \mathscr{N}_4$ and $\forall^*\exists \cap \mathscr{N}_5$ are unsolvable.

Proof. By the three lemmata of this chapter and the Krom and Horn Prefix Theorem, the class of all schemata in $\forall\exists\forall^*$ with matrices of the form $(A_0 \wedge \neg A_1) \vee (A_2 \wedge \neg A_3)$ is unsolvable. Now let $F = \forall y \exists x \forall z_1 \ldots \forall z_m((\hat{\alpha}_0 \wedge \neg\hat{\alpha}_1) \vee (\hat{\alpha}_2 \wedge \neg\hat{\alpha}_3))$ be such a schema, where each α_i is a vector of dimension n. Let w, x', y' be new variables and let $\underset{\sim}{u}_0, \underset{\sim}{u}_1, \underset{\sim}{u}_2$ be new (n+2)-tuples of variables. Then let

$$G = \forall \underset{\sim}{z}\,\underset{\sim}{u}_1\underset{\sim}{u}_2\underset{\sim}{u}_3 w y x y' \exists x'((\hat{\beta}_0 \vee \hat{\beta}_1 \vee \hat{\beta}_2) \wedge (\neg\hat{\beta}_3 \vee \neg\hat{\beta}_4)) ,$$

where $\beta_0, \ldots, \beta_4$ are as follows:

$$\beta_0 = \omega(x')^\frown\omega_{n+1}(y', \ldots, y')^\frown\omega(x)^\frown\omega_{n+1}(y, \ldots, y)$$

$$\beta_1 = \omega(w, w)^\frown\alpha_0{}^\frown\omega(w, w)^\frown\alpha_1$$

$$\beta_2 = \omega(w,w)^\frown\alpha_2{}^\frown\omega(w,w)^\frown\alpha_3$$

$$\beta_3 = \omega(\underset{\sim}{u}_0)^\frown\omega(\underset{\sim}{u}_1)$$

$$\beta_4 = \omega(\underset{\sim}{u}_1)^\frown\omega(\underset{\sim}{u}_2) \quad .$$

Let f be the monadic indicial correlate of x, and let g be the indicial function sign for x' in G. Define a mapping $\gamma: D(G) \to D(F)$ by:

$$\gamma(\ddagger) = \ddagger$$

$$\gamma(g(t_1 \cdots t_n)) = f(\gamma(t_1))$$

and let γ^{-1} be the following inverse of γ:

$$\gamma^{-1}(\ddagger) = \ddagger\,, \qquad \gamma^{-1}(f(t)) = g(\gamma^{-1}(t) \cdots \gamma^{-1}(t)) \quad .$$

For any vector α such that $V(\alpha) \subset D(G)$, let $\gamma(\alpha)$ be the result of replacing each term t by $\gamma(t)$ in α; $\gamma^{-1}(\alpha)$ is defined similarly.

(F $\triangleright$ G) If F is satisfiable, then let $\theta(\iota) \models E(F)$, and let η be the minimal completion of

$$\sum_t \sum_{\delta_1 - \delta_2 \subset \theta} \omega(t,t)^\frown\delta_1{}^\frown\omega(t,t)^\frown\delta_2$$

$$+ \sum_{\substack{t \neq f(s) \\ t' = f(s')}} \omega(t')^\frown\omega_{n+1}(s',\ldots,s')^\frown\omega(\underset{\sim}{t})^\frown\omega(s,\ldots,s) \quad .$$

Then $\eta(\gamma) \models E(G)$. For let $(\hat{\beta}'_0 \vee \hat{\beta}'_1 \vee \hat{\beta}'_2) \wedge (\neg\hat{\beta}'_3 \vee \neg\hat{\beta}'_4)$ be an Herbrand instance of G. If $\eta(\gamma) \models \neg\hat{\beta}'_0$ then $\gamma(\beta'_0) = \omega(t')^\frown\omega_{n+1}(s',\ldots,s')^\frown\omega(t)^\frown\omega_{n+1}(s,\ldots,s)$, where $t = f(s)$ (and also $t' = f(s')$). Then for some Herbrand instance $(\hat{\alpha}'_0 \wedge \neg\hat{\alpha}'_1) \vee (\hat{\alpha}'_2 \wedge \neg\hat{\alpha}'_3)$ of F, and some $r \in D(F)$, $\gamma(\beta'_1) = \omega(r,r)^\frown\alpha'_0{}^\frown\omega(r,r)^\frown\alpha'_1$ and $\gamma(\beta'_2) = \omega(r,r)^\frown\alpha'_2{}^\frown\omega(r,r)^\frown\alpha'_3$, so $\eta(\gamma) \models \hat{\beta}'_1 \vee \hat{\beta}'_2$. Also $\eta(\gamma) \models \neg\hat{\beta}'_3 \vee \neg\hat{\beta}'_4$, directly from the definition of η.

(G $\triangleright$ F) If G is satisfiable, then let $\theta(\iota) \models E(G)$. For each term $t \in D(G)$, consider that Herbrand instance in which t is the substituent for each y-variable. It follows that $\omega(g(t\cdots t))^\frown\omega_{2n+3}(t,\ldots,t) \subset \theta$ for each $t \in D(G)$, whence from the second conjunct it follows that $\omega_{n+2}(\underset{\sim}{t})^\frown\omega(g(t\cdots t))^\frown\omega_{n+1}(t,\ldots,t) \subset \theta$ for each $\underset{\sim}{t}$ and t. Then by the first conjunct again

$$(*) \qquad \theta(\iota) \models (\hat{\beta}_1 \vee \hat{\beta}_2)[w/\ddagger,\ y/t,\ x/g(t\ \ldots\ t),\ \underset{\sim}{z}/\underset{\sim}{s}]$$

for every t and $\underset{\sim}{s}$.

Now let η be the minimal completion of $\Sigma\delta_1$, the sum over all δ_1 such that $\omega(\ddagger, \ddagger)^\frown\delta_1{}^\frown\omega(\ddagger, \ddagger)^\frown\delta_2 \subset \theta$ for some δ_2. Then $\eta(\gamma^{-1}) \models E(F)$. For let $(\hat{\alpha}'_0 \wedge \neg\hat{\alpha}'_1) \vee (\hat{\alpha}'_2 \wedge \neg\hat{\alpha}'_3)$ be any Herbrand instance of F, and for $i = 0, \ldots, 3$ let $\alpha''_i = \gamma^{-1}(\alpha'_i)$. Let

$$\beta'_1 = \omega(\ddagger, \ddagger)^\frown\alpha''_0{}^\frown\omega(\ddagger, \ddagger)^\frown\alpha''_1$$

$$\beta'_2 = \omega(\ddagger, \ddagger)^\frown\alpha''_2{}^\frown\omega(\ddagger, \ddagger)^\frown\alpha''_3 .$$

Then $\hat{\beta}'_1 \vee \hat{\beta}'_2$ is an Herbrand instance of $\hat{\beta}_1 \vee \hat{\beta}_2$ of the form shown in (*) above, so either $\beta'_1 \subset \theta$ or $\beta'_2 \subset \theta$. Moreover, by the second conjunct of G^M, if δ_1 and δ_2 are such that $\omega(\ddagger, \ddagger)^\frown\delta_1{}^\frown\omega(\ddagger, \ddagger)^\frown\delta_2 \subset \theta$, so that $\delta_1 \subset \eta$, then $-\omega(\ddagger, \ddagger)^\frown\delta_2{}^\frown\omega(\ddagger, \ddagger)^\frown\delta_3 \subset \theta$ for every δ_3, so that $-\delta_2 \subset \eta$; it follows that $\eta(\gamma^{-1}) \models E(F)$.

This completes the proof of the Few Atomic Subformula Theorem. ∎

<u>Historical References</u>. The unsolvability of $\mathcal{N}_5$ is due to Lewis and Goldfarb (1973); that of $\mathcal{N}_4$ to Goldfarb (1974).

LIST OF OPEN PROBLEMS

1. Improvement of the known results on the classes determined by extended prefix and similarity type (p. 137).
2. $\forall\exists\forall\exists^n \cap$ Krom, for each $n \geq 1$.
3. Conservative unsolvability for $\forall\exists\forall\forall \cap$ Krom.
4. Classes of Krom formulas determined by prefix and similarity type.
5. Signature subclasses of $\mathcal{D}_2$, the two-disjunct class.
6. $\mathcal{N}_3$ (schemata with three atomic subformulas); and $\mathcal{N}_4 \cap \forall^*\exists^*$.
7. The $\exists^*\forall\forall\exists^*$ class with identity.
8. The conjecture on p. 176.

REFERENCES

Aanderaa, Stål O., 1966. A New Undecidable Problem with Applications in Logic. Ph.D. thesis, Harvard University.

__________, 1971. On the decision problem for formulas in which all disjunctions are binary. Proceedings of the Second Scandinavian Logic Symposium, North-Holland Publishing Company, Amsterdam, pp. 1-18.

Aanderaa, Stål O. and Harry R. Lewis, 1973. Prefix classes of Krom formulas, Journal of Symbolic Logic 38, pp. 628-642.

__________, 1974. Linear sampling and the $\forall\exists\forall$ case of the decision problem. Journal of Symbolic Logic 39, pp. 519-548.

Ackermann, Wilhelm, 1928. Über die Erfüllbarkeit gewisser Zählausdrücke. Mathematische Annalen 100, pp. 638-649.

__________, 1954. Solvable Cases of the Decision Problem. North-Holland, Amsterdam.

Behmann, Heinrich, 1922. Beiträge zur Algebra der Logik, inbesondere zum Entscheidungsproblem. Mathematische Annalen 86, pp. 163-229.

Berger, Robert, 1966. The Undecidability of the Domino Problem. Memoirs of the American Mathematical Society, No. 66.

Bernays, Paul and Moses Schönfinkel, 1928. Zum Entscheidungsproblem der mathematischen Logik. Mathematische Annalen 99, pp. 342-372.

Börger, Egon, 1971. Reduktionstypen in Krom- und Hornformeln. Inauguraldissertation, Westfälische Wilhelms-Universität, Münster.

__________, 1974. Beitrag zur Reduktion des Entscheidungsproblems auf Klassen von Hornformeln mit kurzen Alternationen. Archiv für mathematische Logik und Grundlagenforschung 16, pp. 67-84.

__________, 1976. A new general approach to the theory of the many-one equivalence of decision problems for algorithmic systems. Schriften zur Informatik und Angewandten Mathematik, no. 30, Rheinish-Westfälische Technische Hochschule Aachen.

Büchi, J. Richard, 1962. Turing-machines and the Entscheidungsproblem. Mathematische Annalen 148, pp. 201-213.

Chang, C.C. and H.J. Keisler, 1962. An improved prenex normal form. Journal of Symbolic Logic 27, pp. 317-326.

Church, Alonzo, 1951. Special cases of the decision problem. Revue philosophique de Louvain 49, pp. 203-221; correction, ibid. 50, pp. 270-272.

__________, 1956. Introduction to Mathematical Logic, Volume I. Princeton University Press, Princeton, N.J.

Denton, John, 1963. Applications of the Herbrand Theorem. Ph.D. thesis, Harvard University.

Dreben, Burton, 1961. Solvable Surányi subclasses: An introduction to the Herbrand theory. Proceedings of a Harvard Symposium on Digital Computers, 3-6 April 1961; Annals of the Computation Laboratory of Harvard University 31, Harvard University Press, Cambridge, Massachusetts, pp. 32-47.

Dreben, Burton S. and W. D. Goldfarb, 1979. The Decision Problem: Solvable Classes of Quantificational Formulas. Addison-Wesley Publishing Co., Reading, Mass.

Dreben, Burton, A.S. Kahr, and Hao Wang, 1962. Classification of AEA formulas by letter atoms. Bulletin of the American Mathematical Society 68, pp. 528-532.

Fischer, Patrick C., 1966. Turing machines with restricted memory access. Information and Control 9, pp. 364-379.

Friedman, Joyce, 1963. A semi-decision procedure for the functional calculus. Journal of the Association for Computing Machinery 10, pp. 1-24.

Gödel, Kurt, 1930. Die Vollständigkeit der Axiome des logischen Funktionenkalküls. Monatshefte für Mathematik und Physik 37, pp. 349-360; English translation in van Heijenoort (1971), pp. 582-591.

__________, 1932. Ein Spezialfall des Entscheidungsproblems der theoretischen Logik. Ergebnisse eines mathematischen Kolloquiums 2, pp. 27-28.

__________, 1933. Zum Entscheidungsproblem des logischen Funktionenkalküls. Monatshefte für Mathematik und Physik 40, pp. 433-443.

Goldfarb, Warren D., 1974. On Decision Problems for Quantification Theory. Ph. D. thesis, Harvard University.

Goldfarb, Warren D. and Harry R. Lewis, 1975. Skolem reduction classes. Journal of Symbolic Logic 40, pp. 62-68.

Gurevich, Yuri, 1965. Ekzistential'naya interpretatsiya. Algebra i Logika 4, pp. 71-84.

__________, 1966a. Problema razresheniya dlya yzkogo ischisleniya predikatov. Doklady Academii Nauk SSSR 168, pp. 510-511; English translation in Soviet Mathematics-Doklady 168, pp. 669-670.

__________, 1966b. Ob effektivnom raspoznavanii vipolnimosti formul UIP. Algebra i Logika 5, pp. 25-55.

__________, 1966c. K probleme razresheniya dlya chistogo uzkogo ischisleniya predikatov. Doklady Akademii Nauk SSSR 166, pp. 1032-1034; English translation in Soviet Mathematics-Doklady 166, pp. 217-219.

__________, 1969. Problema razresheniya dlya logiki predikatov i operatsiy. Algebra i Logika 8, pp. 284-308; English translation in Algebra and Logic 8, pp. 160-174.

__________, 1970. Mashini Minskogo i sluchay $\forall\exists\forall \& \exists^{\infty}$ problemy razresheniya. Matematicheskie Zapiski, Uralskiy Gosydarstvenniy Universitet 3, pp. 77-83.

__________, 1971. Problema razresheniya logiki I stupeni. Unpublished manuscript.

__________, 1976. The decision problem for standard classes. Journal of Symbolic Logic 41, pp. 460-464.

Gurevich, Yu. Sh. and T. V. Turashvili, 1973. Usilenie odnogo resul'tata Ya. Shuran 'i. Bulletin of the Academy of Sciences of the Georgian SSR 70, pp. 290-292.

Herbrand, Jacques, 1931. Sur le problème fondamental de la logique mathématique. Sprawozdania z posiedzen Towarzystwa Naukowego Warszawskiego, Wydzial III, 24, pp. 12-56; English translation in Herbrand (1971), pp. 215-271.

__________, 1971. Logical Writings, edited by Warren D. Goldfarb. Harvard University Press, Cambridge, Massachusetts.

Hopcroft, John E. and Jeffrey D. Ullman, 1969. Formal Languages and their Relation to Automata. Addison-Wesley Publishing Company, Reading, Massachusetts.

Hughes, C. E., Overbeek, R., and Singletary, W. E., 1971. The many-one equivalence of some combinatorial decision problems. Bulletin of the American Mathematical Society 77, pp. 467-472.

Joyner, William H., Jr., 1976. Resolution strategies as decision procedures. Journal of the Association for Computing Machinery 23, pp. 398-417.

Kahr, A.S., 1962. Improved reductions of the Entscheidungsproblem to subclasses of AEA formulas. Proceedings of a Symposium on the Mathematical Theory of Automata, Brooklyn Polytechnic Institute, New York.

Kahr, A.S., E.F. Moore, and Hao Wang, 1962. Entscheidungsproblem reduced to the $\forall\exists\forall$ case. Proceedings of the National Academy of Sciences of the U.S.A. 48, pp. 365-377.

Kalmár, László, 1932. Zum Entscheidungsproblem der mathematischen Logik. Verhandlungen des Internationalen Mathematiker-Kongresses Zürich, 2, pp. 337-338.

__________, 1933. Über die Erfüllbarkeit derjenigen Zählausdrücke, welche in der Normalform zwei benachbarte Allzeichen enthalten. Mathematische Annalen 108, pp. 466-484.

__________, 1936. Zurückführung des Entscheidungsproblems auf den Fall von Formeln mit einer einzigen, binären, Funktionsvariablen. Compositio Mathematica 4, pp. 137-144.

__________, 1939. On the reduction of the decision problem, first paper: Ackermann prefix, a single binary predicate. Journal of Symbolic Logic 4, pp. 1-9.

__________, 1950. Contributions to the reduction theory of the decision problem, first paper: Prefix $(x_1)(x_2)(Ex_3)\cdots(Ex_{n-1})(x_n)$, a single binary predicate. Acta Mathematica Academiae Scientiarum Hungaricae 1, pp. 64-73.

__________, 1951. Contributions to the reduction theory of the decision problem, third paper: Prefix $(x_1)(Ex_2)\cdots(Ex_{n-2})(x_{n-1})(x_n)$, a single binary predicate. Acta Mathematica Academiae Scientiarum Hungaricae 2, pp. 19-37.

Kalmár, László and János Surányi, 1947. On the reduction of the decision problem, second paper: Gödel prefix, a single binary predicate. Journal of Symbolic Logic 12, pp. 65-73.

__________, 1950. On the reduction of the decision problem, third paper: Pepis prefix, a single binary predicate. Journal of Symbolic Logic 15, pp. 161-173.

Kostryko, V. F., 1964. Klass cvedeniya $\forall\exists^n\forall$. Algebra i Logika 3, pp. 45-65.

__________, 1966. Klass cvedeniya $\forall\exists\forall$. Kibernetika 2, pp. 17-22; English translation, Cybernetics 2, pp. 15-19.

Krom, M.R., 1970. The decision problem for formulas in prenex conjunctive normal form with binary disjunctions. Journal of Symbolic Logic 35, pp. 210-216.

Lewis, Harry R., 1974. Herbrand Expansions and Reductions of the Decision Problem. Ph.D. thesis, Harvard University, 1974; Center for Research in Computing Technology Technical Report No. TR-18-74.

__________, 1976. Krom formulas with one dyadic predicate letter. Journal of Symbolic Logic 41, pp. 341-362.

Lewis, Harry R. and Warren D. Goldfarb, 1973. The decision problem for formulas with a small number of atomic subformulas. Journal of Symbolic Logic 38, pp. 471-480.

Löwenheim, Leopold, 1915. Über Möglichkeiten im Relativkalkül. Mathematische Annalen 76, pp. 447-470; English translation in van Heijenoort (1971), pp. 228-251.

Maslov, S. Yu., 1964. Obratniy metod ystanovleniya vyvodimosti v klassicheskom ischilenii predikatov. Doklady Akademii Nauk SSR 159, pp. 17-20; English translation in Soviet Mathematics-Doklady 5, pp. 1420-1424.

__________, 1966. Primeneniye obratnogo metod ystanovleniya vyvodimosti k teorii razreshimykh fragmentov klassicheskogo ischisleniya predikatov. Doklady Akademii Nauk SSSR 171, pp. 1282-1285; English translation in Soviet Mathematics-Doklady 7, pp. 1653-1657.

__________, 1967. Obratniy metod ystanovleniya vyvodimosti dlya nepredvarennykh formul ischisleniya predikatov. Doklady Akademii Nauk SSSR 172, pp. 22-25; English translation in Soviet Mathematics-Doklady 8, pp. 16-19.

__________, 1968. Obratniy metod ystanovleniya vyvodimosti dlya logicheskikh ichesleniy. Trudy Matematicheskogo Instituta AN SSSR 98, pp. 26-87; English translation in Proceedings of the Steklov Institute of Mathematics 98, pp. 25-96.

Minsky, Marvin L., 1961. Recursive unsolvability of Post's problem of "tag" and other topics in the theory of Turing machines. Annals of Mathematics 74, pp. 437-454.

__________, 1967. Computation: Finite and Infinite Machines. Prentice-Hall, Inc., Englewood Cliffs, N.J.

Orevkov, V.P., 1968. Dva nerazreshimykh klassa formul klassicheskogo ischisleniya predikatov. Zapiski Nauchnykh Seminarov Leningradskogo Otdeleniya Matematicheskogo Instituta im. V.A. Steklova AN SSSR 8, pp. 202-210; English translation in Seminars in Mathematics, V.A. Steklov Mathematical Institute, Leningrad 8, pp. 98-102.

__________, 1971. O bikon'yunktivnykh klassakh svedeniya. Zapiski Nauchnykh Seminarov Leningradskogo Otdeleniya Matematicheskogo Instituta im. V.A. Steklova AN SSSR 20, pp. 170-174; English translation in Journal of Soviet Mathematics 1 (1973), pp. 106-109.

Pepis, Józef, 1938. Untersuchungen über das Entscheidungsproblem der mathematischen Logik. Fundamenta Mathematicae 30, pp. 257-348.

Post, Emil, 1946. A variant of a recursively unsolvable problem. Bulletin of the American Mathematical Society 52, pp. 264-268.

Robinson, J.A., 1965. A machine-oriented logic based on the resolution principle. Journal of the Association for Computing Machinery 12, pp. 23-41.

Robinson, Raphael M., 1971. Undecidability and nonperiodicity for tilings of the plane. Inventiones Mathematicae 12, pp. 177-209.

Rogers, Hartley, Jr., 1967. Theory of Recursive Functions and Effective Computability. McGraw-Hill Book Company, New York.

Schütte, Kurt, 1934. Untersuchungen zum Entscheidungsproblem der mathematischen Logik. Mathematische Annalen 109, pp. 572-603.

Skolem, Thoralf, 1919. Untersuchungen über die Axiome des Klassenkalkuls und über Produktations- und Summationsprobleme, welche gewisse Klassen von Aussagen betreffen. Videnskapsselskapets Skrifter I. Matematisk-naturvidenskabelig klasse, no. 3.

__________, 1920. Logisch-kombinatorische Untersuchungen über die Erfüllbarkeit oder Beweisbarkeit mathematischen Sätze nebst einem Theoreme über dichte Mengen. Videnskapsselskapets Skrifter, I. Matematisk-naturvidenskabelig klasse, no. 3; Section 1, English translation in van Heijenoort (1971), pp. 251-263.

__________, 1922. Einige Bemerkungen zur axiomatischen Begründung der Mengenlehre. Matematikerkongressen i Helsingfors den 4-7 Juli 1922, Den femte skandinaviska matematikercongressen, Redogörelse (Akademiska Bokhandeln, Helsinki, 1923), pp. 217-232; English translation in van Heijenoort (1971), pp. 290-301.

__________, 1928. Über die mathematische Logik. Norsk matematisk tidsskrift 10, pp. 125-142; English translation in van Heijenoort (1971), pp. 508-524.

__________, 1929. Über einige Grundlagenfragen der Mathematik, Skrifter utgitt av Det Norske Videnskaps-Akademi i Oslo, I. Matematisk-naturvidenskapelig klasse, no. 4, pp. 1-49.

Surányi, János, 1950. Contributions to the reduction theory of the decision problem, second paper: Three universal, one existential quantifiers. Acta Mathematica Academiae Scientiarum Hungaricae 1, pp. 261-270.

__________, 1951. Contributions to the reduction theory of the decision problem, fifth paper: Ackermann prefix with three universal quantifiers. Acta Mathematica Academiae Scientiarum Hungaricae 2, pp. 325-335.

__________, 1959. Reduktionstheorie des Entscheidungsproblems im Prädikatenkalkül der ersten Stufe. Verlag der Ungarischen Akademie der Wissenschaften, Budapest.

__________, 1971. Reduction of the decision problem of the first order predicate calculus to reflexive and symmetrical binary predicates. Periodica Mathematica Hungarica 1, pp. 97-106.

Turing, Alan M., 1937. On computable numbers, with an application to the Entscheidungsproblem. Proceedings of the London Mathematical Society, 2nd series 42, pp. 230-265; correction, ibid., 43, pp. 544-546.

van Heijenoort, Jean, 1971. From Frege to Gödel: A Source Book in Mathematical Logic, 1879-1931. Harvard University Press, Cambridge, Massachusetts.

Wang, Hao, 1961. Proving theorems by pattern recognition II. Bell System Technical Journal 40, pp. 1-41.

__________, 1962. Dominoes and the AEA case of the decision problem. Proceedings of a Symposium on the Mathematical Theory of Automata, Polytechnic Institute of Brooklyn, New York, pp. 23-55.

INDEX OF THEOREMS AND LEMMATA

INDEX OF SYMBOLS

INDEX OF TERMS

THE DECISION PROBLEM

Solvable Classes of Quantificational Formulas

Burton Dreben and Warren D. Goldfarb

Harvard University

Skolem and Herbrand addressed the decision problem by associating with each quantificational formula a set, usually infinite, of quantifier-free formulas. In this book, the authors build on that insight to illuminate the mathematical structures underlying solvability, and provide for the first time a unified treatment of the positive results of the decision problem for quantification theory. The standard solvable classes are generalized, new solvable classes are obtained, and questions of the existence of finite models are investigated. The results of this book, together with those of Lewis's Unsolvable Classes of Quantificational Formulas, delimit the boundary between solvable and unsolvable.

Fall 1979, approx. 320 pp., illus.

Hardbound ISBN 0-201-02540-X